Economics of the Belt and Road Initiative

The Belt and Road Initiative

Interdisciplinary Perspectives

Wenshan Jia
Series Editor

Vol. 2

Chen Yongjun

Economics of the Belt and Road Initiative

PETER LANG
Lausanne • Berlin • Bruxelles • Chennai • New York • Oxford

Library of Congress Cataloging-in-Publication Data

Names: Chen, Yongjun, author.
Title: Economics of the Belt and Road Initiative / Chen Yongjun.
Description: New York : Peter Lang, [2023] | Series: The Belt and Road Initiative, 2689-7970 ; vol. 2 | Includes bibliographical references and index.
Identifiers: LCCN 2022060542 (print) | LCCN 2022060543 (ebook) | ISBN 9781433192951 (hardback) | ISBN 9781636671475 (ebook) | ISBN 9781636671482 (epub)
Subjects: LCSH: Yi dai yi lu (Initiative : China) | China–Foreign economic relations.
Classification: LCC HF1604 .C379 2023 (print) | LCC HF1604 (ebook) | DDC 337.51–dc23/eng/20230524
LC record available at https://lccn.loc.gov/2022060542
LC ebook record available at https://lccn.loc.gov/2022060543
DOI 10.3726/b20617

Bibliographic information published by the **Deutsche Nationalbibliothek.**
The German National Library lists this publication in the German National Bibliography; detailed bibliographic data is available on the Internet at http://dnb.d-nb.de.

Cover design by Peter Lang Group AG

ISSN 2689-7970 (print)
ISSN 2689-7989 (online)
ISBN 9781433192951 (hardback)
ISBN 9781636671475 (ebook)
ISBN 9781636671482 (epub)
DOI 10.3726/b20617

Published by Peter Lang Publishing Inc., New York, USA
info@peterlang.com - www.peterlang.com

This publication has been peer reviewed.

Contents

List of Figures

List of Tables

Preface

Towards a Co-construction of a High-Quality "Belt and Road Initiative": A New Journey

The world today is in the middle of major changes unprecedented in centuries. First, along with the new round of scientific and industrial revolution, the global economic landscape is changing, calling for the global governance system to change in response to the times. Second, the process of economic globalization has been challenged by unilateralism and protectionism, and the international situation is frequently characterized by instability and uncertainty. Third, the outbreak of the new COVID-19 pandemic has hit the already sluggish world economy hard again and set back the development gains of human society by decades. At the same time, after China's economic development enters a high-quality development stage, it strengths to promote common prosperity and the problem of unbalanced and inadequate development needs to be stressed. Therefore, it is urgent for China and the international community to gather the wisdom and strength of all parties in the world to work closely together to address global challenges and inject a strong dose of catalyst to promote practical cooperation among countries and robust inclusive economic growth.

Grounded in this background, China's attempt to build "the Belt and Road" with other countries has entered a high-quality development stage after ten years of practical development, in turn, it has started paving a new journey. It serves as a Chinese solution to assemble the strengths of all countries and address common challenges. It proposes an international regional economic cooperation initiative to foster the growth of the world economy. It is an important practice of the concept of "building a community of a shared future for mankind". Promoting the transformation of "the Belt and Road Initiative" to achieve a high-quality development is not only a practical need to recover the world economy and achieve inclusive growth, but also a natural growth of the Chinese economy into a new development stage.

The "Belt and Road Initiative" (BRI) originated in China and its connotation has evolved in the actualization of the pragmatic context. State Councillor Wang Yi gave a concise summary of the contents of the BRI at the briefing session of the "Second International Summit on the BRI" in April 2019. He noted that "the BRI is an international regional economic cooperation initiative proposed by the Chinese President Xi Jinping, with the core of which is to strengthen all-round connectivity with infrastructure construction as the main line, to tap new momentum for world economic growth and create a new platform for international regional economic cooperation." This statement correctly indicates the general title, content, requirements, tasks, and purpose of the construction of the BRI.

Over the past ten years, the construction of the BRI has adhered to the basic rule of mutual consultation, co-construction and sharing. By focusing on the key issues of concern to all countries and based on the common aspirations for development, it pragmatically promotes numerous landmark cooperation projects to achieve substantive progress. This is evident in the following aspects: (1) in the steady progress of infrastructure construction, with the "Six Corridors, Six Roads, Multiple Countries and Ports" as the main framework; (2) the continued growth of trade exchanges and the steady increase in mutual direct investment; (3) the continued expansion of international consensus, with the concept of building a community of a shared future for mankind that has been gaining popularity and making tangible progress. The BRI has achieved important results in the opening round and successfully completed a "Grand Design" layout of a solidified structure.

Along with the implementation of the BRI, theories on the co-construction of the BRI are also evolving. The BRI follows the trend of the times and the realistic development needs, with the overall goal of building a community of a shared future for mankind and strives to build the BRI into a path of peace, prosperity, openness, greenness, innovation, civilization, integrity, health, growth, poverty reduction and so on. The proposal of these ten paths indicates that the construction of the BRI has not only stepped into a steadily and deeply development stage of a high-quality co-construction in practice, further sharpening and clarifying the focus of cooperation, but also established a more complete system in theory.

The new journey of co-constructing the BRI has set higher requirements in practice. A more demanding practice calls for theoretical guidance that keeps pace with the times. Why does BRI work? How does it work? Has the phased focus of the work changed as we enter the phase of quality co-production? Are there boundaries to the construction of the BRI? How is the optimal investment size captured? A deep doctrinal grasp of these fundamental theoretical issues is essential to lead the work in practice and to boost the morale and confidence of those involved. Therefore, this book focuses on the theoretical system and the actualization path of the BRI, especially revolving around a high-quality development of the Initiative. It is divided into the following four parts:

Part one (from Introduction to chapter II)covers the theoretical basis. The part first rigorously argues for the necessary and sufficient conditions for the co-construction of the BRI, pointing out that the objective matching of supply and demand is a necessary condition, and that the co-construction of a benign interaction model and policy guarantee system based on the five channels is a sufficient condition for the mutual benefit and win-win situation of "the Initiative". Second, it summarizes the cooperation model of the BRI to achieve mutual benefit and win-win results and analyzes the benefit creation and sharing mechanism of "the Initiative". On this basis, finally, a theoretical model of economics is applied to discuss the boundaries of the BRI investment, and a three-country interaction decision model and expansion model is constructed, pointing out that the basis of the BRI for win-win cooperation lies in the highly complementary economic structures of the countries along the routes, and the core is to achieve benefit creation and sharing through project cooperation and construction, and arguing for the existence of the optimal scale of investment.

Part two (from Chapter III to Chapter V) focuses on the practice and systematically summarizes the construction results of the ten years of the BRI. This part first summarizes the fruitful results and effective experiences of the BRI from the perspective of the "Five Links" of policy communication, infrastructure connectivity, smooth trade, financial integration, and civic exchanges. It paints an all-round picture of the thriving construction of the BRI. On this basis, quite a few exemplary landmark projects resulting out of the efforts of the construction of the BRI are then analyzed. Their development history, status and significance are summarized in detail. Finally, in view of the widespread and far-reaching impact of the recent COVID-19 pandemic, a special discussion was presented on how the practice of the BRI can contribute to the world economy to counter the impact of the pandemic. How the "Initiative" can play a stabilizing role for the Chinese economy, the economies of the countries along the routes and the world economy through trade and direct investment.

Part three (including Chapters VI and VII) that takes from a policy perspective and provides a comprehensive analysis of the basic ideas, main requirements, and institutional and mechanism innovations for the co-construction of the BRI towards the stage of a high-quality development. This part begins with a thorough discussion of the basic ideas for the high-quality development of the BRI. The basic requirements for the high-quality development of "the Initiative" are proposed, including strengthening the docking of development strategies and policies, further promoting the interconnection of facilities, continuously expanding two-way trade and investment, and deepening exchanges and cooperation in key areas. In terms of institutional and mechanism innovation, special emphasis was placed on building a new system of higher level open economy, building strategic planning docking system and policy standard linkage platform, creating a four-in-one linkage system of land-sea-sky-internet, improving the construction of domestic supply chain and international industrial chain, improving diversified and standardized investment and financing system, and building a coordinating mechanism of green humanities exchange.

Part four (including Chapters VIII to Chapter XI) focuses on the path to the actualization of the BRI high-quality development. It specifically consists of the following four aspects: investment and financing, risk management, legal enforcement, cultural dissemination, and green development. First, in terms of investment and financing, the investment plan was discussed in terms of the scale, structure, and industrial pattern of investment. On this basis, how to promote the financing for the construction of the BRI was also discussed in terms of the main body of investment and financing, investment and financing methods, investment and financing cooperation and the scale of investment and financing. Second, in terms of risk management, it analyzes the potential risks of co-constructing the BRI and discussed several measures to prevent risks, with a focus on how to manage risks by optimizing project management. Especially in the legal aspect, it explains in detail how to promote the construction of the BRI by constructing and improving the legal system in three aspects: analyzing the legal environment, constructing the legal system, and improving the implementation of the law. Third, in terms of cultural dissemination, the importance of multi-stakeholder participation in promoting the dissemination of the BRI cultures was emphasized. Finally, in terms of green development, it was emphasized that firstly, the requirements should be clarified, and the criteria refined. Furthermore, the focus should be on optimizing energy structure and systemic governance of the industrial structure.

In the concluding remarks, it is pointed out that the BRI is not only a product of China's prudent consideration of the major changes in the international and domestic development environments at the new stage of the world's unprecedented changes across the past several centuries, but also the most important practical innovation and theoretical development of Xi Jinping's Socialist Thought with Chinese Characteristics for a New Era. A full understanding of the ideological and theoretical depth behind the BRI is the most important basis for the efforts to strengthen confidence, continuously improve policies, solidify cooperation, share the fruits of development, and continuously play the strong voice of the new era in co-constructing the BRI.

Acknowledgments

I received generous help from many people in wringing this book and would like to acknowledge them here.

Firstly, I would like to thank my project research team, which is mainly composed of post doctoral researchers, doctoral students, and master students under my guidanced. During the research period of several years, they overcame all kinds of difficulties and completed the basic work of the research project through their excellent work. From data collection, paper writing, to manuscript collation, they have condensed their hard work and sweat. Of course, they also received academic training in project research, improved their scientific research level, and laid a foundation for their future academic career. I would also like to congratulate them on their successful departure or their doctorate and master's degrees. They are Guo Chenglong, Zhang Jihuan, Deng Zhongqi, Gao Tingfan, Luo Lijuan, Yu Wenwen, Li Huanhuan, Chen Jiacheng, etc.

Secondly, I would like to thank the national social science planning office of China, which has given the project research and supported by the National Social Science Foundation of China, and the two universities, Renmin University of China and Guangdong University of Finance and economics, which have provided management services and support during the research period. Guangdong University of Finance and Economics gave me special support during the second half of the project research, which enabled the project research process and the translation and publication of the results to proceed smoothly.

Finally, I would like to thank my old friends, and the translation team and the press. I am deeply grateful to Dr. Jia Wenshan, Dean of Global Engagement Academy at Shandong University (Waihai), China, and Professor of Communication and Global Studies at Chapman University for his enthusiastic encouragement and recommendation to publish this book and include it in the Belt & Road Initiative: Interdisciplinary Perspectivs Book Series edited by him. I would also like to thank Qingdao Tengsheng Global Education and Culture Co., Ltd. for securing the translation of my book manuscript into English. Last, but not least. I thank the editors of Peter Lang who have provided a good guarantee for the final publication of this book.

The research and writing of this book were funded by the National Social Science Foundation of China and the Guangdong University of Finance and Economics with special thanks for the funding of the introduction and employment of high-level talents.

Just as the win-win cooperation inherent in the Belt and Road Initiative has created achievements in project construction, the research, writing and publication of this book are also the results of win-win cooperation among all the parties. I hope that after publication, with the participation of the broad masses of readers, there will be a win-win results just like the construction of the BRI projects. Thanks for reading, everyone.

Introduction

1. The historical meaning of the BRI and its new contemporary connotations

In September and October 2013, Chinese President Xi Jinping proposed the cooperation initiatives of building the "Silk Road Economic Belt" and the "21st Century Maritime Silk Road", or "the Belt and Road Initiative" (abbreviated as the BRI), respectively. (The "Belt and Road Initiative" concept first put forward by Xi Jinping, 2021, Feburary 1) The BRI that is an inheritance of the ancient Chinese Silk Road both on land and on sea renewed its connotations in the new era.

As shown in Figure 1, the "Silk Road Economic Belt" incorporates six international regional economic cooperation corridors, including China-Mongolia-Russia, the New Asia-Europe Continental Bridge, China-Central Asia-West Asia, China-Pakistan, China-Central and South China Peninsula, and Bangladesh-China-India-Myanmar. The "21st Century Maritime Silk Road" starts from the southeast coast of China and goes through Southeast Asia, the Indian Ocean, East Africa, and North Africa to Europe.

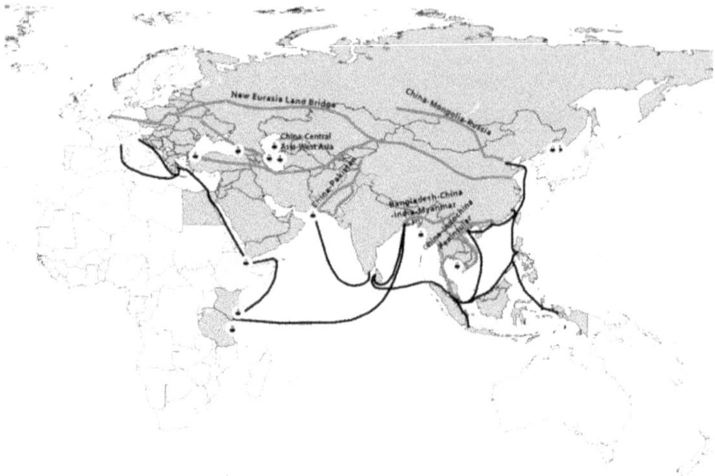

Figure 1 Silk Road Economic Belt and New Maritime Silk Road
Note: The red line is the "Silk Road Economic Belt", the black line is the "New Maritime Silk Road", and the economies marked in blue are countries along the BRI, which may not have signed cooperation agreements with China.
Source: World Bank (2019). *The Economics of the BRI: Opportunities and Risks of Transport Corridors*

(i) The historical Silk Road

The historical Silk Road refers to the overland trade alleyway opened by Zhang Qian on a mission to the West during the Western Han Dynasty (202–8 BC) in China, starting in Chang'an and reaching as far as the West Asian countries. By the time of the Eastern Han Dynasty (25–220), Ban Chao again sent a mission to the West to open the long-deserted Silk Road and extend the route to Luoyang, the capital of the Eastern Han Dynasty, at the eastern end and to Europe (Rome) at the western end. This is the complete historical route of the Silk Road. The Silk Road was named after the silk products that had the greatest impact on the goods shipped westward (China News Network, 2012, November 6).

It should be noted that the overland Silk Road was not a fixed route, but rather a transportation network consisting of several trunk and branch routes, the most prominent of which, starting from Chang'an (modern Xi'an), passed through Lanzhou, Jiuquan, and Yumen, crossed over Anzhi (modern Iran) and Tiaozhi (around modern Iraq), and reached the Great Qin (eastern part of the Roman Empire).

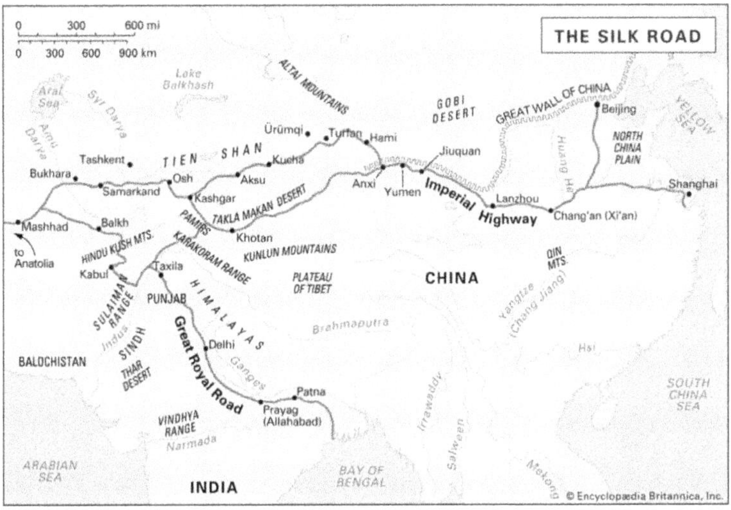

Figure 2 Ancient Overland Silk Road
Source of information: https://www.britannica.com/topic/Silk-Road-trade-route

The Silk Road in the broader sense also includes the Maritime Silk Road. The Maritime Silk Road was a maritime alleyway of transportation, trade and cultural contacts between ancient China and foreign countries. It was also known as the Maritime Ceramic Road and the Maritime Spice Road. The Maritime Silk Road began in the Qin and Han dynasties (221–220 BC), flourished in the Sui and Tang dynasties (581–907), expanded in the Song and Yuan dynasties (960–1368), and reached its peak in the Ming dynasty (1368–1644). The main Maritime Silk Roads trade route consists of the East China Sea trade route and the South China Sea trade route. The South China Sea trade route starts from the Chinese coastal port to Southeast Asia, South Asia, Arabia, and East Africa coastal countries. Its main port has changed over the generations and has shifted among the ancient ports of Xuwen, Guangzhou, Quanzhou, Zhangzhou, etc. The East China Sea trade route starts from the Chinese coastal port to Korea and Japan. Its main port is the ancient port of Qingdao, which can sail to the Liaodong Peninsula, the Korean Peninsula, and the Japanese islands.

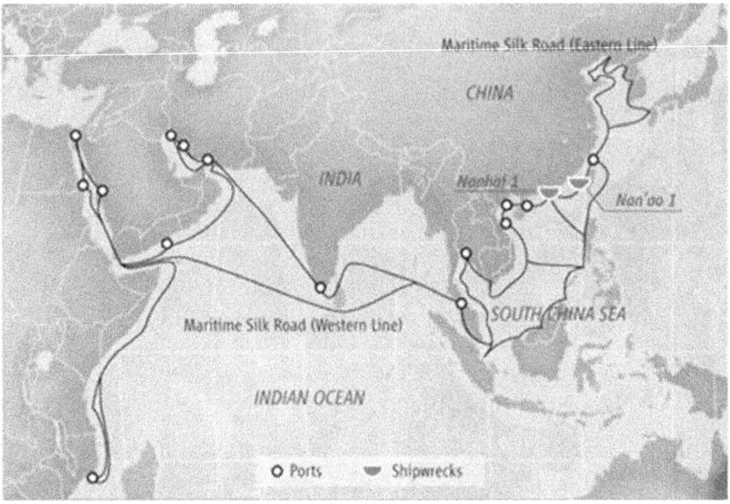

Figure 3 Ancient Maritime Silk Road
From "Unprecedented Excavation Brings Maritime Silk Road to Life" by Li Jiao, https://www.science.org/doi/10.1126/science.328.5977.424

(ii) Historical implications of the BRI

At the end of the 19th Century, German geologist Ferdinand Freiherr von Richthofen called the above-mentioned great alleyway linking East and West the "Silk Road". Based on years of research, the German geographer Hutson wrote a monograph, *The Silk Road* (China News Network, 2012, November 6). Since then, the term Silk Road has gained worldwide recognition for its historical underpinnings, which include the following.

First, the ancient Silk Road pioneered the opening of the Great East-West alleyway and built a large network of world transportation routes for the first time. The ancient Silk Road, which was crisscrossed and connected in all directions, is a miracle in the history of road transport in the world. Both large and small, uncountable Chinese and foreign traffic routes, formed the ancient Silk Road's "blood vein network", resulting in the construction of the basic pattern of the ancient Silk Road, the construction of the ancient Eastern and Western world interconnected transportation network, and became the most convenient alleyway between Asia and Europe.

Second, the ancient Silk Road significantly facilitated the great circulation of commodities, pioneering the interchange of trade and economic exchanges between the East and the West. The Silk Road was the lifeline of trade between

East and West in ancient times. Through the Silk Road, Chinese silk, tea, porcelain, lacquer ware and other commodities were exported to countries along the route incessantly; jewels, herbs, spices, and various crops such as grapes, caraway, walnuts, carrots and caramelis from Central Asia, West Asia and Europe entered China continuously. The change in the European name for "China" from "Seres" to "China", which took about a thousand years, and this is the portrayal of this historical process.

Finally, the ancient Silk Road has facilitated the exchange of diverse cultures and acted as an important link between different countries, races and civilizations of the East and the West, which immersed in and accommodated each other. The ancient Silk Road and the Maritime Silk Road are the road of civilization where different peoples and cultures interact and integrate with each other. The Silk Road spans dozens of countries in Asia, Europe, and Africa, linking and intermingling the ancient civilizations of China, India, Egypt, Persia, Arabia, Greece, and Rome.

(iii) The BRI as the New Silk Road of the present

As the Silk Road of the new era in the context of globalization, the importance of the BRI is at least threefold: 1) it provides a more convenient transportation alleyway for the flow of people, logistics and commerce under the conditions of economic globalization; 2) it provides a new development space for China's economic development and a platform for China's peaceful rise; 3) it provides a new paradigm for the development and mutually beneficial cooperation and win-win results for various countries and regions along the route; and 4) it provides a new way for the re-balancing development of the world economy.

First, the BRI is conducive to promoting the further free flow of production factors and products under economic globalization, strengthening the resource allocation role of the market mechanism, and deepening the globalization process. The manifestation of globalization is the realization of the flow of commerce, logistics and people on a global scale. The essence of which is that the market mechanism works on a global scale, and through the resource allocation and operational regulation of the market mechanism, the free flow of goods, people and capital across borders is constantly realized. Globalization requires individual countries to find their own niche in the global division of labor system. The flow of goods, cultural exchanges, and the movement of people on a global scale has taken on unprecedented proportions. Along with this process, there is a need to open the transport of goods and other factors of production alleyway and to break the real constraints of political factors on the other. This is the deep context in which the BRI was born.

The transport alleyway that needs to be opened contains both the tangible and the intangible. Taking the Eurasian Continental Bridge as an example, it starts from Lianyungang in China in the east, goes through Zhengzhou and Xi'an to Moscow, Warsaw, and Berlin, and terminates in Rotterdam in the Netherlands, with the Pacific Ocean on one end and the Atlantic Ocean on the other, transporting goods from east to west, which is the actual need for economic development. The BRI enables such interconnection. Compared to the ancient Silk Road, although the present BRI has the same course and scope, the development of transport technology has made it possible to build a more efficient and convenient transport alleyway. At the same time, the concept of mutual benefit and a foreign policy of peaceful development has contributed to the formation of a safer environment for the transport of commodities. In essence, this is a major transcendence.

At the same time, the improved international environment has made it possible to overcome political constraints. There were many political constraints in the past, with the Cold War period of the 1950s–1990s, when the Warsaw Pact, formed by the Soviet Union and the socialist countries of Eastern Europe, and NATO, formed by the United States and the capitalist countries of Western Europe, created a full-scale confrontation between the two camps on the European continent. Even if the historical Silk Road had existed, concepts like the BRI would not have been possible. In the new century, with the further development of free circulation in the era of globalization and the demand for global choice and allocation of economic activities, the opening of alleyways such as the Asia-Europe Continental Bridge is back on the agenda, and concepts such as BRI are likely to be proposed and implemented.

Secondly, the BRI, as the Silk Road of the new era, is of great significance to China's development. In terms of China's regional economy, three important strategies have been formed: One is the BRI strategy; another is the "Yangtze River Economic Belt" strategy; and another is the "Beijing-Tianjin-Hebei Cooperative Development" strategy. These three strategies interact and influence each other. The coordinated development of the economies of the three major city clusters in China, namely Beijing-Tianjin-Hebei, the Yangtze River Delta, and the Pearl River Delta, combined with the BRI external strategy, promotes coordinated internal and external development. It makes China's economic layout more rational, promotes more balanced economic development, creates better development momentum, and maintains medium-to-high economic growth.

Unlike the other two strategies that are mainly domestic-oriented, the BRI is the first choice for the new round of China's opening-up strategy because it is a two-pronged strategy both at home and abroad.

Finally, the BRI is conducive to overcoming some of the shortcomings of globalization, ensuring its correct direction of development and helping to achieve long-term balanced growth in the world economy. Important features of the current global economy are as follows: (1) the U.S. and Japanese economies have recovered, and a rising tendency has strengthened; (2) the debt crisis has slowed down and the European economy has grown moderately; (3) emerging markets have become important drivers of global economic growth; and (4) BRI countries have become important drivers of world economic growth. In the future, it remains a question whether the world economy will be able to achieve balanced development in the long term. In his book *Capitalism in the 21st Century* (Piketty, 2014), French economist Thomas Piketty argues that the biggest problem for the future of global development is the growing gap between the rich and the poor, due to the mechanism of return of capital over economic growth rates. Co-constructing the BRI can contribute to the long-term balanced growth of the world economy by reducing transaction costs, enhancing cooperation among countries, providing new growth drivers, and improving economic growth rates.

Therefore, China's BRI is not simply to reopen the historical Silk Road, but to give it more contemporary content in a new context. The great practice of co-constructing the BRI is making a new history.

2. The basic connotation of the BRI

The BRI is a product of China's prudent consideration of the major changes in both the international and domestic development environments at a new stage when the world is entering a century of unprecedented changes. It is the most important practical innovation and theoretical development of Xi Jinping's socialism with Chinese characteristics in the new era. *The Proposal of the Central Committee of the Communist Party of China on The Formulation of the 14th Five-Year Plan for National Economic and Social Development and the Visionary Goals for 2035* (The Central Committee of the Communist Party of China, 2020, November 3), includes a special chapter detailing the BRI high-quality development planning concept. On April 20, 2021, President Xi Jinping delivered an important speech at the Boao Forum for Asia, proposing a new development direction in the process of conveying a strong voice for the BRI, i.e., to make the

BRI a "road to poverty reduction" and a "road to growth", and to make a positive contribution to the common prosperity of mankind. (Xi Jiping, 2021) This prompts the building of the BRI and then leads to the new development stage. Since its inception eight years ago, the BRI has kept pace with the times, enriched its connotations, and demonstrated its vitality and creativity. As for the connotation of the BRI, we can grasp it concisely from Foreign Minister Wang Yi's explanation of the BRI at the briefing session of the Second International Summit on the BRI in April 2019. He noted: "the BRI is an international regional economic cooperation initiative put forward by President Xi Jinping. The core of which is to strengthen all-round connectivity with infrastructure construction as the main line, to tap new momentum for world economic growth and create a new platform for international regional economic cooperation." (Wang Yi, 2019). Wang's statement accurately indicates the general orientation, core content and construction objectives of the BRI.

First, the BRI is an international regional economic cooperation initiative. As the then Chinese Vice Foreign Minister Zhang Yesui pointed out at the China Development High Level Forum 2015, the BRI is a regional economic cooperation initiative, not a geopolitical tool. (Zhang Yesui, 2015) On the one hand, China's economy and the world economy are highly interrelated. Promoting the construction of the BRI is a need for China to further expand and deepen its opening up to the outside world, reflecting its determination to open its doors for construction and integrate deeply into the world economic system; on the other hand, against the background of the current slow global economic growth, rising protectionism and increasing uncertainties in the international environment, the co-construction of the BRI is also a need to strengthen mutually beneficial cooperation with countries along the BRI, aiming to promote a common development. China is willing to assume more responsibilities and obligations within its capacity, making greater contributions to the sustainable and balanced development of the global economy and injecting new growth impetus into the global economy.

Second, an important medium for the BRI cooperation is infrastructure projects. Infrastructure construction projects form the basis for supply-demand matching, with high capital requirements and strong investment-driving capacity for the regional economy. Moreover, according to estimates from international institutions such as the Asian Development Bank and the World Bank, there is a huge demand for infrastructure development in countries along the BRI. The investment gap is also very large. (Belt and Road Initiative Economics: opportunities and risks of Traffic Corridor, World Bank) China's BRI formally targets this priority need. *The Vision and Actions for Promoting the Construction of the Silk Road Economic Belt and the 21st Century Maritime Silk Road* (Ministry of Foreign

Affairs, 2015), jointly released by China's National Development and Reform Commission, the Ministry of Foreign Affairs, and the Ministry of Commerce on March 28, 2015, clearly states that infrastructure connectivity is a priority area for the construction of the BRI. Over the past eight years, the BRI has been promoting with the construction of infrastructure connectivity as its starting point and focus. Many major projects have been built jointly with countries along the routes, with China's state-owned enterprises alone investing in the construction of more than 3,000 major projects, with a considerable amount of investment focused on infrastructure construction. This has effectively improved the infrastructure situation in the relevant countries (Hu Biliang, 2021).

Finally, the purpose of the construction of the BRI is to tap new momentum for the world economic growth and create a new platform for international regional economic cooperation. The globalization led by Western countries since the 1980s, while bringing prosperous development of economic globalization, has also led to many developing countries facing various barriers such as lack of capital, backward technology, and limited market size to successfully integrate into the division of labor system of the global value chain. Although they have integrated into the global value chain division of labor system, they are facing the risk and dilemma of "low-end embedding" or even "low-end locking". The accumulation of the above-mentioned contradictions and problems has led to the current difficulties of sustainability in the further development of economic globalization. Against the backdrop of the urgent need to restructure and optimize global value chain, the BRI construction advocated by China, especially the "promotion of a high-quality development by co-constructing the BRI" was explicitly proposed at the Fifth Plenary Session of the 19th CPC Central Committee. It is committed to building an equal and open international regional economic cooperation platform following the principle of mutual consultation, construction and sharing, actively engaging in substantive docking and cooperation with other countries to co-construct the BRI on the basis of mutual respect and trust, and helping more developing countries to integrate into the global value chain's division of labour system, thereby promoting the healthy and sustainable development of economic globalization (Dai Xiang and Song Jie, 2021).

3. The effect of co-constructing the BRI

Since its inception in 2013 by the Chinese President Xi Jinping, the BRI has gone from scratch, from concept to blueprint, from vision to practice, and from "freehand brushwork" to "meticulous working brush". It has now gone through eight rounds of seasons. Over the past eight years, China has insisted that "though the BRI originated in China, the opportunities and results belong to the world" (*People's Daily*, 2018). This has aroused widespread reactions in the international community and received the active participation and support of many countries and international organizations around the world. It has become the most extensive and largest international cooperation platform and the most popular international public product in both the history of China and perhaps in the history of the world.

First, the international consensus continues to expand. As of August 2021, China signed 206 cooperation documents with 140 countries and 32 international organizations to build the BRI (Data retrieved from the Ministry of Commerce). The friends' circle of the BRI has spread across Asia, Africa, Europe, Oceania， and Latin America. The BRI and its core concepts have been written into relevant documents of the United Nations, the G20, APEC and other regional organizations, covering such areas as connectivity, investment, trade, finance, science and technology, society, humanities, people's livelihood and the sea, and a series of important agreements have been reached to expand the benefits of development achievements. The two BRI international cooperation summits were successfully held, including the second BRI international cooperation summit, which was attended by 38 heads of state and government and more than 6,000 foreign guests from more than 150 countries and 90 international organizations, and resulted in a total of 283 constructive outcomes in six categories.

Second, infrastructure development has been steadily progressing. Focusing on the framework of the "Six Corridors and Six Roads, Multiple Countries and Ports", landmark BRI projects have made substantial progress. The extensive cooperation between China and BRI countriess in the fields of ports, railways, highways, electricity, aviation, and communications has effectively improved the level of infrastructure construction in the countries along the routes. It shortened the transportation time of the countries along the routes, reduced trade costs, released trade and investment potential, and continuously provided new momentum for building a community of a shared future for mankind. The 2019 World Bank research report issued positive comments on this (World Bank, 2019).

Third, trade exchanges continue to grow. The total import and export of goods trade between China and BRI countriess increased from US$1.04 trillion in 2013 to US$1.35 trillion in 2020, with a cumulative total of over US$9.2 trillion in the past eight years; the proportion of China's trade with countries along the routes to the total foreign trade in the same period increased year by year, from 25% in 2013 to 29.1% in 2020.

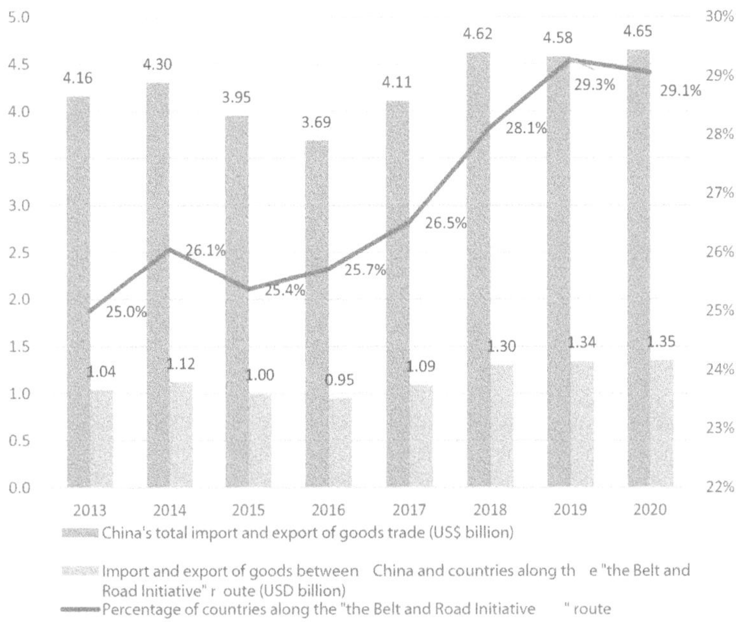

Figure 4 Chinese enterprises' direct total investment to BRI countries, 2013–2020 (unit: USD billion)
Source: National Bureau of Statistics, *China BRI Trade and Investment Development Research Report 2021*

Fourth, mutual direct investment between China and BRI countriess has increased steadily. Chinese enterprises' direct total investment in BRI countriess increased from US$12.63 billion in 2013 to US$22.54 billion in 2020. China's cumulative direct investment in the BRI countries reached about US$140 billion in the eight years; the proportion of China's total investment to the BRI countries in its outward direct total investment during the same period remained roughly stable at 12%; the cumulative number of new enterprises established in China by the BRI countries reached 27,000, and the cumulative actual investment was about US$60 billion.

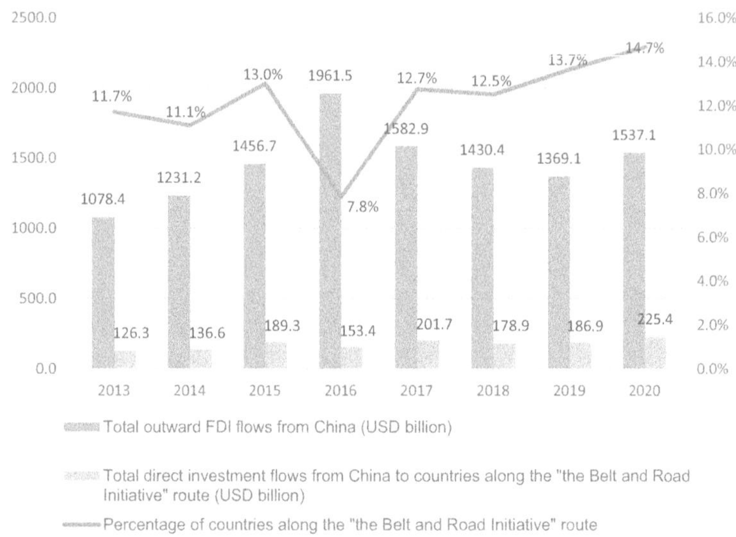

Figure 5 Chinese enterprises' direct total investment to BRI countries, 2013–2020
Source: *Statistical Bulletin on China's Outward Foreign Direct Investment*, 2013–2020

Fifth, promoting the building of a human community of a shared future. Over the eight years of the BRI construction, it has a total investment about US$136 billion and trade in goods of more than US$9 trillion. (CCTV network, 2019, April 29) China's investment in the BRI countries has created new tax and employment channels for these countries and promoted local economic development. By the end of 2020, China had invested nearly US$40 billion in overseas economic and trade cooperation zones built in the BRI countries, paying host countries more than US$4.4 billion in taxes and fees, and creating 330,000 local jobs. (*People's Daily*, 2021) In addition, in the face of the sudden outbreak of the new COVID-19 pandemic, China has done its utmost to prevent and control the pandemic itself, while actively cooperating with many countries and international organizations, including those along the BRI routes, in the fight against the pandemic. Until now, China has provided emergency assistance to 150 countries and 13 international organizations to combat the pandemic. It has exported pandemic prevention materials to more than 200 countries and regions. (China Website, 2021, August 23)

4. Analysis of the theoretical model of win-win cooperation in co-constructing the BRI

Over the past eight years, the theoretical system of the BRI has evolved. Its main contents can be summarized in the following six aspects. (1) Conceptual innovation: we must uphold the principle of mutual consultation, construction and sharing, and adhere to the concept of openness, greenness, and integrity. (2) Clear standards: achieving high standards, benefiting people's livelihoods and sustainability. (3) New approach: rely on the multilateral cooperation mechanisms of China and the countries concerned. It should make use of proven regional cooperation platforms, take the initiative to develop regional economic cooperation partnerships with countries along the routes, and realize the "Five Links", i.e., policy communication, infrastructure connectivity, smooth trade, financial integration, and civic exchanges. (4) Clear path: from free outlining to meticulous brushwork, from realism to depth, to achieve a high-quality development. (5) Clear goal: to jointly build a community of interests, destiny and responsibility with mutual political trust, economic integration, and cultural inclusion, fully demonstrating the main theme of "the world fosters China and China gives back to the world", with the overall goal of building a community of mankind for a shared future. (6) The ten paths: to strive to build the BRI into a path of peace, prosperity, openness, green, innovation, civilization, integrity, health, growth, and poverty reduction.

Looking at the great achievements of the BRI over the past eight years, it is easy to find that a general theoretical framework is embedded in it: to achieve the construction goal of the BRI to tap new momentum and build a new platform, a combination of "necessary" and "sufficient" conditions needs to be achieved in concrete terms. The Party Central Committee, with President Xi Jinping at its core, is the perfect expression of China's wisdom in coordinating the two major international and domestic situations, responding to the trend of regional and global cooperation, meeting the development needs of countries and regions along the routes, and combining the necessary and sufficient conditions based on the present and with a long-term perspective.

The objective match between supply and demand in the economic structure of China and the partner countries is a necessary condition for the co-construction of the BRI to achieve win-win cooperation. From the perspective of China's supply factors, in the new normal economic situation, China's investment and consumption demand growth is relatively slow, and the production capacity is rich and needs new market space; China's manufacturing industry, especially infrastructure project construction technology has reached the world's leading level; the huge foreign exchange reserves can provide relatively abundant funds to guide the search for sound investment projects and opportunities, so it can realize the inherent requirements of China's economic structural transformation through the co-construction of the BRI; and from the perspective of the demand of Asian countries represented by Central and West Asia, the annual demand for infrastructure construction funds in Asia reaches 800 billion U.S. dollars, and financial gap is about 520 billion U.S. dollars per year. (Asian Development Bank, 2017, Feburary 28) Investment in infrastructure has positive spillover effects. The BRI can support and drive regional economic development. It acts as a necessary path to industrialization and modernization in developing countries. China's supply matches exactly the demand of the developing countries. The match between supply and demand in this economic structure is the economic basis for co-constructing the BRI. It is the objective condition for why the BRI is "possible".

The basic model of cooperation can be illustrated by the example of cooperation between China and an Asian country in investment and construction. China provides a low interest rate loan of 10 billion dollars to an Asian country for infrastructure projects. This country uses this low interest rate loan to buy Chinese infrastructure products and labor. Chinese companies receive average profits and achieve rational application of domestic capacity and preservation of foreign exchange reserves. Since infrastructure has a long-term effect on economic development, the participating country can use Chinese funds, technology, and production capacity for infrastructure development. It can use the proceeds from the projects and increased government tax revenue to repay the initial low interest rate loan in the future, while laying the material foundation for future industrialization and modernization. This is a "win-win" situation. This model is equally suitable for China's cooperation with African countries. Infrastructure development can spur the development of numerous developing countries and regional economies in Africa. African countries will benefit greatly from their participation in the co-construction of the BRI.

This basic model of "win-win" for China and other cooperating countries can also be extended to a "multilateral win-win" model with the participation of European and North American countries. As European countries have been facing economic growth pressure in recent years, the major European manufacturing countries represented by Britain, France and Germany can rely on their technological advantages to participate in the co-construction of the BRI. Through participation in the construction of projects to gain access to a wide market in Asia, it has promoted economic growth. At the same time, the "multilateral win-win" model could also be incorporated into North American countries. North American countries, represented by the United States and Canada, can participate in the construction of the BRI through their economic and technological advantages, gaining access to the vast markets of Asia and Africa, relieving the pressure of economic growth, strong market demand and technological spillover to promote the further upgrading of the economies of these countries, benefiting the world's major developed countries. China is uniting the economic and technological advantages of developed countries in project cooperation to continuously improve manufacturing and independent innovation. But this model of cooperation needs to be predicated on good political relations between China and these countries.

Co-constructing the BRI can help re-balance the world economy. The global macro-economy is facing an oversupply and under-demand, with an estimated global demand gap of around one trillion dollars. On August 17, 2016, Xi Jinping pointed out at a central symposium on promoting the construction of the BRI that taking the construction of the BRI as an opportunity to carry out cross-border connectivity. It improves the level of trade and investment cooperation and promotes international cooperation in production capacity. Equipment manufacturing is essentially a way to generate new demand and achieve the goal of re-balancing of the world economy by increasing effective supply (Xi Jinping, 2016, August 17), Especially in the current situation where the world economy continues to be in the doldrums due to the outbreak of the pandemic, it will be conducive to stabilizing the current world economic situation if the huge production capacity and construction capacity formed under the pro-cyclical cycle can be made to go out to meet the urgent needs of countries along the routes to promote industrialization, modernization and improve infrastructure. If China were to invest US$20 billion/year in the future BRI projects, a rough estimate based on a 1:5 investment multiplier could generate US$100 billion in total demand, which is expected to cover about one tenth of the total demand gap in the world economy (Calculated by my research team).

The economic essence of the BRI is to create and realize "dividends". The dividends are manifested at the micro level in win-win cooperation and multilateralism, and at the macro level in contributing to global economic balance. To obtain dividends, we must work together and launch an open cooperation at the national level; at the enterprise level, we must be sincere and united, carefully organized and execute a rigorous and detailed division of labor.

In summary, to achieve the inherent goal of win-win cooperation in building the BRI, it is necessary to provide a series of sufficient conditions to meet the needs of building the BRI, the basic content of which is mainly to form a benign mode of interaction between China and the countries along the routes and an adequate and effective policy guarantee system, so as to realize the combination of the two conditions and make the building of BRI truly "possible". This is the "spine" of what we have done and what we will do in the future.

5. Concrete paths for co-constructing a high-quality BRI in the 14th Five-Year Plan

Based on the above theoretical analysis, further improvement of the sufficient conditions for the construction of the BRI is the central task in promoting a high-quality development of the BRI in the future. Therefore, the *Proposal of the Central Committee of the Communist Party of China on Formulating the 14th Five-Year Plan for National Economic and Social Development and the Visionary Goals for 2035* (hereinafter referred to as the *Proposal*), which was adopted at the Fifth Plenary Session of the 19th CPC Central Committee, pointed out the basic tasks for promoting the BRI high-quality development during the 14th Five-Year Plan period. In accordance with the guiding ideology and working guidelines set out in the *Proposal*, the 14th Five-Year Plan adopted by the National People's Congress this year includes a special Chapter 41 on *Promoting the Co-construction of the BRI and the Co-construction of a High-quality Development*, which provides a specific plan for the high-quality development of the BRI. Three of these efforts are most important.

First, interconnection is a key foundation and a top priority. Infrastructure is the cornerstone of connectivity and a bottleneck to the development of many developing countries. The construction of a high-quality, sustainable, risk-resistant, reasonably priced, inclusive, and accessible infrastructure is conducive to countries giving full play to their resource endowments. Better integrating into the global supply chain, industrial chain, and value chain, and achieving linkage development, is the basis for the international community's recognition and active participation in the BRI and the key to the success of the BRI construction. China should continue its efforts with the cooperating parties to build an interconnection network led by the New Asia-Europe Continental Bridge and other economic corridors. With the China-Europe Train, the New Land and Sea alleyway and other major alleyway and information highways as framing, with railways, ports, and pipeline networks, and with "Six Corridors and Six Roads with Multiple Countries and Ports", it aims to build a global partnership for interconnection to achieve common development and prosperity.

Second, build a good investment and financial platform. It continues to play the role of the BRI special loan, the Silk Road Fund and various special investment funds, develop Silk Road-themed bonds and support the effective operation of the Multilateral Development and Financing Cooperation Centre. While the Asian Infrastructure Investment Bank's investment projects lead the way in the construction of the BRI, multilateral and national financial institutions are welcome to participate in the co-construction of the BRI investment. Financing and cooperation in the third-party market is encouraged to achieve the goal of mutual benefit through the participation of multiple parties.

Thirdly, in-depth theoretical research should be carried out to promote cultural exchanges. The co-construction of the BRI is a comprehensive, strategic, systematic, international, and long-term project. At present, we have numerous major research projects on the construction of the BRI. Combining theoretical and policy research so that project research can form a synergy and provide a higher level of results is a direction for future research.

From the perspective of further providing sufficient conditions for the co-construction of the BRI, the group proposes ten specific policies for the high-quality co-construction of the BRI in the 14th Five-Year Plan, based on the above-mentioned *Proposal* and the *Plan:*

(1) To build a higher-level and open economic system and comprehensively improve the level of opening to the outside world;
(2) to formulate a special national plan for the co-construction of the BRI during the 14th Five-Year Plan, and to systematically deploy the future development of the BRI from the strategic overall perspective;
(3) to build an international industrial chain to promote domestic and international circulation;
(4) it is suggested that the future BRI International Cooperation Summit Forum should set up separate forums in Xinjiang, Fujian and Hainan to promote the formation of the pattern of pluralistic humanistic exchanges;
(5) to strengthen the digital level of connectivity and grasp the innovative dividend of the development of digital economy;
(6) to build an investment and financing platform in line with the international capital market, actively learn from the operational experience of the international capital market, and absorb international capital forces;

(7) to establish an open and transparent mechanism for the release of project construction information to promote the construction of the BRI into a clean road;
(8) to introduce third-party organizations to evaluate the debt sustainability of countries in cooperative construction, in response to the international community's high concern about the debt problems of countries along the routes;
(9) to establish a coordinating organization to form a pluralistic and interactive cultural exchange mechanism to better cooperate with the high-quality development direction of BRI;
(10) to earnestly cultivate the soft power of the BRI and tell a good story about the BRI.

References

China's Belt and Road Initiative Trade and Investment Development Research report, The second "Belt and Road Initiative" Summit Forum on International Cooperation has achieved fruitful results (2019, April 29). CCTV network. http://www.gov.cn/xinwen/2019-04/29/content_5387629.htm.

China provides emergency anti-epidemic assistance to 150 countries and 13 international organizations (2021, August 23). China Website,. http://news.china.com.cn/2021-08/23/content_77709181.html.

Dai Xiang, Song Jie. (2021). 'The optimization effect of global value chain of the "Belt and Road" Initiative: A perspective of the improvement of global value chain division of labor in the participating countries along the route', China Industrial Economics, (06), 99–117.

Foreign investment should not only pursue large-scale "going out", but also go out at a higher level (2021, October 29). People's Daily, https://www.chinca.org/cica/info/21102914052311.

Hu Biliang. (2021). "Eight years of BRI: Strengthening connectivity to achieve win-win cooperation", Guangming Daily, October 22, 2021, p. 12.

"Meeting the Infrastructure Needs of Asia" (2017, Feburary 28). Asian Development Bank (ADB). https://www.adb.org/zh/news/asia-infrastructure-needs-exceed-17-trillion-year-double-previous-estimates.

National Development and Reform Commission, etc. (2015, March 25). Vision and actions of the Silk Road Economic Belt and the 21st Century Maritime Silk Road Ministry of Foreign Affairs, https://www.yidaiyilu.gov.cn/yw/qwfb/604.htm

Piketty, T. (2014). Capital in the twenty-first century. Harvard University Press.

The 5th Anniversary of the Belt and Road Initiative put forward by President Xi Jinping: The Great practice of Building a Human Community with a Shared Future (2018, October 15), People's Daily, http://www.gov.cn/xinwen/2018-10/05/content_5327979.htm.

The Central Committee of the Communist Party of China. (2020). The proposal for the formulation of the 14th five-year plan (2021–205) for national economic and social development and the long-range objectives through the year 2035. Xinhua News Agency, http://www.gov.cn/zhengce/2020-11/03/content_5556991.htm.

The Silk Road. (2012, November 6), In China News Network. https://www.chinanews.com.cn/hwjy/2013/11-06/5471748.shtml.

Wang Yi. (2019, May 1). Start a new journey of high-quality development of "Belt and Road Initiative", Qiushi, http://www.gov.cn/guowuyuan/2019-05/01/content_5388150.htm

World Bank. (2019). Belt and Road economics: Opportunities and risks of transport corridors. The World Bank.

Xi Jinping delivered an important speech (2016, August 17). At the forum on promoting the construction of "Belt and Road Initiative", Presided over by Zhang Gaoli, Xinhua News Agency, http://www.gov.cn/guowuyuan/2016-08/17/content_5100177.htm

Zhang Yesui. (2015, March 25). At the group meeting of the China Development High-level Forum on "Building the Land and Sea Silk Road", Ministry of Foreign Affairs, https://www.mfa.gov.cn/ce/cohk/chn/zt/ydly/t1140481.htm

1

Necessary and Sufficient Conditions for the Actualization of the BRI

Introduction

Over the past eight years, the theoretical system of the BRI has evolved (The 5th Anniversary of the Belt and Road Initiative put forward by President Xi Jinping: the Great Practice of Building a Community with a Shared Future for Mankind, 2018), and its main contents can be summarized in six areas. (1) Conceptual innovation: we must uphold the principle of mutual consultation, construction and sharing, and adhere to the concept of openness, greenness, and integrity. (2) Clear goal: to achieve high standards, benefit people's livelihoods and be sustainable. (3) New approach: rely on the multilateral cooperation mechanisms of China and the countries concerned, make use of proven regional cooperation platforms, take the initiative to develop regional economic cooperation partnerships with countries along the routes, and realize the "Five Links", i.e., policy communication, infrastructure connectivity, smooth trade, financial integration, and civic exchanges (Xi Jinping, 2017, March 22). (4) Clear path: from free outlining to meticulous brushwork, go real and deep to achieve a high-quality development. (5) Clear goal: to co-construct a community of interests, shared future and responsibility with mutual political trust, economic integration, and cultural tolerance, fully demonstrating the main theme of "the world foster China and China gives back to the world". (6) The Ten Paths: to strive to build the BRI into a path of peace, prosperity, openness, green, innovation, civilization, integrity, health, growth, and poverty reduction.

Based on eight years of construction practice, a general theoretical framework can be further refined: to achieve the construction goal of the BRI, to tap new momentum and build a new platform, a combination of "necessary conditions" and "sufficient conditions" is required.

1. Objective matching of supply and demand is a necessary condition for achieving win-win cooperation

The objective match between supply and demand in the economic structure of China and the co-construction countries is a necessary condition for the co-construction of the BRI to achieve win-win cooperation. In terms of China's supply factors, firstly, in the new normal economic situation, China's investment and consumption demand growth has slowed down relatively, and capacity affluence requires new market space. Particularly after more than 30 years of development, China's economy has become the world's second largest economy, and the economy has entered a new development stage. China seeks to transform from its past strategy of "international circulation" to a new development pattern in which the domestic circulation is the mainstay and both international and domestic circulations promote each other (Yu Yongding, 2020) At the same time, due to the development inertia of the "international circulation" strategy, China has formed a huge production capacity. In the strategic push towards a new development pattern of a two-cycle transition period, China's huge production capacity needs to find effective markets and be fully utilized. Secondly, China's manufacturing industry, especially in the construction of infrastructure projects, has reached a world-leading level of technology. For example, China's technologies such as high-speed rail, dynamic trains and nuclear power plant manufacturing all have a global advantage in terms of overall cost performance. In addition, China has rich experience in infrastructure construction, based on which it has cultivated a group of construction and construction teams with both high construction and construction team management capabilities and skilled construction experience, low construction costs and high service quality that make China's infrastructure construction technologies particularly suitable for developing countries with a relatively weak development base, and Chinese infrastructure technologies have a strong competitive advantage. Finally, China's huge accumulated foreign exchange reserves can guide the search for sound investment projects and opportunities by providing relatively abundant funds and can realize the inherent requirements of China's economic structural transformation through the co-construction of the BRI. As China has followed the strategy of "international circulation" and developed an export-oriented economy in the past, a large trade surplus was formed in the process, and with the foreign exchange system policy of compulsory settlement and sale of foreign exchange, China has accumulated a large amount of foreign exchange reserves and has a rich capital element, which matches the high financing demand of infrastructure construction project under the BRI and can serve as a financial precursor basis for the construction of the BRI. Thus, China's objective economic structure lays a solid foundation for China's ability to supply the three elements of abundant capacity, proven technique and abundant capital.

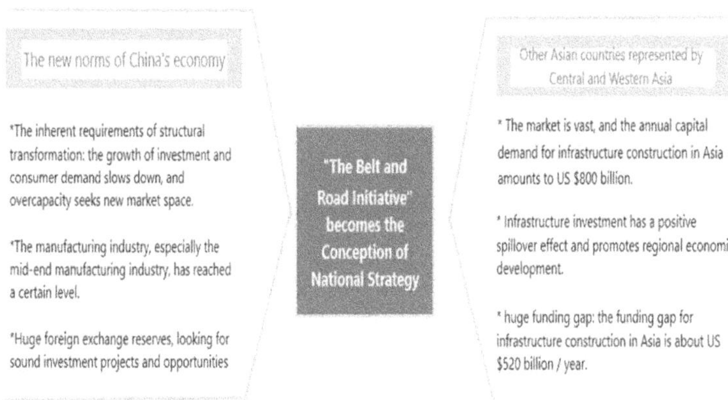

Figure 1.1 Economics supply and demand analysis of the BRI strategy (Chen Yongjun 2016)

Looking at the needs of Asian countries, represented by Central and West Asia, the annual capital needs for infrastructure development in Asia amount to US$800 billion, financial gas about US$520 billion/year (Asian Development Bank, 2017, February 28). The two ends of the Eurasian maritime and land Silk Road economic belt are the developed economic spheres of East Asia and Western Europe, while the middle countries of Central Asia, West Asia and Central and Eastern Europe are mostly in the early stages of industrialization, with weak economic infrastructure and infrastructure development. Investment in infrastructure has positive spillover effects, can support, and drive regional economic development. It is a necessary path to industrialization and modernization in developing countries. In addition, the vision of integrating into the global open economy system, sharing the dividends of globalization, and further unleashing the dynamism of national economic growth has created a huge demand for interconnected infrastructure in a wide range of developing countries. It is worth noting that infrastructure investment is characterized by large investment costs and long investment cycles, and that the funds that can be raised through the World Bank, the Asian Open Bank and these countries' own financing and social funds are limited, leaving a large gap in investment funds, thus creating a huge demand for funds.

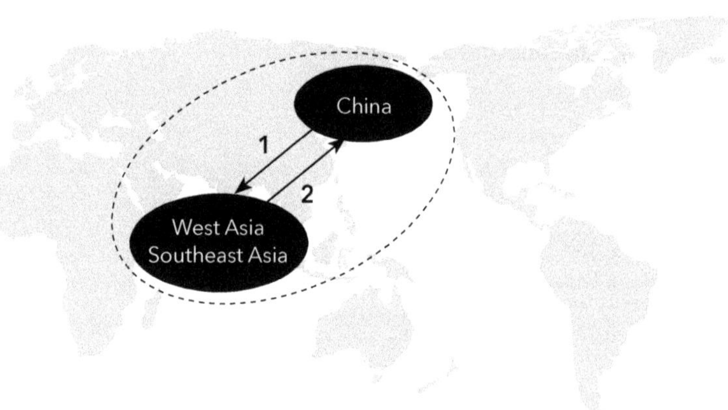

1

- China providing low interest loans to other Asian countries, and carrying out infrastructure construction.
- China's funds mainly come from huge foreign exchange reserves, local version of the Silk Road Fund from provinces along the Belt and Road, and social capital absorbed by creditor's rights and other forms.
- Export of infrastructure construction includes highway, railway, high-speed rail, electric power, and other industries, and will also drive the capacity export of signal communication, construction machinery, and other related domains.

2

- The backward infrastructure construction of West Asian countries seriously obstructs the local economic development and provides broad markets for the mid-end manufacturing industry in China.
- Asian countries, represented by West Asia, drive economic development through infrastructure construction, using it to repay low interest loans.
- Result: achieving "win-win" cooperation.

Figure 1.2 Win-Win economic interaction between China and Asian Countries – Basic Model I (Chen Yongjun 2016)

China's supply matches exactly the demand of developing countries. The match between supply and demand in this economic structure is the economic basis for co-constructing the BRI, and it is the objective condition for why BRI is "possible".

The basic model of cooperation can be illustrated by the example of cooperation between China and an Asian country in investment and construction. China provides low interest rate loan 10 billion dollars to an Asian country for infrastructure projects; this country uses this low interest rate loan to buy Chinese infrastructure products and labor. Chinese companies receive average profits and achieve rational application of domestic capacity and preservation of foreign exchange reserves. Since infrastructure has a long-term effect on economic development, the participating country can use Chinese funds, technology, and production capacity for infrastructure development, and can use the proceeds from the projects and increased government tax revenue to repay the initial low interest rate loan in the future, while laying the material foundation for future industrialization and modernization. This is a "win-win" situation. This model is equally suitable for China's cooperation with African countries. Infrastructure development can spur the development of more than 100 developing countries and regional economies in Africa. African countries can benefit greatly from their participation in the co-construction of the BRI.

This basic model of "win-win" for China and other cooperating countries can also be extended to a "multilateral win-win" model with the participation of European and North American countries. As European countries have been facing economic growth pressure in recent years, the major European manufacturing countries represented by Britain, France and Germany can rely on their technological advantages to participate in the co-construction of the BRI, and through participation in the construction of projects to gain access to a wide market in Asia, which has promoted economic growth. At the same time, the "multilateral win-win" model could also be incorporated into North American countries. North American countries, represented by the United States and Canada, can participate in the construction of the BRI through their economic and technological advantages, and gain access to the vast markets of Asia and Africa to ease the pressure of economic growth. Strong market demand and technology spillovers have driven further economic upgrading in these countries, benefiting the world's leading developed countries. China is uniting the economic and technological advantages of developed countries in project cooperation to continuously improve manufacturing and independent innovation. But the implementation of this model of cooperation needs to be predicated on good political relations between countries.

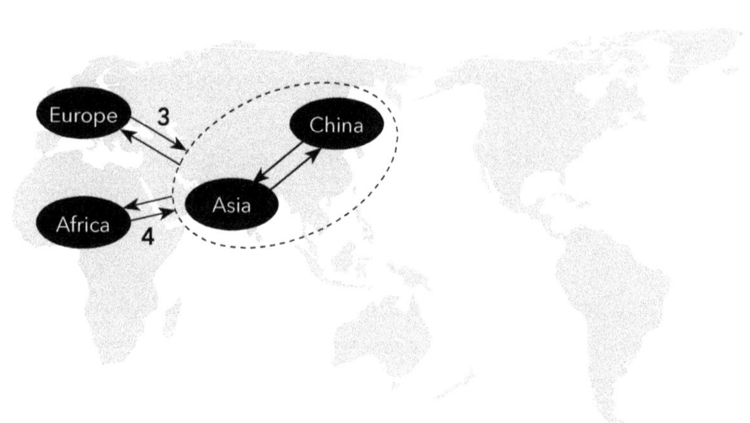

3 **European manufacturing powers represented by England, France, and Germany participate in the Belt and Road strategy with their own technological superiority.**
- All European countries face the pressure of economic growth.
- Making use of their project advantage, European countries obtain broad Asian markets through participating in projects.
- China learns advanced European technology through project cooperation, constantly improving its local manufacturing industry.

4 **African countries will become the beneficiary countries in the next phase.**
- Development of African countries is generally behind other Asian countries.
- Development of regional economy driven by infrastructure construction will bring benefit to all developing African countries.
- Result: multilateral win-win cooperation.

Figure 1.3 China and African and European countries to achieve multilateral win-win-expansion model (I) (Chen Yongjun 2016)

As can be seen from the above two cooperation models, matching supply and demand is necessary to create a win-win cooperation and benefit-sharing mechanism. Along with the high-quality development of the BRI construction, including China, it can achieve win-win and multilateral win-win for countries along the route. China has achieved the goal of strategic transformation under the new development stage, and countries along the route have reduced the infrastructure construction gap, realized interconnection, and shared the globalization dividend of optimizing the efficiency of resource allocation and broad market space, which presents a macroscopic view of injecting vitality into the green and sustainable growth of the world economy and helping to re-balance the world economy.

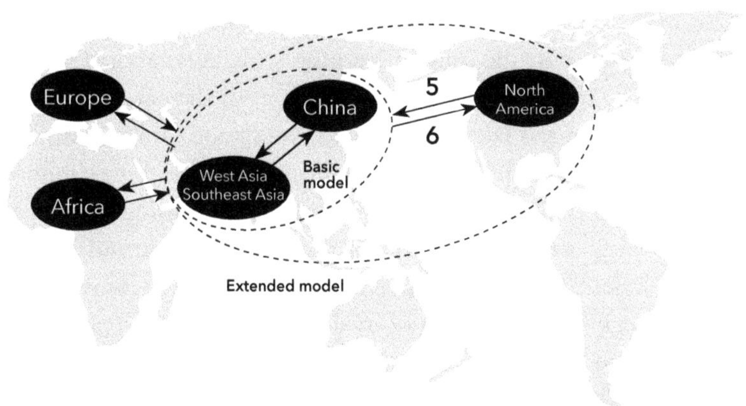

5 **North American countries represented by America and Canada participate in the Belt and Road Strategy with their economic technological advantage.**
- North American countries also face pressure of economic growth.
- Making use of their own economic advantage, North American countries obtain broad Asian markets through participating in project cooperation.
- China unites with the economic technological advantage of North American countries through project cooperation, constantly improving its manufacturing industry and level of independent innovation.

6 **North American countries will become the beneficiaries of the Belt and Road Initiative**
- Development level of North American countries generally higher than Asian countries, and have the capability of participating in Belt and Road project construction.
- Regional economic development driven by infrastructure construction will bring benefit to major developing countries in the world.
- Powerful market demand and technological spillover effect will drive the economy of North American countries to further upgrade.

Figure 1.4 China and African, European, and North American countries to achieve multilateral win-win-expansion model (II) (Chen Yongjun 2016)

First, let us take a Chinese perspective. The full utilization of the rich capacity formed under China's pro-cyclicality helps smooth the transition of China's industrial structure during the transformation and upgrading period and reduce the shocks and pains during the transition period. In the process of actively undertaking the basic construction project of the countries along the route, Chinese enterprises "go out" in a larger and deeper extent. Under the test of fierce competition in the international market, Chinese enterprises have accumulated richer and more advanced technical knowledge, management experience and market experience in the mutual learning with international competitors, which is conducive to the upgrading and iteration of the enterprise as a whole competitive power and has laid a solid micro foundation for the transformation and upgrading of the industrial structure at the meso level and the climbing up of the global industrial chain status. The BRI plays a role in bridging the gap from the "international grand cycle" strategy to the dual domestic and international cycle strategy under the new development pattern, helping China to smoothly achieve the strategic transformation of its development mode, optimize its economic structure and transform its growth momentum.

Second, from the perspective of the countries along the route. The countries along the route can be divided into two categories according to their economic bases: developing countries, represented by Asian and African countries, and developed countries, represented by European and North American countries. In the process of building the BRI, Asian and African countries can enjoy China's low-interest capital support, improve domestic infrastructure development, integrate into the global economic system based on their domestic connectivity, actively participate in cross-border trade and investment, unleash economic growth, and achieve development goals. At the same time, developed countries share the vast market along the BRI route with China, bringing into play their comparative advantage in technology, participating in the competition of the BRI projects, obtaining greater market share and economic profits, and cultivating new growth points for economic growth.

Finally, the co-construction of the BRI can contribute to the re-balancing of the world economy. The global macro-economy is facing an oversupply and under-demand, with an estimated global demand gap of around one trillion dollars. On August 17, 2016, Xi Jinping pointed out at a central symposium on promoting the construction of the BRI that taking the construction of the BRI as an opportunity to carry out cross-border connectivity, improve the level of trade and investment cooperation, and promote international cooperation in production capacity and equipment manufacturing is essentially a way to generate new demand and achieve re-balancing in the world economy by increasing effective supply. Especially in the current situation where the world economy continues to be in the doldrums due to the outbreak of the pandemic, it is conducive to stabilizing the current world economic situation if the huge production capacity and construction capacity formed under the pro-cyclical cycle can be made to go out to meet the urgent needs of countries along the route to promote industrialization, modernization and improve infrastructure. If China were to invest US$20 billion/year in the future BRI projects, a rough estimate based on a 1:5 investment multiplier could generate US$100 billion in total demand, which is expected to cover about one tenth of the total demand gap in the world economy. (Calculated by my research team)

Therefore, the economic essence of the BRI is to create and realize "dividends". The dividends are manifested at the micro level in win-win cooperation and multilateralism, and at the macro level in contributing to global economic balance. To reap the dividends, it is necessary to be open and cooperative at the national level and to work together. At the enterprise level, it is necessary to be sincere and united, carefully organized, precisely constructed, and finely divided.

In summary, to achieve the inherent goal of win-win cooperation in building the BRI, it is necessary to provide a series of sufficient conditions to meet the needs of building the BRI, the basic content of which is mainly the formation of a benign mode of interaction between China and the countries along the route and an adequate and effective policy guarantee system, so as to realize the combination of the two conditions and make the building of the BRI really workable. This is the "main stem" of what we have done and what we will do in the future.

2. Building a virtuous interaction model and policy guarantee system is a sufficient condition for constructing the BRI

Based on the above theoretical analysis, further improvement of the sufficient conditions for the construction of the BRI is the central task in promoting the high-quality development of the BRI in the future. In other words, given the argument that the BRI is a necessary condition, it is a gold mine to be excavated, and the ability to convert the gold mine to be mined into real gold is a sufficient condition.

The improvement of adequate conditions is first and foremost reflected in the implementation of the "Five Links". In September 2013, President Xi Jinping put forward the initiative of jointly building the Silk Road Economic Belt in a speech at the University of Kazakhstan, emphasizing that "we can start with the following areas, leading from point to point, from line to line, gradually forming a regional cooperation". For the first time he proposed the "Five Links" of policy communication, infrastructure connectivity, smooth trade, financial integration, and civic exchanges. (Xi Jinping, 2017, March 22) From 2013 to the present, the BRI has been fruitful, focusing on the improvement of the Five Links.

In terms of policy communication, under the BRI framework, participating members have actively engaged in dialogue on policy docking, established and developed a multi-level, multi-field, and multi-modality policy dialogue mechanism, and built a policy dialogue platform that is widely participatory, effective, and fruitful. As of August 2021, China has signed 206 cooperation documents with 140 countries and 32 international organizations to build the BRI. (Xinhua News Agency, 2021, December 16) The BRI circle of friends has spread across Asia, Africa, Europe, Oceania and Latin America, and the BRI and its core concepts have been written into relevant documents of the United Nations, the G20, APEC and other regional organizations. For example, two Belt and Road Summits on International Cooperation were successfully held, including the second Belt and Road Summit, which was attended by 38 heads of state and government and more than 6,000 foreign guests from more than 150 countries and 90 international organizations, resulting in a total of 283 constructive outcomes in six categories and remarkable policy communication (CCTV network, 2019, April 29).

In road connectivity, focusing on the main framework of the "Six corridors, Six Roads, Multiple Countries and Ports", numerous BRI landmark projects have made substantial progress. The extensive cooperation between China and BRI countries in the fields of ports, railways, highways, electricity, aviation, and communications has effectively improved the level of infrastructure construction in the countries along the route, shortened the transportation time of the countries along the route, reduced trade costs, released trade and investment potential, and continuously provided new momentum for building a human community of shared future. 2019 World Bank Research Report gave this a positive review. As of December 2020, China-Europe Classes, for example, have opened 73 operational routes, connecting 92 cities in 21 countries, and running more than 30,000 trains in total, becoming an important platform for countries along the route to promote interconnection and enhance economic and trade cooperation. (Xinhua News Agency, 2021, January 19) Especially under the impact of the Covid-19 Pandemic, the uncertainty and cost of air, road, port, and other modes of transport have increased, and the China-European train of railway transport stands out by virtue of its low cost, huge capacity, controllable transport time, zero contact, safety and efficiency, etc. It has played an important role during the pandemic to ensure the global supply deployment of anti-pandemic materials and the smooth and stable international supply chain and has injected a strong agent for the global anti-pandemic and economic recovery. In 2020, a total of 12,406 China-EU trains were operated, and the annual number of trains exceeded the "10,000" for the first time, and the number of trains operated in a single month was stable above 1,000, (Xinhua News Agency,

2021, January 19) which played an important role in unblocking the global industrial chain, meeting the consumption needs of the people in the countries along the route and building the "life alleyway" of China-EU to fight against pandemics.

In terms of smooth trade flows, China's trade exchanges with countries along the route have continued to grow. The total import and export of goods trade between China and BRI countries increased from US$1.04 trillion in 2013 to US$1.35 trillion in 2020, with a cumulative total of over US$7.81 trillion in eight years; the proportion of China's trade with countries along the route to the total foreign trade in the same period increased year by year, from 25% in 2013 to 30% in 2020 (Institute of International Trade Regional Economic Cooperation Work). At the same time, trade in services between China and the BRI countries is developing steadily, with rapid growth in tourism services, outstanding cooperation in construction services, and rapid growth in new high value-added areas such as service outsourcing and Chinese medicine services. According to statistics, the number of Chinese tourists visiting BRI countries reaches 25 million each year, and China has become the largest source of tourists for the BRI countries (Xie Cheng Research Team, 2017, May 16).

In terms of currency circulation, direct investment between China and BRI countries has increased steadily, and financial institutions serving the BRI and "project finance" routes have blossomed and developed in various ways. Chinese enterprises' direct total investment to BRI countries increased from US$12.63 billion in 2013 to US$22.54 billion in 2020, with a cumulative total of nearly US$100 billion in 8 years. The proportion of China's direct total investment to BRI countries to foreign direct total investment in the same period remained roughly stable at 12%. The cumulative direct investment in China from BRI countries was about US$60 billion, with more than 27,000 enterprises established. In 2019, the actual amount of foreign investment in China from BRI countries was US$8.42 billion, up 30.6% year-on-year, accounting for 6.1% of China's total actual foreign investment absorption in the same period, and 5,591 new enterprises were established in China, up 24.8% year-on-year (Ministry of Commerce data, 2021, August 24). In terms of financial platform construction, up to now, around the BRI construction of capital financing needs, China has led the establishment of Asian Infrastructure Investment Bank, the BRICS New Development Bank and the Silk Road Fund, the business coverage of these financial institutions, the mode of operation is not the same. As an example, by the end of 2020, the AIIB had invested in 108 projects and approved a cumulative total of US$22.02 billion in infrastructure investments (Asia Infrastructure Investment Bank data, 2022, June 8).

In the area of interpersonal exchanges, humanistic exchanges have continued to deepen, innovative cooperation has been fruitful, the people of countries along the route have shared the fruits of development, and the concept of building a community of a shared future for mankind has taken root in people's hearts. China is actively organizing important humanities exchange activities with countries along the route, such as cultural years, art festivals, film and television bridges, and the Silk Road Book Project, expanding the diverse forms of humanities exchanges and enriching the humanities exchange mechanism. In addition, two-way study is also an important channel for humanities exchanges, and the Chinese government has established the Silk Road Scholarship Program to help cultivate talents and motivate people from countries along the route to study in China. According to statistics, among international students coming to China, the proportion of students from countries along the route has been increasing year by year, reaching 54.6% in 2019 (Cheng & Zhang, 2020). At the same time, science and technology innovation cooperation platforms represented by the BRI Union of National Scientific Organizations, the BRI International Seminar on Science, Technology and Innovation, and the Memorandum Science, Technology and Innovation Cooperation are steadily unfolding, and countries along the route are cooperating to address common challenges such as climate change, natural disasters and life and health, and the BRI is steadily advancing towards the road of innovation and the road of green. In addition, China's investment in BRI countries has created new tax and employment channels for the countries along the route and promoted local economic development, and the countries along the route have shared the fruits and dividends of the BRI construction, bringing tangible benefits to the people of all countries. By the end of 2019, China had invested a total of US$35 billion in cooperation zones built in BRI countries, paid host country taxes and fees of more than US$3 billion and created 330,000 local jobs (Ministry of Commerce, 2020, September 7).

It is worth noting that, along with the gradual deepening of the co-construction of the BRI, the construction of the BRI has entered the high-quality development stage of fine cultivation and deepening. The focus of the work of the BRI sufficient conditions has been adjusted and changed. *The Proposal of the Central Committee of the Communist Party of China on Formulating the 14th Five-Year Plan for National Economic and Social Development and the Visionary Goals for 2035* (hereinafter referred to as the *Proposal*), (Xinhua News Agency, 2020, November 3) adopted at the Fifth Plenary Session of the 19th CPC Central Committee, clearly points out the basic guidelines and tasks for promoting a high-quality development in the BRI during the 14th Five-Year Plan period. Following the guiding ideology and working guidelines put forward in the *Proposal*, the National 14th Five-Year Plan has set up Chapter 41 on *Promoting the Co-construction of the BRI High-quality Development* (Xinhua News Agency, 2020, November 3)which makes specific plans for BRI high-quality development and specifies three important aspects for the BRI high-quality development stage.

First, it is important to grasp the key fundamental position of connectivity. Infrastructure, as the core element of connectivity, is the core driver for many developing countries to recognize and participate in the construction of the BRI. It is an important foundation for promoting the linkage development of BRI countries, which is the key to the success of the BRI. The construction of infrastructure projects required by the BRI high-quality development has shifted to fine-tuning the whole process of project management with the goal of high quality, sustainability, risk resistance, reasonable price, and inclusive accessibility. In practice, under the guidance of the concept of co-business, co-construction and sharing, China should work together with countries along the route to build an interconnection network led by economic corridors such as the New Asia-Europe Continental Bridge, with major alleyway and information highways such as the China-Europe Train and the New Land and Sea alleyway as framing and with railways, ports and pipeline networks, etc., relying on the "Six corridors and Six Roads with Multiple Countries and Ports", to build a global interconnection partnership and achieve common development and prosperity.

Second, build a good investment and financial platform. Funding gap has been a major bottleneck constraining the construction of large-scale infrastructure projects. An efficient and complete, scientifically operated investment and financing system is a strong support for achieving interconnection. In the BRI high quality development stage, we should continue to play the role of co-construction the BRI special loans, Silk Road Fund, and various special investment funds, develop Silk Road-themed bonds, support the effective operation of multilateral development and financing cooperation centers, and develop diversified and effective investment and financing tools. While safeguarding the Asian Infrastructure Investment Bank's leading position in the BRI project construction investment, it welcomes the participation of multilateral and national financial institutions in the co-construction of the BRI investment and financing, encourages the development of third-party market cooperation, and brings together the strengths of multiple parties to achieve the goal of mutual benefit.

Thirdly, we will carry out in-depth theoretical research, promote cultural exchanges and form a pluralistic and interactive pattern of humanistic exchanges.

The co-construction of the BRI is a comprehensive long-term systemic project that requires close international cooperation from a strategic perspective, and theoretical research should be rooted in practice and lead the practice. Therefore, it is necessary to conduct more comprehensive, strategic, systematic, international, and long-term theoretical studies on the co-construction of the BRI. At present, we have numerous major research projects related to the construction of the BRI. Theoretical research and policy research should be combined, so that the project research can form a synergy and provide higher-level results and make the theoretical achievements of the BRI is the direction of future efforts.

3. Conclusions and policy recommendations

The above has demonstrated the necessary and sufficient conditions for the co-construction of the BRI, emphasizing that the full integration is the driving force and realistic guarantee for the high-quality construction of the BRI. Grasping the objective existence of the BRI along the line of the country's supply and demand match, grounding in the fundamental driving force of the existing demand for cooperation and making efforts to jointly build a positive mode of interaction and policy protection system, it is important to elevate the BRI into the high-quality construction phase. Since the necessary conditions for matching supply and demand and for joint development based on the development needs of the BRI countries will persist, the focus of future work will be on further providing sufficient conditions for the co-construction of the BRI. Based on the above-mentioned *Proposal* and the *Plan*, and on the research and practical basis of understanding the national strategic deployment, the subject has made the following ten recommendations on specific policies for the high-quality construction of the BRI in the 14th Five-Year Plan:

First, building a higher level of open economic system and raising the level of openness to the outside world in general. Based on a good start and fruitful results in the "front field", the co-construction of BRI should be made into a historic and innovative project of international cooperation, which requires not only the internal necessary conditions for achieving win-win cooperation, but also the external sufficient conditions for building a new system of higher-level open economy and improving the level of external opening. Specifically, this includes joining the RCEP, promoting membership in the CPTPP, building a free trade port in Hainan, and hosting the Shanghai Import Expo. In the 14th Five-Year Plan period, we should play a good "midfield" to promote the high-quality development of the BRI and realize the transformation of the BRI from the "grand design" of planning and layout to the "brush painting" of delicacy to make the construction of the BRI more stable and durable.

Secondly, it is recommended that a special national plan for the construction of the BRI be formulated for the 14th Five-Year Plan, and that the future development of the BRI be systematically deployed from a strategic overall perspective. It is imperative to implement the important spirit of the *Proposal*, systematically summarize the development of the BRI over the past eight years, objectively assess the progress of the implementation of the BRI, scientifically coordinate and plan the key directions of future efforts and modes of cooperation and strive to implement the "brushwork" in detail. It is proposed that a special plan for the construction of the BRI in the 14th Five-Year Plan period be drawn up, setting out the basic blueprint for development, key tasks, and action plans. It is proposed that the total scale of non-equity investment in the construction of the BRI during the 14th Five-Year Plan period should be limited to about US$100 billion, with investment in digital economy and green development projects accounting for more than 40%. Instead of expanding the scale, we will consolidate the project and maintain its steady progress to produce stronger and better brand and demonstration projects, forming a virtuous cycle and lay the foundation for future victories by allowing smooth advancement of the "halftime match" and in turn achieve success with the whole match.

Third, form an international industrial chain to help promote a dual domestic and international cycle. The dual domestic and international circulation in the new development pattern and the BRI are complementary. The co-construction of the BRI can also help to build the dual domestic and international circulation, and in turn the formation of the dual domestic and international circulation also provides a new platform and a new mechanism for the broader and more solid work of the co-construction of the BRI. Facing the new situation of entering the new development stage, establishing new development concept, and constructing new development pattern, the urgent task of planning the new development of the BRI construction is to build an international industrial chain and promote the formation of a new development pattern of domestic and international double cycle. To promote the operation of the double cycle, it is necessary to carry out supply-side reforms, expand consumption and promote innovation at home; at the international level, it is necessary to establish an international political environment suitable for the development of a new type of globalization, so as to create a favourable development condition for the co-construction of the BRI and the building of a community of shared future for mankind.

Fourth, it is proposed that the future BRI International Cooperation Summit Forum establish regional sub-forums in Xinjiang, Fujian, Hainan, and other regions. Section 2 of the 14th Five-Year Plan, "Promoting Infrastructure Connectivity", mentions promoting the construction of the BRI core zone in Fujian and Xinjiang. At the 2020 Boao Forum for Asia, President Xi delivered a keynote speech on global governance and the BRI, pointing out the important role of Hainan's development in contributing to the overall development of the BRI. Therefore, the establishment of Xinjiang, Fujian, and Hainan summit sub-forums to promote the formation of a diversified pattern of humanities exchanges helps to accelerate the transformation of the BRI practical experience into theoretical results, which in turn, is conducive to the promotion of the BRI high-quality development.

Fifth, the level of digitization of interconnectivity construction should be enhanced. Innovation is the key to achieving China's goal of initial socialist modernization in 2035 in the new era. Technological efficiency changes and organizational changes are the key to achieving the target of about 5% annual GDP growth. Among them, the digital economy is the core of innovation, and the BRI digital construction of connectivity is specifically expressed in the construction of the Digital Silk Road and the Innovation Silk Road.

Sixth, it needs to build an investment and financing platform that is in line with international capital market. The *Proposal* has made clear the need to build a diversified investment and financing platform. The successful experience of projects such as the BOT approach adopted for the construction of power plants in Pakistan and the participation of cartels in the operation of project companies should be drawn upon, and multilateral and national financial institutions should be welcomed to participate in the co-construction of the BRI investment and financing, and cooperation in the third party market should be encouraged to achieve the goal of mutual benefit through the participation of multiple parties.

Seventh, it should establish an open and transparent mechanism for releasing information on project construction. In June 2019, the research report of *Painting the BRI Cartography – Attracting International Private Capital to Infrastructure Development in Countries Along the Route* (KPMG China, the National Development and Reform Commission and the China Chamber of Foreign Contracting projects, 2019, May) noted that the main challenge facing national infrastructure projects in the BRI is the perceived lack of opportunities for financially viable, risk-controlled, reasonably priced and profitable projects by investors in these markets. To address this issue, Chinese and foreign governments, enterprises, international multilateral organizations and Chinese and foreign professional institutions need to cooperate to strengthen institutional capacity building, build risk management systems, innovate and broaden financing channels, and establish effective government-enterprise cooperation platforms to promote the financing and implementation of the BRI projects, so as to build a platform that can help attract international private capital to participate in infrastructure construction and operation in countries along the route. It is important to incorporate the reasonable parts of these recommendations and to introduce specific policies.

Eighth, the introduction of third-party organizations to participate in assessing debt sustainability in cooperation-building countries. On April 25, 2019, the Ministry of Finance officially released the BRI Debt Sustainability Analysis Framework during the Second BRI International Cooperation Summit Funding and Access Sub-Forum. This analytical framework is a debt sustainability analysis tool based on the IMF Petroleum Fund and World Bank debt sustainability analysis frameworks for low-income countries, with the actual situation of the BRI countries taken into consideration. It is encouraged to be used on a voluntary basis by financial institutions and international institutions in China and co-construction BRI countries. It is proposed to introduce the third-party assessment to make it more credible and to give a strong rebuttal to the propaganda of individual countries that BRI has caused a "debt crisis".

Ninth, the establishment of a coordinating organization to form a pluralistic and interactive humanities exchange mechanism. The BRI covers a wide range of sectors and ministries are actively exploring humanities exchanges, but there is no coordination mechanism. The education system has taken the lead in humanities exchanges with the BRI, and Xiamen University Malaysia has set a precedent for Chinese universities to set up overseas branches. Renmin University of China also set up a Sino-France college in Suzhou, contributing to the cultural exchange of the BRI. It is recommended that a coordination mechanism for humanities exchanges be established for the co-construction of the BRI, so as to form a multi-faceted synergy and better match the direction of high-quality development for the co-construction of the BRI.

Tenth, effective strengthening of the BRI soft power. Soft power is an important booster for co-constructing the BRI. It is necessary to do well in the three major tasks of theoretical research, public opinion propaganda and cultural dissemination in the practice of building the BRI, among which public opinion propaganda has assumed a particularly important task. It is proposed that a booklet, about 100,000 words, entitled *Face to Face with BRI* will be published, in which issues of concern of the public, such as why the BRI can achieve win-win cooperation, why China proposes to co-construct the BRI, what role the construction of the BRI have and the prospects for its development, will be published in an easy-to-understand booklet. It is to be disseminated to the staff, entrepreneurs and grass-roots cadres involved in the construction, so as to tell the story of BRI from a global perspective. At the same time, it should continue to play the role of a national think tank and advocate research institutions, private citizens, experts, and entrepreneurs' cooperative study to further coalesce good suggestions for the central government relevant department to consider.

References

2016 BRI Outbound Travel Annual Report, domestic online travel service company Ctrip (2017, May 16). Xie Cheng Research Team, http://www.100ec.cn/detail-6397253.html.

Asian Development Bank, Meeting Asia's Infrastructure Development Needs (2017, February 28). https://www.adb.org/zh/news/asia-infrastructure-needs-exceed-17-trillion-year-double-previous-estimates.

Chen Yongjun. (2016). Research on the Theoretical Model of the "the Belt and Road" Strategy. Journal of the Party School of Guizhou Provincial Committee of the C.P.C. No.161(01):29–35. DOI:10.16436/j.cnki.52-5023/d.2016.01.005.

Cheng Weihua, Zhang Binhai. (2020). "Opportunities, Challenges and Reflections on the Development of Graduate Education in China in the New Era", Graduate Education Research, 000(002):27–33.

China Academy of International Trade and Economic Cooperation, Ministry of Commerce, (2020, September 7), China BRI Trade and Investment Cooperation under the BRI 2020.

China has signed more than 200 cooperation documents on jointly building "Belt and Road Initiative" with 145 countries and 32 international organizations (2021, December 16). Xinhua News Agency. http://www.gov.cn/xinwen/2021-12/16/content_5661337.htm.

China's Belt and Road Initiative Trade and Investment Development Research report, The second "Belt and Road Initiative" Summit Forum on International Cooperation has achieved fruitful results (2019, April 29). CCTV network. http://www.gov.cn/xinwen/2019-04/29/content_5387629.htm.

Institute of International Trade Regional Economic Cooperation Work (2021). "China's 'Belt and Road' Trade and Investment Development Report (2021)".

Ministry of Commerce data, Ministry of Commerce, (2021, August 24). http://fec.mofcom.gov.cn/article/fwydyl/zgzx/202108/20210803190898.shtml.

Painting the BRI Cartography – Attracting International Private Capital to Infrastructure Development in Countries Along the Route, jointly written by KPMG China, the National Development and Reform Commission and the China Chamber of Foreign Contracting projects. https://home.kpmg/cn/zh/home/insights/2019/05/charting-the-course-of-belt-and-road-cooperation-together.html.

The Central Committee of the Communist Party of China. (2020). The proposal for the formulation of the 14th five-year plan (2021-205) for national economic and social development and the long-range objectives through the year 2035. Xinhua News Agency. http://www.gov.cn/zhengce/2020-11/03/content_5556991.htm.

Xi Jinping talking about the "Belt and Road" Asia Infrastructure Investment Bank data, see AIIB website (2022, June 8). https://www.aiib.org/en/projects/summary/index.html.

Xi Jinping. (2017, March 22). What are Belt and Road Initiative's Five links. People's Network.http://theory.people.com.cn/GB/n1/2017/0906/c413700-29519635.html

Yu Yongding. How to realize the transformation from "international cycle" to "double cycle"? (2020, August 18). Sina Finance, https://finance.sina.cn/zl/2020-08-18/zl-iivhvpwy1713127.d.html?vt=4&cid=79615&node_id=79615.

Zhao Chenxin. (2021, January 19). Increase by 50%! In 2020, 12406 China-Europe trains will be opened throughout the year. Xinhua News Agency, http://www.gov.cn/xinwen/2021-01/19/content_5581186.htm.

11

Analysis of the Investment Scale Boundary of the BRI: A Benefit Creation and Sharing Mechanism Perspective

Introduction

President Xi Jinping (2018b) stressed at the symposium on the fifth anniversary of the construction of the BRI that "the co-construction of BRI responds to the inherent requirements of the change of the global governance system, manifests the awareness of building a community of a shared future for mankind with shared power and responsibility, and provides new ideas and solutions for improving the change of the global governance system." The BRI has the domestic goal of promoting China's economic transformation and upgrading, promoting balanced regional development, and deepening its opening-up, and the international goal of promoting the development of the world economy, as well as reforming and building the global governance system. It provides a new direction and platform for the deepening development of economic globalization and a new round of world economic growth, and it is a great feat of landmark significance in the development process of economic globalization.

The economic basis for the BRI to achieve win-win cooperation is the high degree of complementarity and compatibility between China and the countries along the route in terms of economic structure. The core is the benefit creation and benefit sharing mechanism embedded in the initiative, which constitutes the necessary condition for the development of the BRI. The construction of a good mode of interaction and an adequate and effective policy guarantee system for the countries along the route is a sufficient condition for the BRI to achieve cooperation and win together. This is also the focus of the BRI construction work.

It is worth noting that the economic basis of the BRI for win-win cooperation is based on the matching of supply and demand between China and the countries along the route. The core of which is reflected in the matching of supply and demand in infrastructure construction investment. Although infrastructure development is a necessary part of economic development, it continues to follow the general law of investment, namely the law of marginal income diminishing. The progressive development of infrastructure is accompanied by the gradual satisfaction of demand gaps and diminishing economic development returns from additional infrastructure projects. Thus, theoretically, there is a scale boundary for investment in infrastructure projects under the BRI, and there is an optimal scale point for investment. In addition, in practice, due to the complex political environment, weak legal and regulatory system of investment support, diverse social and cultural environment, sensitive ecological security, many uncertainties and high investment risks in BRI countriess, all of these risks should be considered and prevented comprehensively when making infrastructure investment. The aimless pursuit of investment scale brings potential risks. Therefore, from the perspective of enterprise practice, compared with blindly spreading out investment projects, the fine tuning of each project is conducive to guaranteeing the return on investment for enterprises at the micro level and promoting the refined BRI at the macro level.

Since the investment scale boundary of the BRI is essentially a theoretical issue, based on this logic, and based on the preceding full combination of arguments, this chapter constructs a three-country interaction decision model. It uses mathematical theory analysis technique to argue the rationality and inevitability of China's proposed and pioneered construction of the BRI. It identifies two key elements affecting the investment scale of the host country project: the risk attitude of the cooperating countries providing funds and the project's value for money and reveals the essence of the BRI to promote the mutual benefit and win-win situation of the participating parties. Overall, the significance of this chapter is that it not only provides a theoretical explanation of the nature of the economics of benefit creation and sharing in the BRI, but also provides theoretical support to improve the construction of the relevant policy system, which is conducive to the quality development of the BRI in practice.

1. A three-state interactive decision model

(i) Model environment

Suppose the world economy consists of three countries A, B and C. The economic characteristics of the three countries differ as follows: (1) Country A is in the primary stage of industrialization and its degree of domestic infrastructure construction seriously does not meet the needs of economic development. While the limited capacity of capital accumulation and lack of capital and tight production capacity make it difficult for it to carry out large-scale infrastructure construction independently, and it needs to choose a partner in B and C for co-construction through international capacity cooperation. Country A could correspond to the late developing BRI countries. (2) Country B is an emerging economy with a relatively rapid industrialization process. It has large foreign exchange reserves and national savings, but faces problems such as homogenization of investment channels and low investment income rates. While its equipment manufacturing and other industries capacity profile is surplus, the cost-effective supply of production capacity, experienced in infrastructure construction and operation, have a strong demand for expanding international markets. Moreover, country B is more familiar with the political and economic environment of country A and has some regional advantages. Country B can be contrasted with China. (3) Country C is a traditional industrial powerhouse, with a leading international position in financial services and a comparative advantage in science and technology, but there is a "hollowing out" of the manufacturing sector and no surplus capacity to export. Country C may correspond to some developed countries.

(ii) Baseline model setting

Suppose that country A (host country) has limited own capital at the initial moment and proposes to build infrastructure projects through international capacity cooperation, which consists of two aspects: first, financing, and second, construction (or construction + operation), both of which are done uniformly by one of countries B and C through bidding. (This new form of international cooperation, the BRI, can only be in line with the principle of co-business, co-construction and sharing if it is tendered without any conditions and in accordance with internationally accepted commercial rules, in order to achieve mutual benefits and win-win results. Although China has overcapacity problems, surplus production capacity is also high-quality capacity, so the combination of capacity with competitive power and loan interest rates to develop the markets of countries along the route is an important aspect of the construction of the BRI.). During the cooperation, the successful bidder (*i* = B or C), as the cooperating country, provides a loan of the size and is responsible for the construction of the project. At the end of the project, host country A repays the principal and interest on the loan and shares the benefits generated by the project with the cooperating country. Assume that the total financing size of the project between country A and country *i* is E_i. The financing structure can then be broken down into credit b_i from cooperating countries and host country self-financing $(E_i - b_i)$. The above setting can be further interpreted to indicate that the E_i represents the scale of energy cooperation between the two countries and b_i as the scale of capital cooperation between the two countries. The allocation structure between the cooperating country and host country regarding the project benefits is $(\beta_i, 1-\beta_i)$, within which $\beta_i \in (0,1)$ is the share of benefits held by the cooperating country (noted as benefit partition factor). The allocation structure is determined through negotiation between the cooperating country and host country (assuming Nash bargaining issues). The project with a scale of q_i (q_i indicates denoting the natural volume of infrastructure supplied by the country *i*, measured in terms of the number of railroad miles operated, the number of inland waterway miles, port throughput, etc.) for host country A and the cooperating country is net income and, respectively π_A and π_i:

$$\pi_A = (1-\beta_i)H(\lambda_i q_i)^{\alpha} - b_i r_i (1-m) - \sigma(E_i - b_i) \tag{1}$$

$$\pi_i = \beta_i H(\lambda_i q_i)^{\alpha} + b_i(r_i - \sigma) - b_i r_i m \tag{2}$$

$H(\lambda_i q_i)^{\alpha}$ in equations (1) and (2) denote the output expected to be created by a project of size q_i. α denotes the output elasticity of the project in country A. $0 < \alpha < 1$ describing the realistic phenomenon of reduced output efficiency and diminishing marginal output when the project is too large. H denotes the productivity of country A (influenced by the level of science and technology and the level of business and management), which can be interpreted as the ability of country A to absorb and transform the item into economic output. λ_i denotes infrastructure performance (quality), which is related to factors such as the quality of capacity and construction experience in the country i, where $\lambda_i \geq \lambda_{\min}$, as $\lambda_{\min}$ denotes the performance threshold, which is the minimum requirement for infrastructure performance in country A. The project construction unit price is denoted by p_i, where $p_i \le p_{max}$, as p_{max} denotes the price threshold, i.e., the maximum unit price that country A is willing to pay for the construction of the project denotes the credit rate, cost rate for host country external financing, and also rate of return for lending funds by cooperating countries. r_i denotes the international market risk-free rate of return of capital, and therefore denotes the risk-free return of own capital invested in international capital market of size host country. σ represents the risk-free rate of return of capital in the international market, so $\sigma(E_i - b_i)$ represents the risk-free return of the host country's own capital invested in the international capital market of $(E_i - b_i)$. m is an exogenous variable indicating the interest preference granted by the cooperating country to host country on the loan in accordance with the international practices.

It might as well make $\tau_i \equiv \lambda_i / p_i$ the cost-effective ratio of the project, then $\tau_0 = \lambda_{min} / p_{max}$ represents the minimum requirement of country A for the cost-effective ratio of the project. Without losing generality, $\tau_0 = 1$, the cost-effective ratio of infrastructure provided by partner countries must meet $\tau_i \ge 1$, otherwise it is impossible to win the bid. Due to the public bidding of the project, at the initial stage of the project, the host country has known that the construction unit price of the project supplied by the partner country is p_i and the performance is λ_i, which means that the project performance-to-price ratio τ_i is an exogenous observable variable for the host country. Because E_i represents the total investment of the construction scale q_i, so $E_i = p_i q_i$, then formulas (1) and (2) can be rewritten as:

$$\pi_A = (1-\beta_i) H(\tau_i E_i)^\alpha - b_i r_i (1-m) - \sigma(E_i - b_i) \tag{3}$$

$$\pi_i = \beta_i H(\tau_i E_i)^\alpha + b_i (r_i - \sigma) - b_i r_i m \tag{4}$$

(iii) Analysis of decision-making issues

1. Consider the decision-making behavior of host country A. According to the reverse induction, when the benefit distribution rate β_i is given, the host country maximizes its net income by adjusting the total investment scale E_i and credit scale b_i of the project.

$$\max_{E_i, b_i} \pi_A = (1-\beta_i) H(\tau_i E_i)^\alpha - b_i r_i (1-m) - \sigma(E_i - b_i) \tag{5}$$

By referring to Bardhan (1973), Chatterjee et al. (2003), Addessi and Saltari (2012), Meidunju and Gong Liutang (2011), the assumptions in this paper that r_i are determined by the cooperating countries and are a function of the size of host country external financing b_i and the size of total project investment E_i.

$$r_i = \sigma + \delta_i \frac{b_i}{E_i} \tag{6}$$

$\delta_i > 0$ is the risk premium coefficient, which is related to the degree of capital surplus and resilience of the cooperating country, host country political stability, and the degree of mutual trust in connectivity between the two countries. The international credit level $\delta_B < \delta_C$ is due to the large foreign exchange reserves of country B and the fact that country B is more aware of the political and economic environment in country A and more resilient to risk compared to country C. Equation (6) denotes the credit rate r_i as the increasing function of the share of external credit (b_i / E_i), which implies that the larger the share of external credit, the higher the host country repayment pressure, the higher the default risk and therefore the higher the risk premium (higher credit rates increase the risk compensation of the cooperating country). Also, equation (6) indicates that the international credit rate is higher than the risk-free rate σ, otherwise no country would be willing to lend in the absence of subsidiary condition. Making $\gamma_i \equiv b_i / E_i \geq 0$ indicates the proportion of external credit from the partner country i, when $\gamma_i = 1$,the project is fully funded by the partner country, and the loan interest rate is higher. Insert formula (6) into formula (5) to obtain:

$$\max_{E_i, b_i} \pi_A = (1-\beta_i) H(\tau_i E_i)^\alpha - E_i\left[\sigma - \sigma m \gamma_i + \delta_i (1-m)\gamma_i^2\right] \tag{7}$$

In order to solve the above optimization problem and observe the structure of formula (7), with reference to the research ideas of Addessi and Saltari (2012), this paper defines the average project cost of the host country as $c_{A,i}$:

$$c_{A,i} = \sigma - \sigma m \gamma_i + \delta_i (1-m)\gamma_i^2 \tag{8}$$

The above formula indicates that the average project cost of the host country is a quadratic function of the proportion of external credit γ_i, so the optimal credit structure to minimize the average project cost is:

$$\gamma_i^\star = \left(\frac{b_i}{E_i}\right)^\star = \frac{\sigma}{2\delta_i}\frac{m}{1-m} \tag{9}$$

Furthermore, by substituting $\gamma_i^\star$ into formula (7) and finding the first-order condition of E_i, the optimal financing scale of the host country project can be obtained:

$$E_i^\star = \left[\frac{\alpha(1-\beta_i)H\tau_i^\alpha}{\sigma\left(1-\sigma\frac{m^2}{4\delta_i(1-m)}\right)}\right]^{\frac{1}{1-\alpha}} \tag{10}$$

At the same time, credit rates and the decision-making behaviour of host country and cooperating countries are available:

$$r^\star = \frac{\sigma}{2}\frac{2-m}{1-m}, b_i^\star = E_i^\star\frac{\sigma}{2\delta_i}\frac{m}{1-m},$$

$$\pi_A^\star\left(E_i^\star, r_i^\star, b_i^\star\right) = \sigma E_i^\star\left(1-\sigma\frac{m^2}{4\delta_i(1-m)}\right)\left(\frac{1}{\alpha}-1\right) \tag{11}$$

In practice, the project loan size $b_i^\star$ can not exceed the total project financing scale $E_i^\star$, so $\sigma m \leq 2\delta_i(1-m)$, in other words, the loan interest preferential subsidy rate m can not be too high. The optimal credit structure $\gamma_i^\star$ indicates that when the proportion of external credit of the host country is low, that is, $\gamma_i < \gamma_i^\star$, the unit external financing cost is lower, which stimulates the host country to increase the external financing scale b_i from the partner country, thus promoting γ_i to reach γ_i finally. When γ_i is higher, that is, $\gamma_i < \gamma_i^\star$, it can be seen from formula (6) that the higher proportion of external credit in the host country leads to higher credit interest rate, that is, higher external financing cost, which forces the host country to reduce the scale of external credit, so that γ_i will decrease and finally reach $\gamma_i^\star$, and at the same time make the interest rate tend to equilibrium credit rate $r^\star$. Since this paper assumes that the loan interest preferential subsidy rate m is determined by international practice and is constant for two competing countries (B and C), formula (11) shows that the optimal credit rate r^* obtained by the optimal credit structure γ_i^* is a constant, that is, for the same host country, the loan interest rates provided by different cooperative countries tend to be the same, which is more in line with the "one-price law" of the international credit market.

From equations (9) to (11), in the model setting, $\partial\gamma_i^* / \partial\delta_i < 0 \partial b_i^* / \partial\delta_i < 0 \delta_i$, suggests that for host country, the risk premium coefficient δ_i plays the role of screening and selecting international credit cooperation recipients:

Proposition 1: host country prefers to apply for loans from countries with smaller risk premium coefficients δ_i and thus enjoy lower interest rates on loans for the same share of external credit, and such a credit option is a way for host country to address big scale funding shortages.

This study assumes that the risk premium coefficient δ_i is correlated with the degree of capital surplus and risk resistance of the cooperating country, host country political stability, and the degree of mutual trust in connectivity between the two countries, so all other factors being equal. Proposition 1 also suggests that countries with a high degree of capital surplus and a more knowledgeable and accommodating host country political and economic environment (e.g., China) are more likely to achieve big scale credit cooperation with host country because such countries δ_i are smaller.

From equations (10) and (11), $\partial E_i^* / \partial\tau_i > 0 \partial\pi_A^* / \partial\tau_i > 0$, this shows that host country the total project financing size and net income is increasing function of the value for money of the cooperating country in supplying infrastructure, which leads to the following proposition τ_i.

As can be seen from expressions (10) and (11), $\partial E_i^* / \partial\tau_i > 0 \partial\pi_A^* / \partial\tau_i > 0$, this shows that the total financing scale and net income of the host country is an increasing function of the performance-to-price ratio τ_i of infrastructure provided by the partner country, from which the following propositions can be obtained:

Proposition 2: The better the value for money of the cooperating country in providing infrastructure projects, the greater the incentive host country to scale up the total project financing and obtain more net income.

The role of value for money in Proposition 2 has been neglected in previous studies. As shown in equations (3) and (4), the difference in value for money confers heterogeneity in unit financing and it also points to an important direction of competition between B and C – improving value for money. Of course, there can be a variety of ways to improve value for money; when there is no difference in the performance (quality) of the services provided, price competition is the traditional way to compete; when the unit price of project construction is the same, improving service performance is more likely to prevail in the competition. Thus, Proposition 3 follows:

Proposition 3: The host country selects the partner country based on high performance-to-price ratio in order to create more net income.

2. Consider the decision-making behavior of partner countries. Given the credit needs of the host country, the partner country and the host country negotiate to determine the benefit distribution rate β_i. It is assumed that the negotiation on β_i is a Nash bargaining (Nash Bargaining) problem. Under the condition that the Nash axiom is satisfied (Individual rationality, Pareto strong validity, symmetry, invariance of equivariant profitability descriptions, and independence of irrelevant choices.), there is a unique solution to the problem, that is, to solve the maximization of Nash product (Nash, 1950).

$$\max_{\beta_i} \pi_A^{*}\left(E_i^{*}, r_i^{*}, b_i^{*}\right)\pi_i\left(E_i^{*}, r_i^{*}, b_i^{*}\right) = \sigma E_i^{*}\left(1 - \frac{\sigma m^2}{4\delta_i(1-m)}\right)\frac{1-\alpha}{\alpha}\left(\beta_i H\tau_i^{\alpha}\left(E_i^{*}\right)^{\alpha} - E_i^{*}\frac{\sigma^2}{4\delta_i}\frac{m^2}{1-m}\right) \tag{12}$$

Where E_i^{*} equation (10) is satisfied. Solving equation (12) yields:

$$\beta_i^{*} = \frac{1}{2}\frac{(3\alpha-1)\sigma m^2 + 4\delta_i(1-\alpha)(1-m)}{4\delta_i(1-m) - (1-\alpha)\sigma m^2} \tag{13}$$

Under the given conditions $\sigma m \le 2\delta_i(1-m)$, it can be proved $0 \le \beta_i^{*} \le 1$. Further, taking the derivative of β_i^{*} to δ_i can be obtained:

$$\frac{\partial \beta_i^{*}}{\partial \delta_i} = \frac{2\alpha m^2\sigma(m-1)(\alpha+1)}{\left[4\delta_i(1-m) - (1-\alpha)m^2\sigma\right]^2} < 0 \tag{14}$$

From equation (14), the following proposition follows:

Proposition 4: Under the "finance+build" model of international cooperation, there is an alternative relationship between the benefit-sharing mechanism and the credit mechanism for the participation of the cooperating country in the project $\left(\partial\beta_i^{*} / \partial\delta_i < 0\right)$. When the risk premium on interest rates is high, cooperating countries have higher returns to credit and will lower the benefit sharing rate during the negotiation process to concede to host country; conversely, when the risk premium is low, cooperating countries have lower returns to credit and will raise the benefit sharing rate during the negotiation process to compensate for their own returns.

(iv) Discussion of balanced results

Substituting equation (13) into equations (10), (11) and (12) gives the equilibrium result when country A cooperates with country *i*.

$$\hat{E}_i = \left[\frac{2(1-m)\delta_i \tau_i^{\alpha} H\alpha(1+\alpha)}{4\delta_i\sigma(1-m)-(1-\alpha)m^2\sigma^2}\right]^{\frac{1}{1-\alpha}},$$

$$\hat{b}_i = \left[\frac{2(1-m)\delta_i \tau_i^{\alpha} H\alpha(1+\alpha)}{4\delta_i\sigma(1-m)-(1-\alpha)m^2\sigma^2}\right]^{\frac{1}{1-\alpha}} \frac{\sigma}{2\delta_i}\frac{m}{1-m},$$

$$\hat{\pi}_A = \sigma\left[\frac{2(1-m)\delta_i \tau_i^{\alpha} H\alpha(1+\alpha)}{4\delta_i\sigma(1-m)-(1-\alpha)m^2\sigma^2}\right]^{\frac{1}{1-\alpha}}\left(1-\sigma\frac{m^2}{4\delta_i(1-m)}\right)\left(\frac{1}{\alpha}-1\right),$$

$$\hat{\pi}_i = \frac{1}{2}(1-\alpha)H\tau_i^{\alpha}\left[\frac{2(1-m)\delta_i \tau_i^{\alpha} H\alpha(1+\alpha)}{4\delta_i\sigma(1-m)-(1-\alpha)m^2\sigma^2}\right]^{\frac{\alpha}{1-\alpha}} \tag{15}$$

All the above indexes are the increasing function of performance-to-price ratio τ_i. The understanding of E_i as the scale of production capacity cooperation between the two countries and b_i as the scale of credit cooperation between the two countries means that the cooperative countries can improve the cost-effectiveness of infrastructure, which can not only enhance the demand of the host country for investment in infrastructure projects, and expand the scale of capacity cooperation between the two countries, but also increase the scale of credit cooperation between the host country and the partner country, and expand the capital export of the partner country. In addition, the improvement of performance-to-price ratio is conducive to the realization of larger-scale net income and mutual benefit between the two countries in the process of cooperation, based on which proposition 5 can be obtained:

Proposition 5: The more cost-effective the infrastructure supplied by the cooperating country, the more conducive it is to the "going out" of capital and capacity, and to the expansion of the scale of benefit creation and benefit sharing between the two countries.

Observing the expression $\hat{\pi}_A$ in equation (15), the δ_i change gives $\hat{\pi}_A$ both positive and negative effects. On the one hand, the higher the risk premium host country the higher the interest rate paid on the loan, so host country net income is impaired; on the other hand, it follows from Proposition 4 that the higher the risk premium host country the higher the proportion of the benefit distribution process, so net income increases. The result of these two effects when they cancel each other out is $\partial\hat{\pi}_A / \partial\delta_i$, i.e very small (this is also evident from Figure 3–1 later) and therefore will not be discussed in this paper. As mentioned in the previous hypothesis, country B capacity profile has a high level of surplus, and another layer of capacity profile surplus is generally considered by academics to be a low capacity utilization factor and capacity profile exceeding the effective demand of society, with a state of oversupply, which in economic terms leads to lower product prices, so that country B has lower prices relative to country C's projects when providing infrastructure projects with the same performance. At the same time, B Domestic Energy is highly skilled and experienced in infrastructure construction and can deliver higher performance infrastructure project (Lin, 2010). Therefore, country B is able to supply cost-effective infrastructure projects in international cooperation, i.e., $\tau_B > \tau_C$. Further, it follows from Proposition 3 and Proposition 5 that country A prefers country B as the project partner country. The above inference is in line with China's cost-performance advantage in mid- to high-end manufacturing in recent years and the landing of the BRI project.

Therefore, when country A selects country B as the project cooperation country, optimal rate of interest, credit structure, benefit sharing rate, scale of credit cooperation, scale of capacity cooperation, and net income between the two countries are:

$$r^* = \frac{\sigma}{2}\frac{2-m}{1-m}, \gamma_B^* = \frac{\sigma}{2\delta_B}\frac{m}{1-m}, \beta_B^* = \frac{1}{2}\frac{(3\alpha-1)m^2\sigma + 4\delta_B(1-\alpha)(1-m)}{4\delta_B(1-m)-(1-\alpha)m^2\sigma},$$

$$\hat{E}_B = \left[\frac{2(1-m)\delta_B(\tau_B)^\alpha H\alpha(1+\alpha)}{4\delta_B\sigma(1-m)-(1-\alpha)m^2\sigma^2}\right]^{\frac{1}{1-\alpha}},$$

$$\hat{b}_B = \left[\frac{2(1-m)\delta_B(\tau_B)^\alpha H\alpha(1+\alpha)}{4\delta_B\sigma(1-m)-(1-\alpha)m^2\sigma^2} \right]^{\frac{1}{1-\alpha}} \frac{\sigma}{2\delta_B} \frac{m}{1-m},$$

$$\hat{\pi}_A = \sigma \left[\frac{2(1-m)\delta_B(\tau_B)^\alpha H\alpha(1+\alpha)}{4\delta_B\sigma(1-m)-(1-\alpha)m^2\sigma^2} \right]^{\frac{1}{1-\alpha}} \left(1-\sigma\frac{m^2}{4\delta_B(1-m)}\right)\left(\frac{1}{\alpha}-1\right),$$

$$\hat{\pi}_B = \frac{1}{2}(1-\alpha)H(\tau_B)^\alpha \left[\frac{2(1-m)\delta_B(\tau_B)^\alpha H\alpha(1+\alpha)}{4\delta_B\sigma(1-m)-(1-\alpha)m^2\sigma^2} \right]^{\frac{\alpha}{1-\alpha}} \tag{16}$$

Through the above analysis, it can be found that in the case of limited capital accumulation capacity, lack of capital, capacity shortage, and difficulty in achieving large-scale infrastructure construction by itself, through the international cooperation model of "loan + contract construction", country A has found country B, which can provide cost-effective infrastructure and low risk premium coefficient, as a partner country, thus bridging the infrastructure gap and achieving economic net income of scale $\hat{\pi}_A$. In the process, the credit cooperation between the two countries is scaled $\hat{b}_B$ and this loan can be spent with the foreign reserves and national savings of country B, raising the funds utilization factor. Country B used the loan as a "seed" fund to facilitate the construction of infrastructure projects and the "going out" of related domestic industries with a total financing size of $\hat{E}_B$.This could allow for the efficient use of domestic capacity in country B and create space for industrial upgrading. In this way, a "win-win" situation is achieved for both partners, which proves the economic logic of the BRI in terms of value creation and benefit sharing to achieve mutual benefits in the implementation process.

2. Numerical analysis of the baseline model and its application in the BRI framework

The previous section demonstrates the economic logic of the BRI win-win situation. This section discusses the impact of the changes in infrastructure value for money and capital surplus in country B, the political environment in country A on countries A and B and their strategic responses, as well as the accompanying changes in benefit creation and benefit sharing mechanisms. To show the response process more clearly, this paper uses the numerical modeling approach to analyze the economic effectiveness implied by these changes and focuses on the optimal response taken by country B (China). Of course, the purpose of this paper numerical modeling is not to fit realistic data, but to enrich the qualitative conclusions of the model.

Regarding parameter calibration, in studies by scholars such as Carlstrom and Fuerst (1997), Fukunaga (2002), Meidunju and Gong Liutang (2011), the risk premium coefficient is generally in the value taken $[0,0.4]$. In this book, the chosen basic parameter $\delta_B = 0.07$. In the study of Aschauer (1989), Addessi and Saltari (2012), Qiu Junfeng (2015) and other scholars, the project output elasticity in taking the value $[0.1,0.8]$, which is selected $\alpha = 0.4$ in this paper. Referring to the study of Chatterjee et al. (2003). The risk-free rate takes the value of $\sigma = 0.06$. In Asiedu and Villamil's study (2002), the interest subsidy rate m in some countries is taken in a range of values $[0.21,1]$.This paper chose $m = 0.5$. Use the level of technology in country A as a reference, i.e., set the productivity of country A to be $H = 1$. Let the value for money of infrastructure in country B be $\tau_B = 2$.

The respective indicators identified under the baseline model are: $\hat{\pi}_A = 1.8570$, $\hat{\pi}_B = 1.3901$, $\hat{b}_B = 9.9041$, $\hat{E}_B = 23.1096$, $\hat{E}_B - \hat{b}_B = 13.2055$, $\gamma_B^\star = 0.4286$, $\beta_B^\star = 0.3321$, $r^\star = 0.09$. The economic implications of this result in the BRI cooperation framework can be understood as follows: in the baseline environment, when China participated in the host country project through the "loan + contract construction" cooperation model, the size of the credit cooperation with host country was 9.9041 units, which contributed to host country demand for Chinese infrastructure of 23.1096 units, in which host country self-financed 13.2055 units and host country external credit financing accounted for 42.86%. China achieved 1.3901 units of net income and host country achieved 1.8570 units of net income. At the same time, the interest rate for the credit $r^\star = 0.09$ was determined by both countries and the mechanism $(0.3321,0.6679)$ for the distribution of the benefits of the project. From the point of view of the benefit distribution mechanism alone, it is more favorable to host country, which reflects the righteousness and benefit concept of "giving more and taking less" that China upholds in the BRI project.

Figure 2.1 shows the changes in interest distribution ratio $\beta_B^\star$, external credit ratio $\gamma_B^\star$, project demand scale $\hat{E}_B$, credit scale $\hat{b}_B$, host country self-raised capital scale $\left(\hat{E}_B - \hat{b}_B\right)$ and net income $\hat{\pi}_A$ and $\hat{\pi}_B$ when country B risk premium factor δ_B changes. With the increase of δ_B, the interest rate level $r^\star$ remains unchanged, but it can be found that the financing structure of the host country has changed greatly, the external financing scale $\hat{b}_B$ decreases significantly, and the total project financing scale $\widehat{E}_B$ decreases slightly, so the proportion of external credit to $\gamma_B^\star$ of the host country decreases, which increases the pressure of internal financing $\left(\widehat{E}_B - \hat{b}_B\right)$ of the host country, forcing the host country to open up more internal financing channels. Due to the reduction of debt pressure, the host country's net income $\hat{\pi}_A$ has a very weak upward trend, but for countries with limited production capacity and capital, this may crowd out inputs from other production sectors, and high internal financing needs may not be met, so it is necessary to continue to find partners for cooperation. In addition, the increase of δ_B leads to the change of benefit distribution mechanism and the decrease of benefit distribution rate $\beta_B^\star$, which indicates that cooperative countries will stimulate the host country to strengthen credit cooperation on a larger scale and achieve larger benefits by reducing the proportion of project income. This reflects the substitution effect between the benefit distribution mechanism and the credit mechanism (see proposition 4).

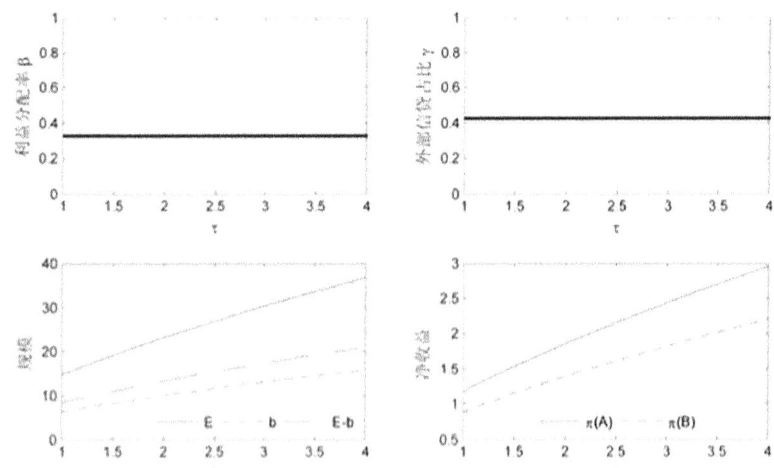

Figure 2.1 Change in equilibrium outcome due to change in risk premium coefficient δ_B

Figure 2.2 shows the impact of the change in the performance-to-price ratio τ_B of country B's supply infrastructure on interest distribution ratio $\beta_B^\star$, external credit ratio $\gamma_B^\star$, project demand scale $\widehat{E}_B$, credit scale $\hat{b}_B$, host country self-raised capital scale $\left(\widehat{E}_B - \hat{b}_B\right)$ and net income $\hat{\pi}_A$ and $\hat{\pi}_B$ of the two countries.

As shown in Figure 2.2, when τ_B increases, the benefit distribution ratio $\beta_B^{\star}$ and the proportion of external credit $\gamma_B^{\star}$ remain unchanged, but the project demand scale $\widehat{E}_B$, external credit scale $\widehat{b}_B$ and internal financing scale $\left(\widehat{E}_B - \widehat{b}_B\right)$ of the host country increase synchronously, indicating that the improvement of infrastructure performance-to-price ratio stimulates the host country's demand for larger-scale project cooperation, and the larger cooperation increases the net income of the two countries. Therefore, the expansion of the "win-win" effect has been realized.

Combined with the analysis in Figures 2.1 and 2.2, the changes in risk premium coefficients δ_B and infrastructure price/performance τ_B are rich in relevance and reflect the changes faced by multiple economic environments in international cooperation in the BRI. This paper considers the following four possible scenarios and analyzes the strategic options for the cooperating country (China) to cope with the changes.

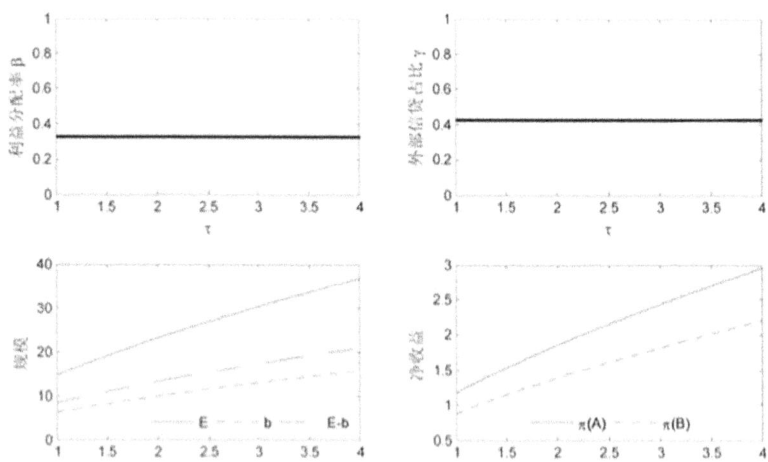

Figure 2.2 Change in equilibrium results due to change in τ_B price/performance ratio

Scenario 1: The size of China's foreign exchange reserves and national savings far exceed normal range and lack effective investment channels. In this context, the large foreign exchange reserves and national savings are to some extent a sign of economic inefficiency for China, leading to a decline in capital marginal income. As mentioned earlier, the risk premium coefficient is related to the degree of capital surplus in the country negative correlation and δ_B will therefore fall in this scenario. Combined with the simulation results in Figure 3.1, the total size of project cooperation between the two countries increases slightly and the size of credit cooperation increases significantly. β_B^* and $\hat{\pi}_B$ rise at the same time, which means that China realizes its own interests to a greater extent by increasing its share of benefits in project cooperation. The decline in δ_B results in little change in the $\hat{\pi}_A$ of the host country's net income and is always positive. Therefore, under the circumstances, the "Belt and Road Initiative" can still ensure a win-win situation for both the host country and China.

Scenario 2: Host country political environment is stable and political risks are declining δ_B. In such a scenario, the same decline occurs and a change due to the scenario occurs, net result remains mutually beneficial. Similarly, the situation is similar when the degree of mutual trust in China's interconnection with host country is enhanced, both reducing the degree of risk premium. Conversely, the instability of the host country political and economic environment is not conducive to mutual benefit and win-win situation, so maintaining a relatively stable host country political and economic environment is a prerequisite and guarantee for the construction of the BRI.

Scenario 3: Chinese technological innovation leads to technological improvements that allow China to offer better quality capacity at the same price (cost). At this point, the cost effectiveness τ_B of the infrastructure provided by China improves, giving host country an expanded economic benefit per unit of infrastructure construction. Combined with the previous analysis of Figure 3.2, equation (16), and Proposition 3, it is clear that host country spontaneously introduces more infrastructure supplied by China and increase the scale of credit cooperation, and eventually both countries' net incomes are enhanced to achieve more big scale benefit creation and benefit sharing.

Scenario 4: China's capacity surplus expands. On the one hand, capacity prices fall due to oversupply, and with no change in infrastructure construction performance, the cost-performance (τ_B) advantage further increases, host country will expand demand for China-provided infrastructure in the hope of more big scale projects, at which point the indicators change in line with scenario 3. On the other hand, overcapacity also implies a lack of robust investment opportunities in the country, which leads to a decline δ_B, and the indicator movement is consistent with scenario one. Therefore, scenario 4 is a compound process that results in host country restructuring of credit and cooperating countries restructuring the distribution of benefits in bargaining, ultimately leading to a win-win situation.

In short, by adopting different coping strategies in the face of the changing economic environment, the cooperating countries can effectively alleviate the pressure of double excess capital and capacity in their own countries, and it also bring greater economic return for host country, thus achieving more big scale value creation and more lucrative benefit sharing in the process of cooperation between the two countries. Scenarios 1 to 4 can also be considered as robustness tests of the baseline model. Combining the results in Figures 3.1 and 3.2, the findings of the baseline model are relatively robust, indicating that the BRI project helps create economic benefits and achieve a "win-win" situation for China and project host country. Further, comparing the baseline model with the four scenarios above, the paper leads to two corollaries that can inform decision-making by cooperating countries:

Corollary 1: Cooperating countries can create space for capacity exports and financial flows and realize greater benefits by fully grasping the changes in the economic environment and making reasonable use of the credit rationing mechanism and benefit-sharing mechanism.

Corollary 2: Cooperating countries can improve the technical content and quality of their production capacity, which can enhance the international competitive power of their production capacity, which is the way to fundamentally solve the problem of overcapacity.

3. Extended model

Equation (16) shows that while A and B enter a cooperative relationship, both countries' gains will increase if host country capacity to absorb and transform infrastructure projects into economic output (H) increases. Using this finding, the paper can extend the benchmark model to introduce developed countries other than host country and China. Developed countries can participate in the project by virtue of their comparative advantages in science and technology, etc., leading to more big scale benefit creation and benefit sharing. Specifically, according to the previous assumptions, country C is a traditional industrial powerhouse, and its direct participation in project investment suffers from lower cost effectiveness and excessive capital opportunity cost and is therefore not advantageous in the project bidding process. However, country C has comparative advantages in the field of financial services and science & technology. After the cooperative relationship between countries A and B, country C participates in project construction from the technical and management aspects, which can enhance the transformation capacity H of country A. Let's assume that the degree of improvement is 1 zero, and country C gets a portion of the $\mu(\mu<1)$ of the project income in return. Make $\theta = (1-\mu)(1+\nu)$, then the net income functions faced by countries A, B and C are:

$$\pi_A = (1-\beta_B)\theta H(\tau_B E_B)^\alpha - b_B r_B (1-m) - \sigma(E_B - b_B) \tag{17}$$

$$\pi_B = \beta_B \theta H(\tau_B E_B)^\alpha + b_B(r_B - \sigma) - b_B r_B m \tag{18}$$

$$\pi_C = \mu(1+\nu)H(\tau_B E_B)^\alpha \tag{19}$$

To simplify the analysis, the value μ assumed are determined exogenously by the three countries through negotiations. μ satisfies the condition $\theta > 1$, i.e., the participation of country C would have a positive effect on countries A and B. If $\theta < 1$, the participation of country C causes losses to country A and B, the two countries will therefore reject the participation of country C. Such assumptions will avoid deviating from the center of this text in the follow-up process due to the problem of taking μ values discussed.

Solving the maximization problem for equations (17) to (19) in the previous game framework yields the following equilibrium results:

$$r_1^* = \frac{\sigma}{2}\frac{2-m}{1-m}, \gamma_{B,1}^* = \frac{\sigma}{2\delta_B}\frac{m}{1-m}, \beta_{B,1}^* = \frac{1}{2}\frac{(3\alpha-1)m^2\sigma + 4\delta_B(1-\alpha)(1-m)}{4\delta_B(1-m)-(1-\alpha)m^2\sigma},$$

$$\hat{E}_{B,1} = \left[\frac{2(1-m)\delta_B(\tau_B)^\alpha H\theta\alpha(1+\alpha)}{4\delta_B\sigma(1-m)-(1-\alpha)m^2\sigma^2}\right]^{\frac{1}{1-\alpha}},$$

$$\hat{b}_{B,1} = \left[\frac{2(1-m)\delta_B(\tau_B)^\alpha H\theta\alpha(1+\alpha)}{4\delta_B\sigma(1-m)-(1-\alpha)m^2\sigma^2}\right]^{\frac{1}{1-\alpha}}\frac{\sigma}{2\delta_B}\frac{m}{1-m},$$

$$\hat{\pi}_{A,1} = \sigma\theta^{\frac{1}{1-\alpha}}\left[\frac{2(1-m)\delta_B(\tau_B)^\alpha H\alpha(1+\alpha)}{4\delta_B\sigma(1-m)-(1-\alpha)m^2\sigma^2}\right]^{\frac{1}{1-\alpha}}\left(1-\sigma\frac{m^2}{4\delta_B(1-m)}\right)\left(\frac{1}{\alpha}-1\right),$$

$$\hat{\pi}_{B,1}=\frac{1}{2}(1-\alpha)H\theta^{\frac{1}{1-\alpha}}(\tau_B)^{\alpha}\left[\frac{2(1-m)\delta_B(\tau_B)^{\alpha}H\alpha(1+\alpha)}{4\delta_B\sigma(1-m)-(1-\alpha)m^2\sigma^2}\right]^{\frac{\alpha}{1-\alpha}},$$

$$\hat{\pi}_{C,1}=\mu(1+v)H(\tau_B)^{\alpha}\left[\frac{2(1-m)\delta_B(\tau_B)^{\alpha}H\theta\alpha(1+\alpha)}{4\delta_B\sigma(1-m)-(1-\alpha)m^2\sigma^2}\right]^{\frac{1}{1-\alpha}} \tag{20}$$

Among them, the subscript "1" distinguishes the extended model from the benchmark model. By comparing the above results with the benchmark model, Table 1 can be obtained. As shown in Table 1, because $\theta > 1$, country C participates in the project by virtue of its advantages in science and technology and management services, which increases the net income of both countries A and B by $\theta^{1/(1-\alpha)}$, and they also win the benefits of $\hat{\pi}_{C,1}$ for themselves. In addition, the participation of country C has expanded the scale of project cooperation and credit cooperation between countries A and B by $\theta^{1/(1-\alpha)}$ times, which can effectively improve the utilization rate of funds and production capacity in country B, and solve the problem of insufficient funds in country A. Accordingly, the extended model realizes the breakthrough from "win-win" to "win-win" of the benchmark model, which proves the "expansion path" put forward in theory in the third part of this paper, and Corollary 3 can be obtained.

Table 2.1 Impact of third-country participation in the project on pre-existing cooperative relationships (Chen Yongjun et al., 2019)

	Profitability of the 3 countries			**Project size**	**Credit size**	**Other parameters**
	Country A	Country B	Country C	E	b	$r^{*}, \gamma_B^{*}, \beta_B^{*}$
Base model	$\hat{\pi}_A$	$\hat{\pi}_B$	0	$\hat{E}_B$	$\hat{b}_B$	unchanged
Extended model	$\theta^{\frac{1}{1-\alpha}}\hat{\pi}_A$	$\theta^{\frac{1}{1-\alpha}}\hat{\pi}_B$	$\hat{\pi}_{C,1}$	$\theta^{\frac{1}{1-\alpha}}\hat{E}_B$	$\theta^{\frac{1}{1-\alpha}}\hat{b}_B$	

Corollary 3: Based on the existing international cooperation, third party countries can create more benefits and achieve Pareto improvement by participating in the construction of the BRI projects with their comparative advantages such as technology, while not changing the original benefit distribution pattern of A and B.

4. Conclusions and policy recommendations

Regarding the logic of win-win cooperation in the BRI, it is based on the highly complementary economic structure of China and the countries along the route. The key is to actualize the creation and sharing of benefits through project cooperation. The analysis of the baseline and extended models in this chapter shows that, first, international cooperation in the BRI is not limited to one or two countries along the route, but each country can participate with its own comparative advantage, resulting in more big scale benefit creation and sharing. Second, with a reasonable international credit mechanism and benefit-sharing mechanism, China's originally relatively abundant capacity profile and foreign exchange reserves can be transformed into comparative advantages, providing favorable conditions for it in the process of international project bidding and construction. Host country can seize the opportunity of the BRI construction, using the rich production capacity and foreign exchange reserves of the investing countries to make up for the double gap of capital and technology and expand the construction of infrastructure projects needed for its own development. In addition, China's understanding and inclusiveness of the host country political and economic environment is an important advantage for China to lead the construction of the BRI and to influence the scale of investment through credit rationing and benefit-sharing mechanisms. Finally, scientific, and technological progress can improve the cost effectiveness of supply projects in partner countries, which is an important means to achieve more economic advantage and increase the scale of effective investment, so continuously improving the technical content and quality of production capacity is the way to fundamentally solve the problem of overcapacity in China.

Based on the above findings, this section makes the following policy recommendations: (1) Improving the quality and technological content of production capacity is an important means for China to achieve greater gains in international cooperation in the BRI. Modern economic development requires more than just infrastructure stock, but also demands higher quality requirement. Infrastructure quality improvement has a positive contribution to host country economic development. Therefore, in the construction of the BRI to improve the quality of China's production capacity supply, can enhance the core competitive power through production capacity "go out" to develop a broader market. (2) Rationalization of international credit policies and business incentives is an important guarantee for the successful implementation of the BRI. Reasonable international credit and investment incentives can be developed to combine national and corporate advantages in a "bundled" approach to going global. Using international credit as a "lead" to stimulate the demand for project construction in countries along the route and pave the way for Chinese enterprises to "go global", thus meeting the demand for infrastructure construction in countries along the route and breaking through development bottlenecks, while also creating space for China's economic transformation and upgrading. (3) Strengthening the study and understanding of the political, economic, cultural, and legal aspects of the countries along the route is an important asset for the better promotion of the BRI. Political and cultural proximity and tolerance are important advantages that distinguish China from European and American countries in cooperation along the route. Increased understanding of host country political culture, among other things, can enhance inclusiveness and strengthen anticipation of cooperation risks, thereby reducing risk premiums. At the same time, the construction of the BRI in accordance with international practice and legal norms is an important condition for ensuring the smooth route of the BRI economic cycle and the smooth realization of the benefit creation and benefit sharing mechanism, as well as an important measure for risk avoidance in the construction of the BRI. Therefore, in the next phase of the BRI policy design, the improvement of laws and regulations should occupy an important position.

References

Addessi, W. and E. Saltari, (2012). "The perverse effect of debt tax benefits on firm investment decisions", Economic Notes, 41(3):101–114.

Aschauer, D. A. (1989). Is public expenditure productive?. Journal of Monetary Economics, 23(2):177–200.

Asiedu, E., and Villamil, A. P. (2002). Imperfect enforcement, foreign investment, and foreign aid.Macroeconomic Dynamics, 6(4):476–495.

Bardhan, P. K. (1973). "On terms of foreign borrowing", American Economic Review, 63(3):458–461.

Carlstrom and Fuerst, (1997). "Agency costs, net worth, and business fluctuations: A computable general equilibrium analysis", American Economic Review, 1997, 87(5):893–910.

Chatterjee, S., G. Sakoulis. and S. J. Turnovsky, (2003). "Unilateral capital transfers, public investment, and economic growth", European Economic Review, 47(6):1077–1103.

Chen Yongjun, Deng Zhongqi and Zhang Jihuan. (2019). An economic analysis of the Win-Win cooperation of the Belt and Road Initiative: Based on the benefit creation and sharing mechanisms. Journal of Xiamen University (Arts & Social Sciences). (5)15: 83–97.

Fukunaga (2002)."Financial accelerator effects in Japan's Business Cycles", Bank of Japan Working Paper, No. 02–6, 2002.

Justin Yifu Lin, Wu Hemao, and Xing Yiqing. (2010). "'Surge phenomenon' and the formation mechanism of excess capacity.",[林毅夫, 巫和懋, & 邢亦青. (2010). "潮涌现象" 与产能过剩的形成机制. 经济研究, 10(4), 19.]

Meidunju and Gong Liutang (2011). "Currency mismatches, exchange rate appreciation and economic volatility", Quantitative Economics Technical Economics Research, 28(6), 37–51. [梅冬州, & 龚六堂. (2011). 货币错配, 汇率升值和经济波动. 数量经济技术经济研究, 28(6), 37–51.]

Meidunju and Gong Liutang (2011). "Exchange rate regime options for emerging market economies", Economic Research, (11), 73–88. [梅冬州, & 龚六堂. (2011). 新兴市场经济国家的汇率制度选择. 经济研究, (11), 73–88.]

Nash, J. F. , & Jr. (1950). "The bargaining problem", Econometrica, 18(2):155–162.

Qiu Junfeng (2015). "International reserve accumulation, physical and asset price inflation and monetary policy independence", International Currency Review, 87. [裘骏峰. (2015). 国际储备积累, 实物与资产价格通胀及货币政策独立性. 国际货币评论, 87.]

III

Basic Information on the Eight Years of the Construction of the BRI

Introduction

Since 2013, when President Xi Jinping proposed the construction of the Silk Road Economic Belt and the 21st Century Maritime Silk Road. The achievements of the BRI in the past eight years have been remarkable. From infrastructure connectivity to policy and interpersonal communication, from trade exchanges to financial flows, the results of the BRI have benefited the world and have gained widespread trust and praise.

In terms of policy communication, international consensus continues to expand. As of August 2021, the Chinese Government had signed 206 intergovernmental cooperation agreements with 172 countries and international organizations (Ministry of Commerce data, 2021, August 24). The scope of the commercial visa has spread to relevant countries located in Asia and Europe, Africa, Latin America, South Pacific, Western Europe, and other regions. It has continued to formulate numerous cooperation plans jointly with countries with a solid foundation for cooperation, a large volume of cooperation and a strong will to cooperate. It has held the BRI International Cooperation Summits I & II to strengthen the international consensus on the co-construction of the BRI, which has been expanded in scale, enriched in content, increased in the number of participating countries and growing in results year by year.

In the area of facility connectivity, substantial progress has been made on landmark projects. Over the past eight years, infrastructure connectivity has been built around the blueprint of " the Six corridors, Six roads, Multiple Countries and Ports" ("The Six Corridors, Six Roads, Multiple Countries and Ports", 2021), and a number of landmark projects in railways, ports, aviation, energy and other areas have made substantial progress. At the same time, as the backbone of international logistics land transport, China-European Liner is becoming more and more active with its advantages of short distance, fast speed, safety, green and environmental protection. As of the end of December 2020, the cumulative number of China-EU train rides had been 33,631, including 12,406 China-EU train rides in 2020, trasnsporting 1.135 million TEUs, up 50% and 56% respectively year-on-year, with an integrated heavy container rate of 98.4%. The number of annual departures exceeded 10,000 for the first time, and the monthly departures all remained stable at more than 1,000 (National Rail Group, 2020).

In terms of smooth trade flows, economic, trade and investment cooperation has been expanding. Over the past eight years, China's trade with the BRI countries has increased year by year as a proportion of total foreign trade. In 2020, the total imports and exports of goods with the BRI countries reached US$1.35 trillion, with 0.7% year-on-year growth (Ministry of Commerce data, data retrieved from Xinhua, 2021). Meanwhile, China has signed bilateral investment agreements with numerous countries along the routes, promoting a continuous increase of two-way investment. As the world's first import-themed large-scale national exhibition, the China International Import Expo has been successfully held for three sessions, including the third session held in 2020 both online and offline and attracted more than 3,800 enterprises from more than 180 countries, regions, and international organizations worldwide, with a cumulative intended turnover of US$72.62 billion, an increase of 2.1% over the previous session (China International Import Expo, 2020.). At the same time, China also actively promotes the construction of overseas cooperation parks, and all kinds of Chinese enterprises follow the market-oriented legal principle to independently go to the BRI countriess to build cooperation parks, creating new sources of tax revenue and employment channels for the BRI countries.

In terms of financing, a diversified investment and financing system is being built. Over the past eight years, China's financial institutions have focused on promoting the construction of a long-term, stable, sustainable, and risk-controlled diversified financing system to provide sufficient and secure funding for the BRI construction project. The approval of the *BRI Financing Guidelines*, (China Belt and Road Initiative Wesbsite, 2017, May 16) the publication of the BRI Debt Sustainability Analysis Framework and the establishment of the Multilateral Development Finance Cooperation Centre helps to improve the science of investment and financing decisions of all parties involved in the BRI and strengthen the debt management capacity of countries along the route. The establishment of the Silk Road Fund, the launch and establishment of the Asian Infrastructure investment bank and the assistance of China Export Credit surety company have jointly promoted various banks and insurance institutions in countries along the route to provide financial support for the BRI construction project. By the end of June 2019, China had achieved a cumulative insured amount of USD 770.4 billion in the BRI countries and paid USD 2.87 billion in claims (China Export Credit surety company official website, 2016). As of the end of December 2019, the Silk Road Fund invested the amount of capital as much as about USD 12.3 billion (Silk Road Fund, 2021). In the area of interpersonal exchanges, humanistic exchanges and cooperation have continued to deepen. Since the introduction of the BRI, China and the BRI countries have held art festivals, film festivals, music festivals, exhibitions of cultural relics, book exhibitions and many other activities, and cooperated in the creation and mutual translation and broadcasting of books, radio, film and songs. Over the past eight years, it has actively promoted the construction of the BRI press cooperation alliance, whose board consists of 182 mainstream media from 86 countries (the BRI Press Cooperation Allianc, 2004). With 352 member organizations from 72 countries and over 400 livelihood projects and activities, the Silk Road CSO Cooperation Network has become an important platform for promoting civil friendship and cooperation. The BRI countries ISO alliance, initiated by CAS in 2018, aims to promote close cooperation between countries and regions in science, technology, innovation, and capacity building. It has held three BRI countries ISO alliance council meetings until 2020. The interpersonal relationship is also reflected in China's active promotion of foreign aid. During the three-year period from 2017 to 2020, China provided CNY 60 billion in aid to developing countries and international organizations participating in the construction of the BRI to build numerous livelihood projects. It provided CNY 2 billion in emergency food aid to developing countries along the BRI, and it increased its capital to the South-South Cooperation Assistance Fund by USD 1 billion. China has also implemented 100 "happy homes", 100 "love and help" and 100 "rehabilitation and medical help" projects in countries along the routes (China BRI website, 2021).

Over the past eight years, the current good pattern of regional cooperation has been gradually formed around innovative cooperation models in five areas: policy consultation, infrastructure connectivity, smooth trade, financial integration, and civic exchanges, leading from point to point and from line to line. Among them, policy consultation is an important guarantee for the construction of the BRI, infrastructure connectivity is a priority area, investment and trade cooperation is a key element, financial integration is an important support, and people's civic exchanges is a social foundation.

As the construction of the BRI has moved from top-level design to the stage of promoting practical cooperation, the "Five Links" have been expanded into eight specific areas of cooperation, including infrastructure, international trade, cross-border investment, resource development, financial cooperation, humanistic exchanges, ecological and environmental protection, and maritime cooperation to facilitate operational implementation. After eight years of efforts, the BRI and its core concepts have been written into the relevant documents of the United Nations, the G20, APEC and other regional organizations, and have promoted mutual benefits and win-win results for the BRI countries through co-consultation, co-construction and sharing, focusing more on nodal countries and important projects for the benefit of people from the BRI countries, injecting new momentum into world economic development and making new contributions to global governance, and becoming a practical platform for promoting the building of a human community of a shared future.

1. Eight years of practice in co-constructing the BRI: Policy consultation

China has actively carried out policy consultation with international organizations and countries along the routes. It achieved obvious results by seeking support for the co-construction of the BRI, promoting the effective docking of the BRI with the development strategies and general planning of countries along the routes, and promoting the joint planning and construction of economic corridors. As of August 2021, the BRI had received positive responses and support from many countries and international organizations, including the United Nations and the Security Council. China has signed the BRI cooperation agreements with 140 countries and regions and 32 international organizations.

(i) Positive developments in bilateral cooperation

The BRI has actively dovetailed with the development plans of Russia's Eurasian Economic Union, Europe's "Juncker Plan" ("What is the meaning behind "International production capacity Cooperation" and "Juncker Plan"?, 2015, July 3), Kazakhstan's Bright Road ("Bright Road" in Kazakhstan, 2017, April 20), Mongolia's Steppe Road (Mongolia's "Prairie Road", China net, 2017, April 20) and Indonesia's Global Maritime Pivot (Aufiya, 2017). It has become an important element of bilateral and multilateral international cooperation, embodying the principles of mutual consultation, co-construction and sharing. Under the framework of the BRI, participating countries and international organizations have fully exchanged views on economic development plans and policies and negotiated the formulation of regional economic cooperation plans and measures, based on the principle of seeking common ground while preserving differences.

On May 8, 2015, in Moscow, China and the Russian Federation issued a joint statement on dovetailing cooperation between the *People's Republic of China and the Russian Federation on the Construction of the Silk Road Economic Belt and the Construction of the Eurasian Economic Union* (2017), announcing the launch of negotiations on an agreement between China and the Union in economic and trade cooperation. On October 25, 2019, Chinese Premier Li Keqiang, and the prime ministers of the member states of the Eurasian Economic Union issued a Joint Statement on the entry into force of the Agreement on Economic and Trade Cooperation between the People's Republic of China and the Eurasian Economic Union, signed on May 17, 2018, announcing the formal entry into force of the Agreement. The Agreement is an important early harvest of the implementation of the consensus between President Xi Jinping and the leaders of the member States of the Eurasian Economic Union on promoting the docking of the BRI construction and EEU. It is the first institutional arrangement in the economic and trade field between China and EEU, as well as the first institutional arrangement for the docking of the BRI construction and EEU, marking a new stage of institutional leadership from project cooperation between China and the member states. The member countries of the Eurasian Economic Union are all located along the Silk Road Economic Belt and are important partners of China in the construction of the BRI, as well as key areas for China to promote road infrastructure, trade and investment cooperation and production capacity cooperation, and Central Asian countries can function as important barricades for China's westward security. Chinese economy can complement the economies of the member states of the Union and portends for a large potential for trade cooperation. China has become EEU's number one trade partner and a major source of investment. By intensively promoting all-round economic and trade cooperation with the countries of the region, the two sides are striving to dovetail the construction of the Silk Road Economic Belt with that of the Eurasian Economic Union, ensuring sustained and stable regional economic growth, strengthening regional economic integration, and maintaining regional peace and development.

On December 14, 2015, the European Bank for Reconstruction and Development (EBRD) issued a statement accepting China as a shareholder of the bank, which has expanded its investment objectives from Eastern Europe at its inception to Central Asia, Central Europe, and Southeast Europe in recent years. It has many geographical overlaps with China's BRI, making it logical for the two sides to deepen their cooperation. On January 15, 2016, the State Council decided to accede to the *Agreement on the Establishment of the European Bank for Reconstruction and Development (EBRD)* (Yicai net, 2016, January15) and to accept the Resolution on China's Membership adopted by the Board of Governors of the EBRD. Since the EBRD's investment areas are highly compatible with the countries covered by the BRI, joining the EBRD helps China fill the gap in financial cooperation with the above-mentioned areas. It provides strong support for Chinese enterprises and financial institutions to "go global". In addition, the investment experience and model of the EBRD in the process of economic transformation is also worth learning from. The EBRD is one of the active participants in Europe's Juncker Plan, into which China can participate. China's membership in the EBRD gives a strong impetus to the docking of the BRI with the European Investment Plan, promoting cooperation between the official and private sectors, and providing broad space for China and the Bank to carry out project investments and deepen cooperation and exchanges in the fields of industry and technology in Central and Eastern Europe, the eastern and southern Mediterranean and Central Asia and North Africa.

As the neighbor whose border line with China is the longest, Kazakhstan is a friendly neighbor and a comprehensive strategic partner of China. The two countries have a good basis for political, economic, and cultural cooperation. On September 7, 2013, in his speech at Nazarbayev University, President Xi Jinping proposed for the first time the co-construction of the Silk Road Economic Belt, to which the Kazakh side responded positively, willing to strengthen economic, transport and humanistic connectivity with the Chinese side and jointly build a new Silk Road. In November 2014, President Nazarbayev proposed the New Economy policy called the "Bright Road", which was also fully recognized by the

Chinese side. In October 2016, China and Kazakhstan jointly released the *Silk Road Economic Belt Construction and Bright Road New Economy Policy Matching Cooperation Plan* (hereinafter referred to as the *Plan*) (National Development and Reform Commission, 2016, October, 24). This is the first bilateral cooperation plan signed and published under the BRI framework, which is an important step in guiding policy consultation at the national level under the BRI framework. Kazakhstan is the the country for China to annouce the BRI and one of the key pivot countries in the construction of the BRI. Strengthening cooperation with Kazakhstan is an important model for the comprehensive promotion of the construction of the BRI. As the first bilateral cooperation plan signed and released under the framework of the BRI, the China-Kazakhstan Plan is not only a top-level design to strengthen the docking of development strategies and deepen practical cooperation between China and Kazakhstan, but also an important step towards the construction of the China-Central Asia-West Asia international regional economic cooperation corridor. The *Plan* is a very useful exploration of the alignment of China's BRI with the development strategies of the countries along the route, and it can be said to have set an example.

(ii) Important progress in multilateral cooperation

China has also made important progress in multilateral cooperation around the BRI. On the one hand, it makes full use of existing multilateral cooperation mechanisms, such as the Shanghai Cooperation Organization, the China-Central and Eastern Europe 16+1 cooperation mechanism, the China-ASEAN "10+1", APEC, Boao Forum for Asia, the Asia-Europe Meeting, the Asia Cooperation Dialogue, the Asia Trust Conference, the Forum on China-Africa Cooperation, the Greater Mekong Sub-region regional economic cooperation, the Central Asian region regional economic cooperation, the Lancang-Mekong Cooperation Mechanism, the China-GCC Strategic Dialogue and other mechanisms. Through strategic dialogues, China has strengthened communication and deepened cooperation with relevant countries and regions, leading to the participation of more countries and regions in the construction of the BRI. On the other hand, it continues to play a constructive role in the relevant international forums and exhibitions in the BRI regions and sub-regions, such as the China-ASEAN Expo, China-Asia-Europe Expo, China-South Asia Expo, China-Arab States Expo, China West International Expo, China-Russia Expo, Qianhai Cooperation Forum, Eurasian Economic Forum, China International Fair for Investment and Trade, and other platforms.

The BRI is included in important documents of international organizations. The BRI and its core concepts have been incorporated into the relevant documents of the United Nations, the G20, APEC and other regional organizations. In July 2015, the Shanghai Cooperation Organization issued the Ufa Declaration of the Heads of State of the Shanghai Cooperation Organization, supporting the initiative on the construction of the Silk Road Economic Belt. In September 2016, the Communiqué of the G20 Leaders' Summit in Hangzhou, Zhejiang Province adopted an initiative to establish the Global Alliance for Infrastructure Connectivity. In November 2016, 193 Member States of the United Nations adopted a consensus resolution, welcoming economic cooperation initiatives such as the BRI and calling on the international community to provide a safe and secure environment for the construction of BRI. In March 2017, the UN Security Council unanimously adopted Resolution 2344, calling on the international community to strengthen regional economic cooperation through the BRI construction and enshrining the concept of "the community of mankind for a shared future" for the first time.

The number of countries and international organizations that have signed the BRI intergovernmental cooperation document has been increasing year by year. The BRI has received enthusiastic responses from relevant countries and international institutions. At the G20 Hangzhou Summit, China, as the presidency, actively promoted the possibility of dovetailing cooperation with other G20 members in regions such as the Middle East, Central Asia, and climate issues. Under the BRI, China has built a free trade area with ASEAN countries and expanded cooperation in the areas of finance, tourism, investment, and agriculture through concession of tariffs, the removal of non-tariff barriers and other measures. The cooperation between China and European countries is also deepening, with the EU's "Juncker Plan" dovetailing with China's BRI and the rapid development of economic and trade exchanges between China and Europe and countries and regions along the BRI routes, with frequent trade and logistics exchanges. Under the framework of the BRI, participating countries and international organizations have fully exchanged views on economic development plans and policies and negotiated the formulation of regional economic cooperation plans and measures, based on the principle of seeking common ground while preserving differences.

The co-construction of the BRI professional areas of docking cooperation in an orderly manner. The construction of the Digital Silk Road has become an important component of the co-construction of the BRI, China and Egypt, Laos, Saudi Arabia, Serbia, Thailand, Turkey, the United Arab Emirates and other countries jointly launched the *BRI Digital Economy International Cooperation Initiative*, and signed cooperation documents with 16 countries to strengthen the construction of the Digital Silk Road. China released the *Action Plan for Standard Linkage to Build the "One Belt, One Road" (2018–2020)* and signed 85 standardization cooperation agreements with 49 countries and regions. The long-term mechanism of the BRI tax cooperation is maturing. China has organized a conference on the BRI tax cooperation and issued the *Astana BRI Tax Cooperation Initiative*, extending the cooperation network of tax agreements to 111 countries and regions. China and 49 of the countries along the routes jointly issued the *Joint Declaration on Furthering Practical Cooperation on Intellectual Property Rights in the BRI Countries*. China organized the BRI International Forum on Rule of Law Cooperation and issued the BRI Statement of the Co-Chairs of the International Forum on Rule of Law Cooperation. China organized the BRI energy ministers' meeting, and 18 countries jointly announced the establishment of the BRI energy partnership. China released *The Vision and Actions for Jointly Promoting Agricultural Cooperation in the Construction of the BRI* and *The Vision for Maritime Cooperation in the Construction of BRI*, etc. China has promoted the establishment of an international commercial court and a "one-stop" mechanism for the diversified settlement of international commercial disputes.

(iii) Steady progress in the construction of economic corridors

The six international regional economic cooperation work corridors, including the New Asia-Europe Continental Bridge, China-Mongolia-Russia, China-Central Asia-West Asia, China-Central South Asia, China-Pakistan, and Bangladesh-China-India-Myanmar, link the Asian economic sphere with the European economic sphere and play an important role in establishing and strengthening the partnership of connectivity among countries and building an efficient and smooth Asia-Europe joint market.

In the nine years since the introduction of the BRI, regional cooperation in the New Asia-Europe Continental Bridge Economic Corridor has become increasingly deep, taking the open and inclusive, mutually beneficial partnership to a new level and giving a strong impetus to economic and trade exchanges between the two continents. China, Mongolia, and Russia are actively promoting the formation of a network of cross-border infrastructure links, mainly by rail, road, and border crossings. The three countries reached a *Planning Outline for the Construction of the China-Mongolia-Russia Economic Corridor*, promoting the official implementation of the first multilateral regional economic cooperation corridor of the BRI. In the eight years since the BRI was launched, the China-Central Asia-West Asia Economic Corridor has deepened cooperation in the areas of energy cooperation, facility connectivity, economic and trade and capacity cooperation. Over the past eight years, the China-Central South China Economic Corridor has made positive progress in infrastructure connectivity and the construction of cross-border regional economic cooperation zones. The layout of cooperation in the China-Pakistan Economic Corridor focusing on energy, transport infrastructure, industrial park cooperation and Gwadar port was determined for implementation. The China-Pakistan Economic Corridor is opening the third-party cooperation, and more countries are already or have the will to participate. The four parties of Bangladesh, China, India and Myanmar have jointly promoted the construction of the corridor under the framework of the Joint Task Group, and they have developed and planned a number of key projects in the areas of mechanism and institution building, infrastructure connectivity, trade and industrial park cooperation, international financial opening and cooperation, humanities exchanges and livelihood cooperation.

(iv) Summary

When discussing the "Five Links" model of the Silk Road Economic Belt, President Xi Jinping put policy communication at the top of the list, considering it as the basis and guarantee for infrastructure connectivity, smooth trade, financial integration, and civic exchanges. *The Vision and Action for Promoting the Construction of the Silk Road Economic Belt and the 21st Century Maritime Silk Road*, which is the top-level design of the BRI plan, points out that strengthening policy communication is an important guarantee for the construction of the BRI. With smooth policy communication, countries along the route can fully exchange and dovetail on economic development strategies and countermeasures, gradually forming a mechanism for dovetailing economic development strategies, macroeconomic policies, and major planned project among governments, forming a convergence of strategies, decisions, policies, and rules, and forming a more consolidated "community of a shared future". In the eight years since the BRI was proposed, China has established policy communication channels with countries along the route at different levels, in different areas and in different ways, laying a good foundation for further development in other areas.

1. Breadth: Increasing number of cooperating organizations

The construction of the BRI involves dozens of countries and regions in Central, West, South-East and South Asia and Europe, with a population of more than 3 billion, many of which are less developed regions. The core value of the BRI has made the countries along the routes realize that China's BRI "benefits the world". As more and more bilateral and multilateral agreements are signed and major projects are put in place, the BRI countries have fewer doubts about the BRI and more consensus is reached at the national leadership level. As it can be seen, as policy communication continues to deepen, the scope of cooperation becomes broader and the number of countries and organizations cooperating continues to increase.

2. Height: Increasing strategic alignment

At the beginning of the implementation of the BRI, more interdepartmental agreements were signed with the BRI countries, such as the customs agreement on customs clearance facilitation with relevant countries, and the Ministry of Commerce signed a memorandum on its entry with many countries on trade. In October 2016, China and Kazakhstan jointly released the *Silk Road Economic Belt Construction and the Bright Road New Economy Policy Matching Cooperation Plan*. Since then, many countries have begun to seek strategic alignment with the BRI, such as Russia's "Eurasian Economic Union", Mongolia's "Steppe Road" strategy, Europe's "Juncker Investment Plan', Vietnam's "two corridors and one circle" and so on (Bai Jie, 2018). With the success of the strategic alignment, the level of policy communication under the BRI framework is also rising.

3. Substantive effects: Widening areas of cooperation

Under the leadership of the BRI, as the breadth and depth of policy communication continues to expand, China has made efforts to dovetail with the development aspirations of the BRI economies, giving rise to a series of regional economic cooperation, major projects, capacity cooperation, foreign trade and investment and other areas of cooperation are also expanding.

2. Eight years of practice in co-constructing the BRI: Facility connectivity

China is actively promoting the construction and connection of infrastructure such as roads, railways, water transport, air transport, energy, and communications in BRI countries, as well as the docking of relevant regulatory systems with countries along the route. In the eight years since the launch of the BRI, new progress has been made in promoting the effective connection of various facilities.

(i) Transport infrastructure

Since the introduction of the BRI, the construction of the BRI has grown from scratch, from concept to action, from a national initiative to a global consensus, and has made remarkable achievements. The lack of infrastructure investment is a bottleneck for the economic development of developing countries, and the construction of transport infrastructure is a key area of cooperation in the Belt and Road. Numerous international and domestic Belt and Road key infrastructure projects are being carried out in an orderly manner and accelerating the construction of facility links is a key area and core content of the co-construction of the Belt and Road.

1. Railway facilities

Railways are the much-anticipated BRI infrastructure sector, with key railway projects including the China-Laos Railway, China-Thailand Railway, Hungarian-Serbian Railway, Mokha High Speed Railway, Muse-Mandalay Railway, Yawan High Speed Railway, Mak Mak High Speed Railway, East Coast Railway of Malaysia, Southern Malaysia Railway, Akha Railway of Nigeria, Yaji Railway, Kenya Monel Railway, Nigeria Coastal Railway, China-Pakistan Railway, China-Jiwu Railway, etc., as well as related railway facilities construction project and cooperation mechanisms such as the electrification of the Tehran-Mashhad Railway of Iran, the renovation of the Belgrano Freight Railway of Argentina, the T9 Tunnel of the Georgian Railway Modernization Project, the Padma Bridge Railway Link and the China-Europe Class Train.

China-Laos Railway. The Sino-Lao Railway is a railway that passes between China and Laos and is an important component part of the Central Line of the Trans-Asian Railway. The China-Laos Railway starts from Yuxi City in Yunnan Province, China in the north, and passes through Pu'er City, Xishuangbanna, the Sino-Laos border crossing of Mochan, and through the famous tourist destination of Luang Prabang in Laos to Vientiane, the capital of Laos. Construction of the China-Laos Railway began in December 2016 and began full scale construction in mid-2017, with completion and opening scheduled for the end of 2021. The project is designed, constructed, and operated in accordance with Chinese railway standards, with a design speed of 160 km/h. The Chinese section, the main line of the Yuma Railway, is 508.53 km long. Among them, the section from Yuxi to Xishuangbanna is a double line, and the section from Xishuangbanna to Mochan is a single line. The total investment is US$50.5 billion. The Lao section, the railway from Phnom Penh to Vientiane, is 418 km long. The total investment is US$37.4 billion. The project contractors are China Railway Fifth Bureau, China Railway International, China Railway Eighth Bureau, Hydropower International, Electric Construction International, China Railway Second Bureau, etc.

Mocha High Speed Rail. China, Mongolia, and Russia are actively promoting the formation of a network of cross-border infrastructure links, mainly by rail, road, and border crossings. The Mokha High Speed Rail project is an important component of Russia's Transport Development Plan 2030, issued in 2008, with the line starting from Kurskoye Station in Moscow, the capital of Russia, in the west, going east through Moscow, Moscow Region, Vladimir Region, Nizhny Novgorod Region, Chuvash Republic, Mari El Republic, and east to Kazan, the capital of the Republic of Tatarstan. With a total length of 770 km and a maximum design speed of 400 km/h, it is regarded as the first Chinese high-speed rail "going out". The work on the Chinese side of the Sino-Russian Tongjiang-Nizhniy Leninskoye Boundary River Railway Bridge was completed in October 2018. Construction of the Heihe-Bragoveshchensk border river vehicular bridge is progressing well. The Russia-China consortium of companies has basically completed the sketch design of the Mocha high-speed railway project. When the project is completed, the travel time from Moscow to Kazan will be reduced from the current 14 hours to 3.5 hours.

Yawan High Speed Rail. Yawan high-speed railway is a high-speed railway from Jakarta to Bandung, Indonesia, with a total length of 142 km and a maximum design speed of 350 km/h. It is a major project dovetailing China's BRI and Indonesia's maritime pivot point strategy. It is also the first single of China's high-speed railway moving outside of China in a comprehensive manner, with a total investment of US$5.135 billion. The Yawan High Speed Rail is being built and operated by Kereta Cepat Indonesia China (KCIC), a joint venture between China and Indonesia. The joint venture was formed by a consortium of Chinese companies led by China Railway main office and Indonesian state-owned enterprises led by Vika Indonesia. Other members of the Chinese consortium include China Railway (design and construction), CEC (civil construction), China CNR (rolling stock), and China Tongxiang (train control system).

Kenya's Monel Railway. The Mombasa-Nairobi Standard Gauge Railway, which runs from the eastern Kenyan port of Mombasa in the east to the capital Nairobi in the west, is about 480 km long and was opened to traffic on May 31, 2017, with a total investment of US$3.8 billion. It is the first modern railway built in Kenya in 100 years and is the flagship project of Kenya's Vision 2030.

The construction of the railway alone has boosted the local economy by 1.5% and has been hailed as one of the "world's most rewarding" railway travel routes. As a landmark and exemplary project of China-Kenya and China-Africa cooperation, the Mombasa-Nairobi Standard Gauge Railway is of far-reaching significance in promoting Kenya's economic and social development, accelerating industrialization, and advancing connectivity and integration in East Africa, and has been hailed as the "Road of Friendship", "Road of Cooperation" and "Road of Win-Win" for China-Africa in the new era. According to the forward plan, the railway connects the East African countries of Kenya, Uganda, Rwanda, Burundi, and South Sudan.

China-Europe Class Train. The China-Europe Class Train refers to the container international railway intermodal transport train between China and Europe and the BRI countries in accordance with the fixed number of trains, lines, and other conditions. The China-Europe Class Train has initially explored the formation of a multinational collaborative international train operation mechanism. Seven railway companies from China, Belarus, Germany, Kazakhstan, Mongolia, Poland, and Russia signed the Agreement on Deepening Cooperation on China-Europe Class Trains. As of the end of December 2020, the cumulative number of China-EU train rides was 33,631, including 12,406 China-EU train rides in 2020, sending 1.135 million TEUs, up 50% and 56% respectively year-on-year, with an integrated heavy container rate of 98.4%. The number of annual departures exceeded 10,000 for the first time. The number of single-month departures all remained stable at more than 1,000.

2. Road facilities

China has signed 18 bilateral and multilateral international transport facilitation agreements, including the *Intergovernmental Agreement on Facilitation of International Road Transport among the Member States of the Shanghai Cooperation Organization*, with 15 countries along the route. China-Mongolia-Russia, China-Jiwu, China-Russia (Dalian-Novosibirsk) and China-Vietnam international direct road transport preliminary operation activities have been successfully held. In February 2018, international road transport between China, Kyrgyzstan and Ukraine became a regular operation. The second bridge over the Beilun River highway between China and Vietnam was opened to traffic. China has formally acceded to the International Road Transport Convention (TIR Convention). Positive progress has been made in the implementation of the Greater Mekong Subregion Agreement on Facilitating the Cross-border Movement of Goods and People.

On the whole, it seems that nine years since the BRI was launched, China's BRI road facilities have been fully connected, with some major projects having been successfully started and others in the planning stage. In the direction of China-ASEAN, two railway crossings (Pingxiang-Tongdeng Railway Crossing and Hekou-Laojie Railway Crossing) connecting China and ASEAN have been formed. There are eight main highways basically completed or planned for construction. Yunnan's China-Laotai Highway, China-Vietnam Highway, China-Myanmar Highway, and most of the domestic section of the China-India Highway have been completed as expressways, while the completion of the Chongzuo and Jingxi-Longbang Expressway in Guangxi at the end of 2016 opened the land route alleyway to Southeast Asian countries from western Guizhou, eastern Yunnan and southern Qianjiang in China. In the China-Central and West Asia direction, China has formed a tri-continental transport alleyway with Central Asian countries, including six cross-border highways. As the flagship project of the BRI construction, the second phase of Karakorum Highway and Karachi Expressway in Pakistan started construction. The Karachi-Lahar Expressway Project and the Karakorum Highway Upgrade Project, constructed by China Construction and China Communications Construction respectively, have contracts worth US$2.89 billion and US$1.32 billion respectively. Taking the key projects of the China-Pakistan Economic Corridor as an example, the Raikot-Islamabad Road project is about 487 km long, connecting the first phase of the Karakoram Highway to Kashgar in Xinjiang, China, via the Hongqilaf Pass, and is the only land alleyway to China in the northern part of Pakistan.

3. Port facility

The port is an important infrastructure to serve the 21st Century Maritime Silk Road. The maritime industry is the main mode of transportation of the Maritime Silk Road, which also plays an important role in supporting and promoting the formation and development of the Silk Road Economic Belt. In terms of maritime transport, which is connected to the BRI facilities, Chinese enterprises have participated in the construction and operation of many ports, and maritime transport services cover many coastal countries along the route. The Gwadar port in Pakistan opened a regular container liner trade route, and the supporting facilities in the starting area have been completed, attracting more than 30 enterprises to the park. Sri Lanka's Hambantota Port special economic zone has completed preliminary work such as industrial positioning and concept planning for the park. The port of Piraeus, Greece, was completed as an important transit hub. The construction of the third phase of the port is nearing completion. The second phase of the Khalifa Port container terminal in the UAE was officially opened in December 2018. China Ningbo Shipping Exchange has continuously improved the "Maritime Silk Road Shipping Index" and released the 16+1 Trade Index and Ningbo Port Index. In the past eight years, Chinese port enterprises have actively participated in the BRI, deepened international cooperation, demonstrated their strength on the international stage, and brought mature operating experience to BRI countries.

4. Aviation transportation facilities

Since the past eight years, China has expanded air traffic rights arrangements with a number of countries, signed bilateral governmental air transport agreements with a number of countries and regions, and achieved direct flights with a number of countries along the Belt and Road, focusing on improving connectivity, promoting infrastructure dockening, expanding civil aviation cooperation platforms, and strengthening cooperation with countries along the Belt and Road, forming a good situation for the whole industry to jointly promote the construction of the Belt and Road.

It aims to accelerate the level of the BRI air connectivity. As of early 2020, the country had signed bilateral intergovernmental air transport agreements with 127 countries and territories. Among them, to serve the construction goal of the BRI, bilateral governmental air transport agreements have been signed with 96 BRI countries. Regular passenger and cargo flights have been maintained with 54 of them. The first regional air transport agreement has been signed with ASEAN. Bilateral air talks and air traffic rights arrangements have been expanded with Russia, Armenia, Indonesia, Cambodia, Bangladesh, Israel, Mongolia, Malaysia, Egypt, and other countries along the route. Meanwhile, as of early 2019, China has achieved direct flights with 45 BRI countries, with about 5,100 flights per week. In the summer and autumn seasons of 2018, a total of 29 Chinese airlines operated regular round-trip flights from 47 cities in China to 81 cities in 37 BRI countries, with 2,849 flights per week, including 2,751 passenger flights and 98 cargo flights. Meanwhile, a total of 90 airlines from 37 BRI countries operated regular flights from 84 foreign cities to 52 domestic cities, with more than 2,346 flights per week, including 2,204 passenger flights and 142 cargo flights (China BRI website).

It aims to contribute to the construction of airports in BRI countries. The airport construction covers many professions, high technology content and long industrial chain, involving engineering design, construction, equipment integration, electronics, information, materials, machinery manufacturing and other related industrial fields. The project has a strong leading role. China's civil aviation has achieved remarkable development and has rich experience in airport construction, which is welcomed and valued by BRI countries. Since the BRI was proposed eight years ago, Chinese enterprises have built many airports construction projects in countries along the "Belt and Road", such as Beijing Urban Construction Group Co., Ltd. has built the expansion project of Velana International Airport in Maldives, China Construction Co. has built the construction of the new airport in Algiers, and China civil engineering work Group Co.

It aims to actively expand the BRI civil aviation cooperation platform. Using mechanisms such as the China Civil Aviation Cooperation Platform for Central Asia, the China-Africa Regional Cooperation Platform, and the China-ASEAN Aviation Regional Cooperation Platform, we have promoted comprehensive cooperation with countries and regions along the BRI route. For example, the first annual China-Europe Aviation Safety Conference was held, which became an important platform for China-Europe civil aviation exchange and dialogue and common development. A letter of intent was signed with international civil aviation organization on BRI cooperation to jointly promote the development of civil aviation in countries and regions along the BRI route. 21 bilateral meetings were held with 19 countries and international organizations at the first Asia-Pacific Ministerial Conference on Civil Aviation. The *Beijing Declaration* was adopted to help promote the implementation of the BRI in the global and regional civil aviation sector (China BRI website). During President Xi Jinping's visit to South Africa, the Civil Aviation Administration signed the *Understanding on Regional Aviation Partnership Memorandum* entry with the civil aviation authorities of Brazil, Russia, India, and South Africa, defining the content and modalities of cooperation among the five BRICS countries in the field of aviation and establishing a cooperation mechanism. Solidly promoting comprehensive aviation cooperation with countries along the route, China's civil aviation provides professional training for countries in Central Asia, Africa, Southeast Asia, and other regions, covering aviation safety, traffic control, airport management, flight standards, safety and security, aviation meteorology and many other fields. A China-ASEAN aviation safety accident/incident investigation cooperation understanding memorandum entry was signed with ten ASEAN countries, strengthening the aviation safe practice cooperation and exchange between China and ASEAN.

(ii) Energy infrastructure

The BRI runs through the continents of Asia, Europe, and Africa, connecting all countries of the world, from the dynamic East Asian economic sphere in the east to the developed European economic sphere in the west, with developing countries with great economic development potential located in the vast hinterland in between, and the world's most important energy producing, consuming and alleyway countries located along the route. The BRI is the heart of the world economy and energy. Energy cooperation is an important foundation and support for the BRI. In May 2017, China released the *Vision and Actions for Promoting BRI Energy Cooperation*, which explains the cardinal rule of international energy cooperation, sets goals and principles for energy cooperation, and advocates strengthening cooperation in seven areas, including policy communication, trade facilitation, energy investment, and energy production capacity. *The Vision* clearly states that it insists on "openness and inclusiveness" All countries and international and regional organizations can participate in the BRI energy cooperation, strengthen dialogue among countries, seek common ground while reserving differences, and build and share together, so that the fruits of cooperation can benefit a wider region.

1. Construction of energy facilities

China has signed a series of cooperation framework agreements and memorandums of understanding with BRI countries. It has carried out extensive cooperation in the fields of electricity, oil and gas, nuclear power, new energy, and coal, etc. It has worked with relevant countries to maintain the safe operation of oil and gas pipeline networks and promote energy optimal allocation of resources between countries and regions. The China-Russia crude oil pipeline and China-Central Asia rock gas pipelines maintain stable operations. The east line of the China-Russia rock gas pipeline was partially connected to gas in December 2019 and will be fully connected to gas in 2024. In the eight years since the BRI was proposed, in addition to active cooperation at the government level, Chinese state-owned energy enterprises, represented by CNPC, Sinopec, State Grid and CGN, have played an important role in the construction of energy facilities in the BRI. These state-owned energy enterprises have actively promoted the connection of oil and gas pipelines and the construction of overseas power grids in countries along the routes, ensuring the security of China's energy resources, advancing the pace of Chinese enterprises going abroad, directly benefiting the development of the local economies of the host countries, and promoting regional connectivity.

2. Cooperation in energy resources

In the eight years since the launch of the BRI, positive progress has been made in the construction of numerous key energy resources cooperation projects. At the first BRI Energy Ministers' Meeting in 2018, China, together with Afghanistan, Bolivia and 17 other countries, issued a *Joint Declaration on the Establishment of the BRI Energy Partnership*, confirming the formal establishment of the partnership in 2019. For both China and the BRI countries, the broad common interests in the energy sector are behind the strong push to get the initiative off the ground. In addition to the continuous promotion of the BRI energy partnership, China has also landed a number of major energy cooperation projects with countries and regions along the route, including important oil and gas pipelines between China-Russia, China-Central Asia and China-Myanmar, as well as a large number of thermal power plants, water power station, nuclear power plants and power grids, such as the Vinh Tan coal-fired power plant in Vietnam, the Deninskaya power station in Russia and the Hinkley Point C nuclear power station in the United Kingdom construction project.

New and renewable energy sources have become key areas of cooperation, helping to build a green and low-carbon global energy governance pattern. Under the framework of the BRI energy cooperation, China's international cooperation in energy has gradually expanded from the initial focus on oil and rock gas to include unconventional oil and gas such as tight oil and gas, oil shale fuel gas and coal-bed methane, and new energy sources, including nuclear energy and other types of renewable energy, such as solar, wind and tidal energy. As for the use of nuclear energy, the Sigma nuclear power project and the Karachi nuclear power project are progressing well. In terms of hydropower utilization, China Three Gorges Group has undertaken the first hydropower project in the China-Pakistan Economic Corridor, the Calot water -driven power station project, which is expected to be completed ahead of schedule. According to the long-term vision of the BRI, it is expected that by 2030, some 60 countries will have built a total of more than 200 nuclear power plants. In addition to nuclear power, China has also cooperated with other countries in the areas of investment, technology, and finance, taking specific situations in the fields of solar energy, wind energy and carbon trading into account.

(iii) Information network facilities

1. Construction of communication network facilities

Improving the level of interconnection of international communication network facilities plays an important role in promoting the popularization of the application of new-generation information and communication technologies, enhancing the level of information services in the region, and promoting the economic and social development of countries and regions along the BRI route. Since the introduction of the BRI, BRI countries have jointly promoted the construction of cross-border fibre-optic cables and other communication networks to improve the level of international communications connectivity. China has built a number of cross-border fiber optic cables and international submarine cables with 12 BRI countries, and has made significant progress in the construction of cross-border fiber optic cable information alleyway between China and Myanmar, China and Pakistan, China and Kyrgyzstan, and China and Russia, and China has signed cooperation agreements with Kyrgyzstan, Tajikistan and Afghanistan on the Silk Road fiber optic cable, and has substantially launched the Silk Road fiber optic cable project (China BRI website). On May 12, 2017, during the BRI International Cooperation Summit, *China and the International Telecommunication Union signed the Letter of Intent on Strengthening Cooperation in the Field of Telecommunications and Information Networks Under the BRI Framework.* Actively promote the participation of basic telecommunications enterprises in the construction of the "China-ASEAN Information Port" and the "Six Vertical and Six Horizontal" backbone fiber optic cables in Africa. In 2015 and 2017, China released the *Vision and Action for Promoting the Construction of the Silk Road Economic Belt and the 21st Century Maritime Silk Road* and the *BRI Construction of Maritime Cooperation Vision* (National Development and Reform Commission, Ministry of Foreign Affairs, Ministry of Commerce, 2015, March), proposing to "Plan and Build Intercontinental Submarine Fiber Optic Cable Projects" (National Development and Reform Commission, State Oceanic Administration, 2017 June 20) and "promote joint planning and construction of submarine fiber optic cable projects to improve international communication interconnection".

2. Digital commerce platform construction

Under the mechanisms of China-Myanmar, China-Pakistan and China-Laos economic corridors, China-India regional economic cooperation strategic dialogue and China-Central and Eastern European countries cooperation, the digital economy has become a common concern and actively explored area of cooperation. The accelerated improvement of digital infrastructure lays a solid foundation for the interconnection of the Digital Silk Road. In the eight years since the BRI was launched, the level of information infrastructure connectivity has been effectively enhanced with the joint participation of all parties, the construction of information highways in countries and regions along the BRI route has been further extended. The construction of broadband network infrastructure, mobile Internet and Internet platforms has been speeding up. On March

3, 2017, at the 2nd World Internet Conference, China launched the *BRI Digital Economy International Cooperation Initiative* (2017) with relevant authorities from Laos, Saudi Arabia, Serbia, Thailand, Turkey, and the United Arab Emirates. China has signed cooperation documents with 16 countries to strengthen the construction of the Digital Silk Road. The construction of the Digital Silk Road has become an important component part of the co-construction of the BRI. The construction of the "Digital Silk Road" continues to make efforts in the development of multi-level regional cooperation. Since the introduction of the BRI, numerous domestic cities, such as Shenzhen, Hangzhou and Chongqing, have established cooperation mechanisms with foreign cities around e-commerce, smart cities and other areas of the digital economy, and cooperation among cities in the BRI has been steadily promoted. Meanwhile, the construction of the China-ASEAN Information Port launched in Guangxi Province and the construction of the Sino-Arab Digital (Online) Silk Road carried out in Ningxia Province have advanced practical cooperation from different dimensions of the digital economy, laying the foundation for multi-level regional cooperation.

(iv) Summary

1. Construction of major projects to build a complex infrastructure network

Since the introduction of the BRI, the four-pronged network of land-sea-air-internet connectivity between China and countries along the Belt and Road has taken shape. China-Lao Railway, Sino-Thai Railway Projects such as the Yavan high-speed railway and the Hungarian-Serbian railway are progressing steadily. Projects such as the Gwadar port, the Hambantota port, the Piraeus port, and the Khalifa port are progressing smoothly. Air Silk Road construction has been accelerated, bilateral intergovernmental air transport agreements have been signed with 126 countries and regions, and China-Russia crude oil pipeline, China-Central Asian Energy projects such as the rock gas pipeline and the Sino-Russian rock gas pipeline east line are progressing in an orderly manner. From roads to railways, from sea to air, from oil and gas pipelines to sea and land fiber optic cables, China and the BRI countries have gradually built up a comprehensive land-sea-air transportation, energy and communication composite interconnection network.

2. Significant increase in the level of infrastructure facilitation

While attaching importance to the construction of hardware facilities, our country also attaches importance to improving the level of infrastructure facilitation through innovations in regulations and cooperation models. As of December

2020, the China-Europe Class Trains operated 73 routes, with a total of more than 30,000 trains, covering 92 cities in 21 countries, becoming an important platform for countries along the way to promote interconnection and enhance economic and trade cooperation. In 2020, a total of 12,406 China-EU trains were operated. The annual number of trains exceeded the "10,000" mark for the first time, and the number of trains operated in a single month was stable above 1,000, which played an important role in unblocking the global industrial chain, meeting the consumption needs of the people in the countries along the route and building the "life alleyway" of China-EU to fight against pandemics. In addition, China has signed 16 bilateral and multilateral transport facilitation agreements with 15 BRI countries. The signing of these agreements has greatly enhanced the level of international transport facilitation for BRI countries.

3. Nine years of practice in co-constructing the BRI: Smooth trade

Smooth trade is an important element of the "One Belt, One Road" initiative, which has achieved initial results in trade and investment facilitation, eliminating trade barriers and creating a favourable business environment to promote smooth trade, with the scale of trade continuing to expand, the structure of trade being optimized, trade in services growing rapidly, cross-border e-commerce flourishing, the level of trade facilitation continuing to rise, trade networks being further improved, trade openness and innovation increasing, and countries continuing to promote the wave of economic globalization through the "One Belt, One Road" initiative.

(i) Increasingly prosperous trade in goods and services with countries along the route

In 2020, China's total exports of goods to BRI countries reaches CNY 1,793.26 billion, an increase of 4% over the previous year. From 2014 to 2019, China's total import and export of goods trade with BRI countries exceeded CNY 44 trillion, accounting for an increasing share of China's total goods trade, with an annual average growth rates of 6.1%, higher than the growth rate of China's foreign trade during the same period.

China's trade in services with countries along the route has grown from small to large and steady. In 2018, China's import and export of services trade with countries and regions along the BRI reached US$121.7 billion, up 24.5%

year-on-year, accounting for 15.4% of China's total services trade, up 1% age point year-on-year. At the same time, new business models and modes such as cross-border e-commerce are becoming an important new force in promoting smooth trade flows. As of June 2020, 105 cross-border e-commerce integrated pilot zone have been established in China (see Table 4.1), and each integrated pilot zone enterprise has built more than 1,200 overseas warehouses with global service coverage. Chinese goods are sold through cross-border e-commerce to more than 100 countries and regions that have signed the BRI cooperation document, including Russia, Israel, South Korea and Vietnam, and the scope of online commerce has expanded from Eurasia to many countries in Africa.

Table 3.1 China's comprehensive cross-border e-commerce pilot zone

TIME	Pilot city
March 7, 2015 (first batch)	Hangzhou
January 6, 2016 (second installment)	Tianjin, Shanghai, Chongqing, Hefei, Zhengzhou, Guangzhou, Chengdu, Dalian, Ningbo, Qingdao, Shenzhen, Suzhou
July 24, 2018 (third installment)	Beijing, Hohhot, Shenyang, Changchun, Harbin, Nanjing, Nanchang, Wuhan, Changsha, Nanning, Haikou, Guiyang, Kunming, Xi'an, Lanzhou, Xiamen, Tangshan, Wuxi, Weihai, Zhuhai, Dongguan, Yiwu
December 24, 2019 (fourth installment)	Shijiazhuang, Taiyuan, Chifeng, Fushun, Hunchun, Suifenhe, Xuzhou, Nantong, Wenzhou, Shaoxing, Wuhu, Fuzhou, Quanzhou, Ganzhou, Jinan, Yantai, Luoyang, Huangshi, Yueyang, Shantou, Foshan, Luzhou, Haidong, Yinchuan
April 27, 2020 (fifth installment)	Xiongan New Area, Datong, Manzhouli, Yingkou, Panjin, Jilin, Heihe, Changzhou, Lianyungang, Huaian, Yancheng, Suqian, Huzhou, Jiaxing, Quzhou, Taizhou, Lishui, Anqing, Zhangzhou, Putian, Longyan, Jiujiang, Dongying, Weifang, Linyi, Nanyang, Yichang, Xiangtan, Chenzhou, Meizhou, Huizhou, Zhongshan, Jiangmen, Zhanjiang, Maoming, Zhaoqing, Chongzuo, Sanya, Deyang, Mianyang, Zunyi, Dehong Dai Jingpo Autonomous Prefecture, Yan'an, Tianshui, Xining, Urumqi

Data source: compiled by the author from BRI.com

(ii) Continuous breakthroughs in the construction of trade cooperation platforms

Various trade cooperation platforms provide important vehicles for the promotion of goods and services, investment, and the fostering of new drivers of growth. First, the role of trade and investment promotion platforms has been highlighted. Since the introduction of the BRI, the quality and influence of national exhibition platforms such as the China International Import Expo, the China Import and Export Fair, the International Trade in Services Fair and the China International Fair for Investment and Trade have been continuously enhanced. They have become important platforms for promoting smooth trade flows between China and relevant countries. Second, the overseas economic and trade cooperation zone has become an important support for industrial cooperation. Chinese enterprises are promoting the construction of overseas economic and trade cooperation zones in BRI countries, which plays a positive role in promoting the economic and social development of host country, promoting the industrialization process and industrial upgrading, and deepening economic and trade cooperation. Third, the role of the platform of the border regional economic cooperation working area and cross-border regional economic cooperation working area has been enhanced. China has established 17 border regional economic cooperation working zones, and cross-border regional economic cooperation working zones such as the China-Harbin Khorgos International Border Cooperation Center and the China-Lao Mochan-Maoding regional economic cooperation working zone have become important platforms for border regions to actively participate in the construction of the BRI and carry out border regional economic cooperation working.

The construction of free pilot zone and free trade ports has been accelerating. From August 2013 to September 2020, China approved the establishment of 21 free trade pilot zone in multiple batches (see Table 4.2), forming a reform and opening-up innovation pattern of "1+3+7+1+6+3: covering the east, west, north, south and middle, providing a new institutional supply, experience, and practice model for continuously improving the new system of open economy and exploring high-standard international economic and trade rules. In April 2018, China decided to support the construction of free trade pilot zone on the whole island of Hainan and explore the construction of a free trade port with Chinese characteristics. On June 1, 2020, the *General Plan for the Construction of Hainan Free Trade Port* (*The State Council*, 2020, June 1) was issued, and the construction of Hainan Free Trade Port opened a new journey. China's free trade pilot zone and free trade port deeply promote institutional innovation and open highland construction. It explores the implementation of high-level trade and investment liberalization, as well as facilitation policies. Closely combined with their own geographical characteristics and cultural characteristics, it strengthens exchanges and cooperation with countries along the BRI, promoting the smooth trade of the "One Belt Road", and the construction of a higher level of open view of the new economic system to make positive contributions.

Table 3.2 Construction of free trade pilot zone in China

Date of approval of establishment	Name of the free trade pilot zone
August 2013	China (Shanghai) Free Trade pilot zone
April 2015	China (Guangdong) Free Trade pilot zone, China (Tianjin) Free Trade pilot zone, China (Fujian) Free Trade pilot zone
March 2017	Free Trade in China (Liaoning) pilot zone, Free Trade in China (Zhejiang) pilot zone, Free Trade in China (Henan) pilot zone, Free Trade in China (Hubei) pilot zone, Free Trade in China (Chongqing) pilot zone, Free Trade in China (Sichuan) pilot zone, Free Trade in China (Shaanxi) pilot zone
September 2018	China (Hainan) Free Trade pilot zone
July 2019	China (Shanghai) Free Trade pilot zone Lingang New Area
August 2019	Free Trade in China (Shandong) pilot zone, Free Trade in China (Jiangsu) pilot zone, Free Trade in China (Guangxi) pilot zone, Free Trade in China (Hebei) pilot zone, Free Trade in China (Yunnan) pilot zone, Free Trade Experimental Zone in China (Heilongjiang)
September 2020	China (Beijing) Free Trade pilot zone, China (Anhui) Free Trade pilot zone, China (Hunan) Free Trade pilot zone, China (Zhejiang) Free Trade pilot zone Extended Area

Data source: Compiled by the author from the China Free Trade Zone Service

The FTA network system is gradually taking shape. China has strengthened policy communication and coordination with countries along the route. It has signed or upgraded free trade agreements with many countries and regions, including ASEAN, Singapore, Pakistan, Georgia, Maldives, Chile, and Mauritius. It signed economic and trade cooperation agreements with the Eurasian Economic

Union. By the end of 2020, China had signed 21 FTAs with 25 countries and regions (see Table 4.3), including 7 FTAs with 13 BRI countries, and was promoting negotiations on 18 FTAs. On November 15, 2020, after eight years and 31 rounds of formal negotiations, representatives of member countries formally signed the Regional Comprehensive Economic Partnership Agreement (RCEP) in the presence of the leaders of 15 countries, marking the successful launch of the world's largest free trade area.

Table 3.3 Selected free trade agreements signed by China

Signatory	**Start-up time**	**Signing time**	**Conclude an agreement**
China – Chile	November 2004	November 2005	*China – Chile Free Trade Agreement*
China-Pakistan	April 2005	November 2006	*China-Pakistan Free Trade Agreement*
China-New Zealand (with upgrades)	November 2004	April 2008	*China-New Zealand Free Trade Agreement* *China-New Zealand Protocol on the Upgrading of the Free Trade Agreement between the Government of the People's Republic of China and the Government of New Zealand*
China-Singapore	October 2006	October 2008	*China-Singapore Free Trade Agreement*
China-Peru	September 2007	April 2009	*China-Peru Free Trade Agreement*
China-ASEAN	November 2002	August 2009	*China-ASEAN Free Trade Area Investment Agreement*
China-Costa Rica	November 2008	April 2010	*China-Costa Rica Free Trade Agreement*
China-Iceland	December 2006	April 2013	*China-Iceland Free Trade Agreement*
China-Switzerland	January 2011	July 2013	*China-Switzerland Free Trade Agreement*
China-Korea	May 2012	Jun2015	*China – Korea Free Trade Agreement*
China-Australia	April 2005	Jun2015	*China-Australia Free Trade Agreement*

Table 3.3 Continued

Signatory	**Start-up time**	**Signing time**	**Conclude an agreement**
China-ASEAN ("10+1" upgrade)	August 2014	November 2015	*Protocol between China and ASEAN on the Revision of the China-ASEAN Framework Agreement on Comprehensive regional economic cooperation Cooperation and Part of the Agreement thereunder*
China-Georgia	December 2015	May 2017	*China – Georgia Free Trade Agreement*
China-Maldives	September 2015	December 2017	*China-Maldives Free Trade Agreement*
China-Singapore (upgrade)	November 2015	112018	*Protocol on the Upgrading of the Free Trade Agreement*
China-Mauritius	December 2017	October 2019	*China-Mauritius Free Trade Agreement*
China-Pakistan Phase II	Mar 2011	April 2019	*Protocol between the Government of China and the Government of Pakistan on Amendments to the Free Trade Agreement*
Chinese Mainland-Hong Kong-Macao			*Mainland and Macao Closer Economic Partnership Arrangement*
China-Chile (upgrade)	November 2016	July 2020	*Protocol between the Government of China and the Government of Chile amending the Free Trade Agreement and the Supplementary Agreement on Trade in Services to the Free Trade Agreement*
China-Cambodia	January 2020	October 2020	*China-Cambodia Free Trade Agreement*
15 Member States	2012	November 2020	*Regional Comprehensive Economic Partnership Agreement* (RCEP)

Source: Compiled by the author from the China Free Trade Zone Service

(iii) Steady improvement in the level of trade facilitation

China has promoted cooperation on "smart customs, smart borders and smart connectivity" among BRI countries. It established a pilot "single window" for international trade and continued to promote port connectivity. In 2019, China signed a total of 198 documents on customs inspection and quarantine cooperation with foreign countries, involving 89 documents from BRI countriess. It has also promoted agricultural secondary product fast customs clearance, and opened seven "green alleyway" border crossings with Kazakhstan, Tajikistan, Kyrgyzstan, Vietnam, Mongolia, and other neighboring countries for agricultural secondary product fast customs clearance one after another, shortening customs clearance time by 90%. Finally, it has dromoted the facilitation of customs clearance for people, and set up 89 BRI people facilitation alleyways at 18 air and land crossings (Port Office, 2021, September 23).

China is deeply involved in the formulation of international tax rules and has established close cooperation with major international organizations such as the Organization for Economic Cooperation and Development (OECD), the International Mombasa-Nairobi Standard Gaugeary Fund, the United Nations Committee of Experts on International Cooperation in Tax Matters, and the Tax Administration Forum. It has established bilateral tax cooperation mechanisms with the tax authorities of 121 countries and regions, including those along the "Belt and Road", and signed tax treaties with 111 countries and regions, including 80 BRI countries and regions. In 2019, China successfully held the first BRI Tax Administration Cooperation Forum, signed the BRI Tax Administration Cooperation Mechanism Memorandum of Understanding with the BRI countries built the BRI Tax Administration Cooperation Mechanism, and formed the BRI Tax Administration Capacity Promotion Alliance. As one of the important initiatives to promote trade and investment liberalization and facilitation, a joint BRI tax collection and administration cooperation mechanism has been established, with 36 Council members and 28 observers by 2020 (China BRI website, 2021).

(iv) Continued improvement in the level of trade openness

China has taken the initiative to open its market to the world, becoming the world's second largest import market for 11 consecutive years, with imports accounting for more than 10% of the world's total imports. Since the international financial crisis in 2008, China's imports have contributed one-sixth of the increase in global imports and are a booster and stabilizer of the global economic recovery. At the same time, the rapid development of foreign trade has promoted the integration of China's economy with the world economy. China is already a major trading partner of more than 120 countries and regions and has increasingly close economic and trade ties and mutually beneficial exchanges with countries around the world. As the world's first import-themed large-scale national exhibition, the China International Import Expo has been successfully held for three sessions, including the third session held in 2020, which was held both online and offline. It attracted more than 3,800 enterprises from more than 180 countries, regions, and international organizations worldwide, with a cumulative intended turnover of US$2.62 billion, an increase of 2.1% over the previous session. China's average tariff level has dropped from 15.3% at the time of its accession to WTO to 7.5% at present; it has signed or upgraded free trade agreements with numerous countries and regions, including ASEAN, Singapore, Pakistan, and Georgia; it has signed economic and trade cooperation agreements with the Eurasian Economic Union. A network system of free trade zones with countries along the route has gradually taken shape. At the same time, China also actively promotes the construction of overseas cooperation parks, and all kinds of Chinese enterprises follow the market-oriented legal principle to independently go to the BRI countries to build cooperation parks, creating new sources of tax revenue and employment channels for the countries along the route.

"China's open door will not be closed, but it will only be opened much wider." China is open to the outside world on all fronts, at multiple levels and in a wide range of fields, and it is committed to developing a higher level of open economy. 21 free trade pilot zone form a "goose array". Hainan free trade port set sail. Foreign trade new business, new models flourish, cross-border e-commerce integrated pilot zone increased to 105. Foreign investment law and regulations to optimize the business environment officially implemented. The national and free trade pilot zone foreign investment access negative list entries were reduced to 33 and 30. Financial industry, manufacturing, services, and other areas to expand open, capital, talent, science and technology and other areas of international cooperation to expand (China Free Trade Zone Service Network). From creating a new platform for opening to the outside world, to building a new system of open economy, to accelerating the construction of a new development pattern, China's new round of high-level opening up to the outside world is increasingly accelerated, breaking a new picture of comprehensive opening up.

(v) Summary

1. Significant increase in the scale and efficiency of trade

In the eight years since the BRI was launched, it has significantly expanded the scale of trade among countries and regions along the route, boosted the economic growth of the countries and improved the livelihoods of the countries along the route. China firmly upholds the multilateral trading system, effectively implements the WTO *Trade Facilitation Agreement,* and promotes the free flow of global commodity factors. China has launched the *BRI Trade Facilitation Cooperation Initiative* (*Ministry of Commerce,*2017, May 15), with the active participation of 83 countries and international organizations. It has established working groups on trade facilitation with 12 countries to quickly solve problems arising from bilateral economic and trade cooperation and promote the development of economic and trade cooperation. Customs inspection and quarantine cooperation has been deepening. Since the first BRI International Cooperation Summit Forum in May 2017, China has signed more than 100 cooperation documents with countries along the route, achieving quarantine access for more than 50 kinds of agricultural products and foods. The construction of the "green alleyway" fast-track customs clearance for agricultural products between China and Kazakhstan, Kyrgyzstan and Tajikistan has been actively promoted, and the customs clearance time for agricultural products has been reduced by 90% (China BRI website, 2021).

2. Trade transformation and effective upgrading

China has seized the historical opportunity of global industrial transfer and rapid development of global trade, actively promoted high-quality trade development, and continuously optimized its trade structure. Positive results have been achieved in optimizing and improving the layout of international markets, domestic regional layout, commodity structure, business entities and trade patterns, and the status of a major trading country has been more consolidated. The layout of the international market was significantly optimized, the layout of domestic regions was more reasonable, and the commodity structure was continuously upgraded. For example, exports of electromechanical products accounted for 58.1%, and exports of high-quality, hi-tech, high value-added products integral circuit and complete automobiles grew by 25.4% and 8.6%, respectively. The contribution of general trade strengthened, with the share of exports increasing by 1.5% age points to 57.8%, adding value to the chain and trade competitive power. At the same time, various types of business entities are energetic.

For the first time, private enterprises overtook foreign-invested enterprises to become China's number one foreign trade entity. The structure of trade patterns was further optimized, and the proportion of general trade imports and exports increased significantly.

4. Eight years of practice in co-constructing the BRI: Financial inclusion

(i) Setting up relevant financial institutions to provide ongoing support

Up to now, around the BRI construction funds financing needs, China has led the establishment of the Asian infrastructure investment bank, the BRICS New Development Bank, and the Silk Road Fund. The business coverage of these financial institutions and the operating model is not the same. The AIIB is a regional multilateral development institution in Asia of an intergovernmental nature that focuses on supporting infrastructure development. The BRICS New Development Bank excludes Central Asian countries. The Silk Road Fund is a form of fund investment that provides financing support for relevant projects with the goal of making a profit. These newly established financial institutions are of even greater strategic importance in promoting the construction of the Silk Road Economic Belt.

The Asian Infrastructure Investment Bank (AIIB), established at our initiative, is dedicated to financing the construction of BRI countries. In October 2014, representatives of the first 22 intended founding members signed the *Preparation for Asia's Infrastructure Investment Bank Memorandum Entry.* In June 2015, representatives of the 50 intended founding members co-signed the *Asia Infrastructure Investment Bank Agreement* (Ministry of Foreign Affairs, 2016, October 13), and 7 other countries subsequently signed the Agreement in succession by the end of 2015. In December 2015, the Asian Infrastructure investment bank Agreement met the statutory conditions for entry into force and the AIIB was officially declared (Asia Infrastructure Investment Bank official website, 2015, December). As a multilateral development bank and the world's first multilateral financial institution involving a Chinese initiative, the AIIB's mission is to finance tomorrow's infrastructure, with the aim of unlocking new capital, technologies, and approaches to address climate change by investing in sustainable infrastructure and building linkages with Asia and the world.

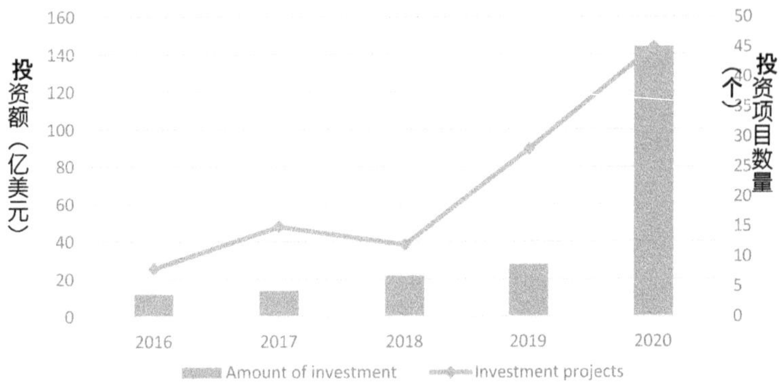

Figure 3.1 Number of AIIB investment projects and investment value by year
+Data source: Asia Infrastructure Investment Bank official website

By the end of 2020, AIIB had invested in 108 projects and approved a cumulative US$22.02 billion in infrastructure investments (see Figure 4.1 for details). To meet the needs of its members, the AIIB is working to enhance its local currency financing capacity while disbursing loans in the US dollar as the main currency. The current lending currencies are mainly the US dollar, the Chinese CNY and the Euro. In terms of sector distribution, the AIIB's investment areas focus on the main purposes and key areas of focus of the AIIB, specifically on energy, transportation construction, water resources, financial inter-mediation, and urban construction, and it also increased investment in public health during the global pandemic of 2020.

The AIIB has now reached key achievements in five areas: (1) rapid growth in global membership, from 57 founding members at the start of its operations to 103 approved members, then only to the World Bank; (2) three major rating agency awarded the highest credit rating and the AIIB now has a AAA rating with a very stable development outlook; (3) the AIIB has been granted permanent observer status in the deliberations of two major development-focused bodies, the UN General Assembly and the Economic and Social Council; (4) governance mechanisms that enhance efficiency and accountability, in line with the principles and development directions of transparency, openness, independence and accountability; and (5) strategies for raising funds for all major infrastructure have been approved and are being implemented.

On July 15, 2014, during the sixth meeting of BRICS leaders, the five BRICS countries signed the Agreement on the Establishment of a New Development Bank, deciding to establish the BRICS Bank, headquartered in Shanghai, China, whose main objective is to provide developing countries with the necessary funds for infrastructure development and thus to promote the sustained economic development of BRICS countries. The BRICS New Development Bank (NDB) was officially opened on July 21, 2015. It became fully operational after the signing of the Headquarters Agreement with the Government of the People's Republic of China and the Understanding memorandum entry with the Shanghai Municipal People's Government on July 27, 2016. By the end of 2020, the BRICS New Development Bank has been approved 67 projects totaling US$24.4 billion. In addition, the BRICS New Development Bank has become an observer to the United Nations General Assembly, laying a solid foundation for active and fruitful cooperation with the United Nations (the BRICS New Development Bank, 2020).

On November 8, 2014, President Xi Jinping announced at the Dialogue on Strengthening Connectivity Partnerships that China would contribute US$40 billion to establish the Silk Road Fund. The Fund is a medium- to long-term development investment fund, jointly funded by the China Foreign Exchange Reserve, China Investment Corporation, The Import-Export Bank of China, and China Development Bank and established in accordance with the Company Law of the People's Republic of China. On December 29, 2014, Silk Road Fund Limited Liability Company was incorporated and officially operated in Beijing. The Fund promotes connectivity with the BRI countries and regions by providing financing support for infrastructure, resource development, industrial and financial cooperation projects, and provides good hardware conditions for enterprises to invest abroad. By the end of 2019, the Silk Road Fund has committed to invest about US$12.3 billion and has signed 34 projects in Central Asia, Russia, South Asia, Southeast Asia, West Asia and North Africa, Europe, North America and other regions and countries, with 70% of the investment funds applied to large-scale international cooperation projects such as power plant development, infrastructure construction, port shipping and high-end manufacturing (Silk Road Fund Ltd. Official website, 2021).

(ii) Continued increase in financial security and diversification of forms of investment cooperation

Over the past eight years, AIIB, BRICS New Development Bank, Silk Road Fund, Chinese banks, and Chinese enterprises have provided financing support for a number of projects along the BRI route, providing financial guarantee for the smooth start and implementation of relevant projects through loans, establishment of special investment funds, direct investment and other forms. For example, the China Development Bank has implemented a special loan of CNY 250 billion for the BRI over a three-year period from 2017 to 2020. The China-Africa Development Fund has identified a cumulative investment of US$4.4 billion in 90 African projects (China Development Bank, 2021).

1. The number of investments continues to expand

In the construction of the BRI, China has continued to increase its investment in BRI countries. In 2020, China invested US$17.79 billion in BRI countries, an increase of 18.3%, accounting for 16.2% of the country's foreign investment; investment in key industries achieved faster growth, with investment in equipment manufacturing, information technology, scientific research and technical services growing by 21.9%, 9.6% and 18.1% respectively (Data from the "Go Global" public service platform). Since 2015, the Ministry of Commerce has been publishing China's investment cooperation with BRI countries (see Figure 4.2), and as can be seen from the publicly available data, the amount of non-financial direct investment in countries along the route has remained steadily increasing over the past six years, while the selection of investment targets has also remained basically stable.

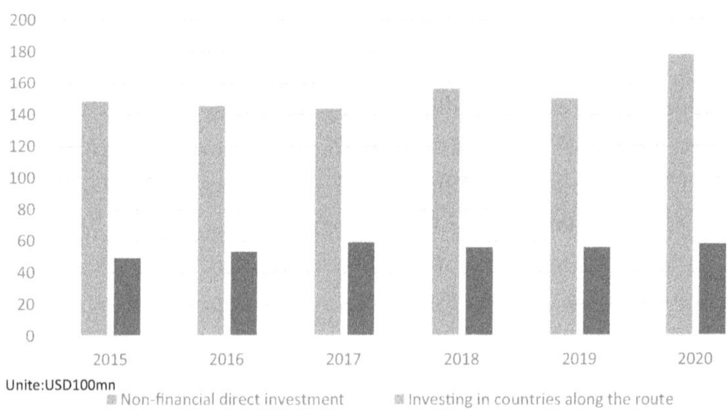

Figure 3.2 China's investment cooperation in BRI countries
Data source: Ministry of Commerce of the People's Republic of China

2. Diversification of financial cooperation

Financial cooperation among the BRI countries is essentially a sub-regional financial cooperation across borders. By promoting the construction of Mombasa-Nairobi Standard Gaugeary cooperation system, investment and financing system, financial service system, credit cooperation system and financial supervision cooperation system, we promote the free flow of funds in the BRI region and enhance the supervision and coordination among countries along the route, so as to improve the efficiency of capital allocation and realize the effective identification, response and early warning of various financial risks in the region. Financial cooperation in the BRI countries not only goes range of action beyond the market and sovereign jurisdiction of a single country, but also its risk regulation and institutional norms need to be carried out across borders. According to the BRI country cooperation degree appraisal report (2018) released by the National Information Center, the average score of the BRI country cooperation degree index in 2018 is 47.12, an increase of 2.01 and 3.57 points compared with 2017 and 2016, respectively. The level of the BRI country cooperation is climbing year by year. In terms of financial cooperation, China has carried out various forms of cooperation with countries along the route in such areas as currency, investment and financing, financial services, credit and financial supervision.

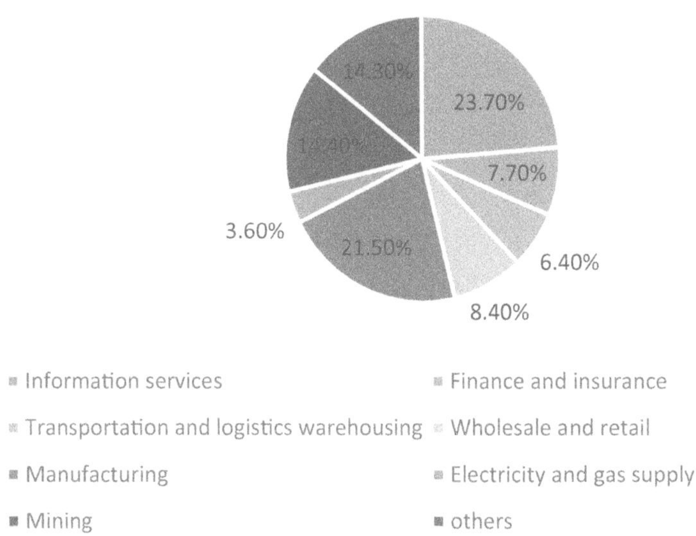

Figure 3.3 Sectoral distribution of China's investment in BRI countries
Data source: EMIS

3. Diversification of investment sectors

As of June 2020, the top five industries in terms of cumulative investment in China along the BRI route were: information services (23.7%), manufacturing

(21.5%), extractive industries (14.4%), wholesale and retail (8.4%), and finance and insurance (7.7%) (see Figure 4.3). In terms of investment amount, the extractive industry, due to its more capital-intensive attributes, accounted for 35.5% of the cumulative investment, accounting for the first place. Manufacturing, information services and professional services accounted for a larger cumulative investment quantity, but the cumulative investment amount accounted for a significantly lower share, at 15.0%, 11.7% and 1.0% respectively. In addition, electricity, gas and water supply, transportation and logistics and warehousing all account for more than 10% of the cumulative investment, although the cumulative number of investments is only 3.6% and 6.4% (see Figure 4.4).

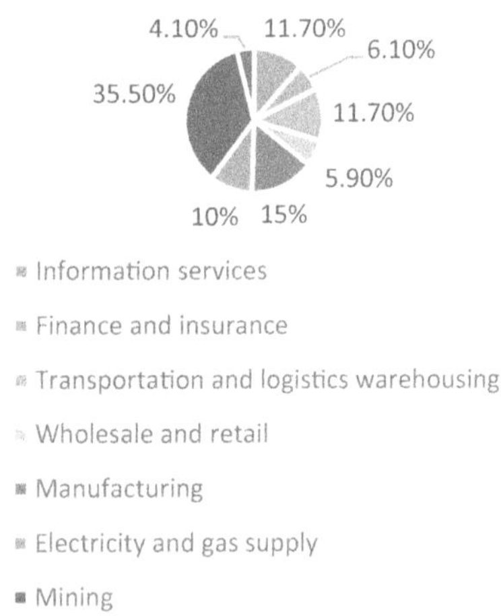

Figure 3.4 Sector distribution of China's investment in BRI countries
Data source: EMIS

(iii) Steady increase in the internationalization of the CNY

In October 2016, the CNY was officially included in the Special Drawing RightSDR (SDR), and in 2017 the IMF officially announced the CNY as a global reserve currency. These advances cannot be separated from the many cooperation opportunities brought by the BRI. Over the past eight years, the acceptance and use of the CNY in the BRI region has been increasing, as evidenced by the outstanding progress in foreign exchange transactions between the CNY and the BRI currencies, the CNY becoming one of the important reserve currency options in the region and the expansion of the CNY clearing network and bank layout.

With the BRI closer economic and trade exchanges, the CNY foreign exchange transactions against the relevant countries' currencies continue to increase. In the national inter-bank foreign exchange market, the CNY realized direct trading against nine relevant national currencies. In the inter-bank regional trading market, China launched foreign exchange trading of CNY against Kazakhstan Tenge, Mongolia Tugrik and Cambodia Riel in Xinjiang, Inner Mongolia and Guangxi respectively, which is of positive significance in reducing transaction costs and exchange rate risks for enterprises and enhancing regional economic and trade exchanges.

With the continuous promotion of the BRI construction and the development of CNY internationalization, China has established bilateral local currency swap arrangements with the Mombasa-Nairobi Standard Gaugeary authorities of 21 relevant countries. The ASEAN "10+3" is considering the inclusion of CNY in the Chiang Mai Agreement arrangement and CNY is beginning to show the characteristics of a regional reserve currency. Relevant countries have included the CNY in their foreign exchange reserves. Russia has even increased its share of CNY reserves to around 15% in the context of "de-dollarization". According to the latest data from the IMF Petroleum Fund organization (IMF), as of the first quarter of 2020, the size of global CNY reserves reached US$221.48 billion, accounting for 2.02% of total international reserves, the highest level ever.

In addition, eight countries related to the BRI have already established CNY clearing arrangements. The CNY cross-border payment system covers more than 60 countries and regions. The Chinese Commercial Bank has become the main force in the promotion of the use of the BRI CNY. In addition, a number of foreign banks are also optimistic about the prospects of BRI and CNY businesses and have increased their resources and financial cooperation.

(iv) Summary

The aim and objective of the BRI is to break the financing bottleneck, establish a new financial mechanism and form an investment and financing development platform that is complementary to the existing multilateral development banks. Strengthen financial regulatory cooperation and improve institutional arrangements for risk response and crisis management. The main initiatives currently being implemented include promoting the construction of the Asian Mombasa-Nairobi Standard Gaugeary stability system, investment and financing system and credit system; building and playing the role of the Asian Infrastructure Investment Bank, BRICS China Development Bank and Silk Road Fund;

building a regional financial risk early warning system and forming an exchange and cooperation mechanism to deal with cross-border risks and crisis management. In the past eight years, the practice of BRI in capital financing is mainly reflected in three aspects, such as the relevant organization structuring, the provision of financial protection and the internationalization of the CNY. As of the end of 2020, the overall results achieved include: the AIIB membership has reached 103, representing 79% of the global population and 65% of the global GDP respectively. The Silk Road Fund has signed 47 various projects with mainly equity investment, with a committed investment amount of US$17.8 billion, covering many BRI countries. Chinese banks have continued their overseas layout, setting up Chinese banks in more than 60 countries/regions affiliated agency, of which 11 Chinese banks have set up a total of affiliated agency 29 in countries and regions along the BRI route. There are 42 direct participants and 1103 indirect participants in the CNY cross-border payment system. In 2019, China's cross-border CNY receipts and payments with BRI countries exceeded CNY 2.73 trillion, accounting for 13.95% of total CNY cross-border receipts and payments in the same period, and by the end of 2019, China had signed local currency swap agreements with 21 BRI countries and established CNY clearing mechanism arrangements in eight BRI countries (China BRI website, 2021).

5. Eight years of practice in co-constructing BRI: Interpersonal contacts

(i) Deepening humanistic exchanges

Over the past eight years, the co-construction of BRI has provided new opportunities for figamong the countries along the route through the establishment of important humanities exchange activities and humanities exchange mechanisms. China and the countries along the route have mutually organized cultural years, art festivals, film and television bridges, the Silk Road Book Project, etc. The BRI humanities exchange and cooperation has achieved positive results. Humanities exchange and cooperation is the root and soul of the BRI. To further implement this major initiative, it is necessary to vigorously promote the exchange and mutual appreciation of civilizations and cultural intermingling and interpersonal contact with countries along the Silk Road.

In terms of two-way study, China is currently the world's largest source country for international students, with about 1.6 million students studying abroad and about 1.4 million currently overseas. In terms of studying in China, the

Chinese government has set up the Silk Road Scholarship Program to help cultivate BRI talents, while building the brand of "Study in China". The proportion of students studying in China with academic qualifications has increased year by year, reaching 54.6% in 2019 (Ministry of Education official website, 2020). In addition, China has continued to strengthen Sino-foreign cooperative education, with 2,282 Sino-foreign cooperative education institutions and projects of various types and levels currently in operation.

The Confucius Institute is an important platform for the world to get to know China. Since 2004, when the first Confucius Institute in the world was signed in Uzbekistan, Confucius Institutes have always adhered to the model of equality and mutual benefit and win-win cooperation between China and foreign countries, starting from language, blending with culture and communicating with the heart, making important contributions to promoting Chinese and foreign cultural exchanges and enhancing friendship between Chinese and foreign people. Since the launch of the BRI, Confucius Institutes in the relevant countries have continued to train language talents, carrying out cultural activities and promoting interpersonal exchanges. It becomes an important force in promoting the construction of the BRI. As of September 2018, 137 Confucius Institutes and 130 Confucius Classrooms have been established in 53 of the 65 countries in the BRI, accounting for about a quarter of the total number of Confucius Institutes worldwide (Economic Times, 2018, September 11). Confucius Institutes have provided a numerous talented persons who are proficient in Chinese language and understand China for the construction of the BRI. According to the data, Confucius Institutes have dispatched 34,000 directors, teachers, and volunteers, given away 4.51 million copies of Chinese language teaching materials, and trained more than 60,000 Chinese language teachers through the Confucius Institute Scholarship and other channels (Economic Daily, 2018, September 11). The Confucius Institute has established a series of Chinese teaching systems from kindergartens to schools and universities, from academic to non-academic education, from "zero starting point" to high-end translation and academic research. It has trained a total of 2.04 million students of all kinds (Economic Daily, 2018, September 11)

(ii) Fruitful cooperation in science, technology, and innovation

Cooperation in science and technology innovation is an important element of the co-construction of the BRI, which provides lasting impetus for the construction of the BRI. At the same time, the construction of the BRI provides a broad space for cooperation in science and technology innovation. Over the past eight years, science, technology, and innovation cooperation has been further developed in countries along the route with the BRI Union of National Scientific Organizations, the BRI International Seminar on Science, Technology and Innovation, and Memorandum entry of Science, Technology, and Innovation Cooperation, etc. All of these have played a key role in addressing common challenges such as climate change, natural disasters and life and health.

In 2017, President Xi Jinping initiated the launch of the *BRI Science and Technology Innovation Action Plan* (2017, May 14) at the BRI International Cooperation Summit Forum, focusing on the implementation of four actions, including science and technology humanities exchanges, joint laboratories, cooperation in science and technology parks, and technology transfer. Under the inspiration of the BRI science and technology innovation cooperation initiative, China had signed intergovernmental agreements on science and technology cooperation with 49 BRI countriess by the end of 2019, and it launched a series of science and technology partnership programs with countries along the route. Over the past eight years, four training courses on the management and construction of science and technology parks and incubators have been held annually, training nearly 100 participants and sharing the experience and models of China's science and technology park development. Delegations from Egypt, South Africa, Iran and other countries and diplomats in China have been arranged to visit Zhongguancun in Beijing, Zhangjiang in Shanghai, Chengdu, and other national high-tech zones to actively dock innovation cooperation. Chinese delegations have visited the Philippines, Iran, South Africa, and other countries to conduct docking and negotiation on science and technology park cooperation; localities are encouraged and supported to cooperate with BRI co-construction countries on science and technology parks through the co-construction of innovation parks, incubators, accelerators, and innovation centers, etc.

(iii) Continued enhancement of foreign assistance

Since the introduction of the BRI, China has steadily increased the scale of its foreign aid funding and further expanded the scope of its assistance. From 2013 to 2018 alone, China's foreign aid amounted to CNY 270.2 billion, including non-reimbursable aid, interest-free loans, and concessional loans. Of this amount, CNY 127.8 billion in non-reimbursable aid, accounting for 47.30% of total foreign aid, is focused on helping other developing countries build small to medium social welfare projects and implement human resources development cooperation, technical cooperation, material assistance, South-South cooperation assistance funds and emergency humanitarian assistance projects. CNY 11.3 billion in interest-free loans, accounting for 4.18% of total foreign aid, is mainly used to help other developing countries build social public facilities and livelihood projects. Foreign aid concessional loans of CNY 131.1 billion, accounting for 48.52% of total foreign aid, are provided to help other developing countries build productive project and large and medium-sized infrastructure with economic and social benefits, providing equipment set, electromechanical products, technical services, and other materials, etc. (Information Office of the State Council of the people's Republic of China, 2021)

In 2015, on the 70th anniversary of the United Nations, China announced that it would provide an initial US$2 billion to establish a South-South cooperation assistance fund to support developing countries in implementing the 2030 Agenda for Sustainable Development. In 2017, at the BRI International Cooperation Summit, the Chinese Government announced that it would provide CNY 60 billion to developing countries and international organizations participating in the construction of the BRI to build livelihood projects, US$1 billion in targeted funds for cooperation in poverty reduction, agriculture, industry and trade promotion, and support countries along the route to implement poverty reduction projects such as "happy homes", "love and help" and "rehabilitation and medical help", while providing CNY 2 billion in food aid to improve the basic living conditions of poor people in countries along the route (China's International Development Cooperation in a New Era white paper). In terms of historical data, after the financial crisis in 2008, China's total aid to BRI countries gradually stabilized to about 50% of China's global aid, and the share of investment stock in countries along the route in China's global investment stock rose rapidly to about 80% after 2013 (Zhang Yuan, 2018).

Over the past eight years, the construction of the BRI has involved numerous livelihood projects. China has implemented subsidized f housing, water supply, medical care, education, rural roads, assistance to vulnerable groups and other livelihood projects in the countries where the BRI is being built, helping to make up for the shortcomings in infrastructure and basic public services. We help countries such as Côte d'Ivoire, Cameroon, Ethiopia, and Djibouti to build water supply systems to solve the problems of difficult drinking water and poor water quality for the people. Hospitals built for Sri Lanka, Senegal, Guinea, Niger, Mozambique, the Democratic Republic of the Congo, South Sudan, Jamaica, Suriname, Dominica, and other countries have been built to improve the level of local medical services and make it easier for people to see a doctor. It helps

Belarus build social security housing and improve the housing and living conditions of vulnerable groups.

(iv) Increasing awareness and action on ecological protection

To promote the construction of green BRI, the International Alliance for Green Development BRI was formally established in April 2019 at the Green Road Sub-Forum of the Second BRI International Cooperation Summit, creating a platform for policy dialogue and communication, environmental knowledge and information, and green technology exchange and transfer for the BRI green development cooperation. The sub-forum also officially launched the BRI ecological and environmental protection data service platform, and released the Green Efficient Refrigeration Action Initiative, the Green Lighting Action Initiative, and the Green "Go Global" Action Initiative. Among them, the establishment of the BRI international alliance for green development, along with the green BRI construction of the depth, and constantly expand China's green circle of friends. By the end of 2019, China had signed nearly 50 bilateral and multilateral ecological and environmental protection cooperation documents with co-construction countries and international organizations (the second BRI International Cooperation Summit Green Roads disclosure data). Countries along the route are actively promoting the construction of platforms and building a solid foundation, such as the launch of the BRI green supply chain platform, the establishment of the Lancang-Mekong River Environmental Cooperation Center, the establishment of the Sino-Cambodian Environmental Cooperation Center with the Cambodian Ministry of Environment, the preparation of the China-Africa Environmental Cooperation Center in Kenya, the preparation of the Sino-Lao Environmental Cooperation Office in Laos and other ecological and environmental cooperation platforms. The opening of the BRI ecological and environmental protection big data service platform, through information sharing and consultation service tools, share the green development concepts and practices of various countries, promote green trade, green investment, green infrastructure construction and green industry technology exchange, and promote the BRI ecological and environmental protection cooperation.

By the end of 2020, 25 typical projects of green development practices in the construction of the BRI have been under construction and completed. 36 cases of water pollution control technology and 52 cases of airborne pollution control have been completed (Zhou Guomei, 2020). The joint BRI has formed ten international alliances for green development, including biodiversity and ecosystems, green energy and energy efficiency, green finance and investment, environmental quality improvement and green cities, South-South cooperation and sustainable development goals, green technology innovation and corporate social responsibility, sustainable transport, global climate governance and green transformation, environmental laws and regulations and standards, as well as the Ocean Destiny Community and marine environmental governance, etc. Hundreds of domestic and foreign partners have been involved in the BRI ecological and environmental protection big data service platform, and five sub-platforms have been built, including the Shanghai Cooperation Organization environmental information sharing platform, the China-ASEAN environmental information sharing platform, the green supply chain, the international think-tank platform for environmental protection technology and the ecological and environmental protection big data service platform of the China-Mongolia Economic Corridor.

(v) Summary

The purpose and goal of the BRI is to build the foundation of diplomacy and public opinion ininterpersonal exchanges. Through humanistic exchanges, cultural cooperation, and friendly interpersonal activities, we promote understanding of public sentiment, communication, and connection between people's hearts, consolidate the foundation of cooperation and provide support for talents. The main initiatives currently being implemented include extensive cooperation in the fields of science, education, culture, sports, media, medicine, health, tourism and civil contacts, and various forms of activities such as fairs, film festivals, cultural festivals, education/training and Chinese studies conferences. Over the past eight years, the practice of the BRI in interpersonal exchanges is mainly reflected in four areas: humanities exchanges, cooperation in science and technology innovation, foreign aid, and ecological protection. By the end of 2020, the effect achievements achieved include: 49 national cultural centers, 550 Confucius Institutes and 1,172 Confucius Classrooms have been built (People's Daily Overseas Edition, 2019, December 10), and 2,629 pairs of friendship cities and provincial and state relations have been established with 136 countries in 2019, (Beijing News, 2019, April 5); and visa-free or visa-on-arrival has been achieved with 57 countries and regions along the route in 2019, (Information Office of the State Council, 2019, April 22). "Two-way study abroad, cooperative education and teacher training programmes along the Silk Road are well under way, and countries and regions along the route have become a net inflow of international tourists.

References

Aufiya, M. A. (2017). Indonesia's global maritime fulcrum: Contribution in the Indo-Pacific Region. Andalas Journal of International Studies (AJIS), 6(2), 143–158.

Bai, Jie. (2018). Xi Jinping met with Vietnamese Prime Minister Nguyen Chun Phu: Speed up the docking of "Belt and Road Initiative" and "two corridors and one Circle" (2018, November 4). Xinhuanet, https://www.yidaiyilu.gov.cn/xwzx/xgcdt/70604.htm.

Beijing News, (2019, April 5). China has relations with 136 countries and 2629 sister cities and provinces and states https://finance.sina.com.cn/china/gncj/2019-04-25/doc-ihvhiewr8186083.shtml

"Bright Road" in Kazakhstan, (2017, April 20), China Web, http://keywords.china.org.cn/2017-04/20/content_40656151.htm.

China BRI website. (2017). BRI Digital Economy International Cooperation Initiative. https://www.yidaiyilu.gov.cn/xwzx/gnxw/38241.htm.

China BRI website. (2018). Outlook of aviation market Made in China has a way out by grasping the "the Belt and Road" https://www.yidaiyilu.gov.cn/.

China BRI website. (2019). Direct flights between China and 45 countries along the "the Belt and Road" Portal. https://www.yidaiyilu.gov.cn/.

China BRI website. (2019) Jointly building the "the Belt and Road" initiative: progress, contributions and prospects. https://www.yidaiyilu.gov.cn/.

China Development Bank. (2021). annual report2020:High Quality Development of Belt and Road Initiative. http://www.cdb.com.cn/gykh/ndbg_jx/2020_jx/.

China Export Credit surety company, (2016). https://www.sinosure.com.cn/gywm/gsjj/xxpl/xxgg/index.shtml.

China International Import Expo. (2020). The accumulative intended turnover of the third China International Fair was 72.62 billion US dollars, up 2.1% from the previous one https://www.ciie.org.

Fang Jiangshan. (2019). Jointly tell stories of struggle, success and happiness in jointly building the "the Belt and Road". China Economic Weekly. https://m.gmw.cn/baijia/2019-04/26/32783126.html

Liu Jin. (2020). China has signed agreements on mutual recognition of academic qualifications and degrees in higher education with 54 countries. Wenhui client. https://wenhui.whb.cn/third/baidu/202009/06/369472.html

Liu Meng. (2019). 2018 "the Belt and Road" memorabilia: these major changes have taken place in the joint construction of the "the Belt and Road" China BRI website, https://www.yidaiyilu.gov.cn/xwzx/roll/76800.htm

Ministry of Commerce data, Ministry of Commerce, (2021, August 24). http://fec.mofcom.gov.cn/article/fwydyl/zgzx/202108/20210803190898.shtml.

Ministry of Commerce. (2017, May 15). BRI Trade Facilitation Cooperation Initiative, Belt and Road Portal, https://www.yidaiyilu.gov.cn/zchj/sbwj/14163.htm.

Ministry of Commerce. (2021). The total volume of goods trade with countries along the "the Belt and Road" in 2020 will be $1.35 trillion. http://chinawto.mofcom.gov.cn/article/e/r/202101/20210103035381.shtml

Ministry of Finance of China et al. (2017, May 16) Guiding Principles on Financing the Development of the Belt and Road The approval of the BRI Financing Guidelines. China Belt and Road Initiative Wesbsite. https://www.yidaiyilu.gov.cn/zchj/qwfb/13767.htm

Ministry of Foreign Affairs. (2016, October 13), Asia Infrastructure Investment Bank Agreement Belt and Road Portal. https://www.yidaiyilu.gov.cn/yw/qwfb/1672.htm.

Mongolia's "Prairie Road", (2017, April 20), China net, http://keywords.china.org.cn/2017-04/20/content_40656252.htm.

National Development and Reform Commission, Ministry of Foreign Affairs, Ministry of Commerce. (2015, March). Promote the vision and action of jointly building the Silk Road Economic Belt and the 21st Century Maritime Silk Road. Belt and Road Portal. https://www.yidaiyilu.gov.cn/wcm.files/upload/CMSydylgw/201702/201702070519013.pdf

National Development and Reform Commission, State Oceanic Administration. (2017, June 20), Vision for Maritime Cooperation under the Belt and Road Initiative Belt and Road Portal. https://www.yidaiyilu.gov.cn/zchj/jggg/16621.htm.

National Development and Reform Commission. (2016. October 24). China and Kazakhstan jointly released the Silk Road Economic Belt Construction and Bright Road New Economy Policy Matching Cooperation Plan. https://www.yidaiyilu.gov.cn/yw/qwfb/2163.htm

National Rail Group. (2020). Concerned that the China-EU train will continue to be popular in 2021? China's Belt and Road Network. https://card.weibo.com/article/m/show/id/2309404594755811016930.

New Development Bank. (2020). NDB Annual Report 2020:Meeting ever-evolving Development challenges https://www.ndb.int/wp-content/uploads/2021/07/NDB-AR-2020_complete_v3.pdf.

Office of the Leading Group for Promoting the Construction of the "the Belt and Road". (2019). Jointly building the "the Belt and Road" initiative: progress, contributions and prospects.The Central People's Government of the People's Republic of China Website, http://www.gov.cn/xinwen/2019-04/22/content_5385144.htm.

People's Daily Overseas Edition, (2019, December 10). There are 550 Confucius Institutes worldwide. http://www.gov.cn/xinwen/2019-12/10/content_5459864.htm

Silk Road Fund Ltd. (2021). Silk Road Fund:About Us. Silk Road Fund Ltd. Official website. http://www.silkroadfund.com.cn/enweb/gywm/gsgk/gsjj/index.html.

The General Administration of Customs of the People's Republic of China. (2021, September 23) National "14th Five-Year Plan" port development plan.Port Office, http://ka.sz.gov.cn/xxgk/qt/tzgg/content/post_9149277.html

The People's Bank of China. (2020). 2020 RMB Internationalization Report.The People's Bank of China.http://www.pbc.gov.cn/goutongjiaoliu/113456/113469/4071737/2020081416503717495.pdf

The State Council. (2020, June 1). The General Plan for the Construction of Hainan Free Trade Port. The Central People's Government of the People's Republic of China Website.http://www.gov.cn/zhengce/2020-06/01/content_5516608.htm

The State Council Information Office of the people's Republic of China. (2019, April 22). The "Belt and Road" initiative has achieved remarkable results in the past five years. Website of the Information Office of the State Council, http://www.scio.gov.cn/xwfbh/xwbfbh/wqfbh/39595/40298/zy40302/Document/1652479/1652479.htm

The State Council Information Office of the people's Republic of China (2021, January). White Paper on China's International Development Cooperation in the New era. http://www.scio.gov.cn/zfbps/32832/Document/1696685/1696685.htm.

Xi Jinping. (2016). Speech at the Opening Ceremony of the Asian Infrastructure Investment Bank. CPC News Network .http://cpc.people.com.cn/n1/2017/0105/c64094-29000774.html

Xi Jinping. (2017, May 14). Jointly promote the construction of the "the Belt and Road",CPC News Network, http://cpc.people.com.cn/n1/2017/0515/c64094-29274601.html.

Yicai net. (2016, January15) China's formal accession to the European Bank for Reconstruction and Development is expected to link up with the shareholder country strategy. Yicai net. https://www.yicai.com/news/4739803.html.

Zhang Xiaohe. (2015). What is the meaning behind "International production capacity Cooperation" and "Juncker Plan"?, People's Daily Online. https://new.qq.com/omn/20200611/20200611A02WV500.html

Zhang Yuan. (2018). "Poverty reduction effect of Chinese aid and OFDI in 'the Belt and Road'", Finance and Trade Economics, 12:111–125.

Zhou Guomei. (2020). 'What efforts has China made to promote the co-construction of green BRI?', China Environment News, 12.[周国梅.推动共建绿色"一带一路"，中国做出了哪些努力? [J].中国环境报, 2020, 12.]

Zhou Mingyang. (2018, September 11).53 of the 65 countries along the "the Belt and Road" have established 137 Confucius Institutes - start with language and communicate with the soul. Economic daily, http://www.ce.cn/xwzx/gnsz/gdxw/201809/11/t20180911_30266354.shtml.

IV Analysis of the Emblematic Projects in the Construction of the BRI

Introduction

Over the past eight years, numerous important projects have been started and developed under the "Belt and Road" project, each with its own unique significance and playing an irreplaceable role in the "Belt and Road". The emblematic projects selected in this book are mainly as follows: (1) the China-Europe train, as the most important landmark project, undertakes the important task of land connections, deepens the economic and trade cooperation of the countries along the way, and enhances the economic vitality of the BRI countries; (2) the Kasim coal-fired power station. It is the starting point for the construction of the China-Pakistan Economic Corridor. It not only brings great benefits to the Pakistani people, but also builds a tough bridge for the friendship between China and Pakistan; (3) the Kenya-Inner Mongolia Railway. The project connects the profound friendship between China and Africa, enabling the East African region and even Africa to achieve connectivity; (4) The Sino-Russian Yamal project. It is the rapid development of Sino-Russian relations and a favorable prerequisite for the joint construction of the "Ice Silk Road" between China and Russia that will effectively supplement China's energy supply chain; (5) The China-Singapore (Chongqing) strategic connectivity demonstration project. It strongly reflects the good trust and in-depth cooperation between the governments of China and Singapore, deepens the bilateral cooperation between China and Singapore, and promotes the economic and trade development of the western region of China; (6) the China-Belarus Industrial Park. The Chinese and Belarusian governments are committed to make it into a "Pearl" along the BRI. It is one of the major industrial parks carried out by China abroad, which has effectively activated the Belarusian economy; (7) Xiamen University in Malaysia, which is an important interpersonal exchange project in the construction of the BRI. It greatly promotes cultural exchanges between China and Malaysia, and it also provides a talent training base for the construction of the BRI project.

Over time, the completion and successful operation of these landmark projects continues to export their huge instrumental force in the future construction of the BRI, becoming one of the benchmarks for China's construction of the BRI and building up a valuable pool of China's multilateral or bilateral project construction experience. Looking ahead, China continues to contribute its powerful Chinese wisdom and unique Chinese solutions to the world economy through the construction of other important projects on the BRI. China's power encourages the world economy to take off in terms of globalization and deep economic cooperation.

1. China-European Train

(i) History of development

The rise of the China-Europe Class train began in 2014. During his visit to the European Union, President Xi Jinping proposed that China and Europe should upgrade their trade cooperation from simple buying and selling to complex economic and trade cooperation with links in various fields and strive to achieve the target of US$1 trillion in annual trade volume at an early date. So, China Railways main office organized the first domestic coordination meeting of the China-European Liner in Chongqing on August 14, 2014. The meeting promulgated the *Interim Measures on the Organization and Management of China-European Liner*, (Liu Xian, 2014) and the representatives of all parties signed the Agreement on the *Establishment of China-European Liner Domestic Transportation Coordination Council Memorandum Entry* above the meeting (Liu Xian, 2014). The memorandum entry involves the future unified brand logo, unified transport organization, unified full price, unified service standard, unified operation team, unified coordination platform, and strengthened mechanism and equipment guarantee for

China-European trains everywhere. On December 16 of the same year, the second domestic transport coordination meeting of the China-EU Liner, hosted by China Railway Container main office, was held in Zhengzhou, Henan Province. At this meeting, the *China-EU Liner Transshipment Consolidation Organization Measures* was agreed upon and signed (Henan Provincial People's Government, 2014, December 17)

In 2020, the China-European Liner operated safely, smoothly, and steadily, with the number of departures growing against the trend, strongly serving the new development pattern and international cooperation in pandemic prevention, with 12,400 China-European Liner train rides operating and 1,135,000 TEUs sent throughout the year, up 50% and 56% respectively year-on-year, with a comprehensive heavy container rate of 98.4%. Regionally, the cumulative number of cities with more than 100 train rides in China increased to 29, with access to more than 90 European cities, involving more than 20 countries, and the scope of service continued to expand. (Hanlin: The Current Development of China-European Liner)

(ii) Status of the project

The China-Europe Class Train refers to the container international railway intermodal transport train that runs in accordance with the fixed number of trains, lines, and other conditions, between China and Europe and countries along the BRI regions. The project has laid out three alleyways in the west and Middle East, which constitute the main operating lines of the China-EU trains: alleyway in the west from the central and western part of China via Alashankou (Horgos), alleyway in the middle from the north of China via Erenhot, and alleyway in the east from the southeastern coastal area of China via Manzhouli (Suifenhe).

In 2021, 3,072 China-EU train rides have been operated, an increase of 82% year-on-year. At present, the number of cities with more than 100 China-Europe train rides has increased to 31. China-Europe class train all the way high at the same time, the western land and sea new alleyway radiation effect continues to appear. From January to February 2021, 702 western land and sea new alleyway train rides were operated, up 175% year-on-year; 38,356 TEUs were sent, up 192% year-on-year. On March 13, 2021, the 21028 heavy machinery China-European train, carrying 11 units of hammer head crane, departed from Nanning International Railway Port in Guangxi to Nur Sultan, Kazakhstan. This is the first heavy machinery China-European train sent from Guangxi, also opened up a new line of "Nanning-Xi'an-Nur Sultan" (Sohu.com, 2021, April 29). On March 24, 2021, the "Xiangxi Europe" international freight train carrying 50 containers departed from the railway logistics yard of Xiangyang Jinying Heavy Industry Logistics Park (Hanlin: The Current Development of China-European Liner) .

(iii) Significance of the project

1. Domestic significance

(1) Active planning for unification of international rail freight standards

China, as the organizer and coordinator of the China-European Train, guides the BRI countriess to develop various forms of cooperation around the train, which may include developing uniform or following uniform international rail freight standards. At present, there is a problem of inconsistent and outdated standards of railway transport infrastructure on the China-European Liner that departs from China and passes through several countries in Central Asia or the CIS before reaching Europe. The different rail gauges in China, Central Asia, the CIS and European countries require stopping midway to change trains, which increases the risk of loss of train cargo. Chinas, in the future, needs to play its role as a Silk Road organizer. The relevant departments of the country can develop and put into production of bigger trains with greater capacity. It also increases efforts to develop new energy trains to stimulate the strength of China's infrastructure innovation to facilitate future trade. This line could elevate China's energy and material security, to a new level in the future.

(2) Effective relief of pressure from maritime transport

It can effectively relieve the pressure from sea transportation and keep multiple routes open to avoid being cut off during special periods when there are problems with sea transportation. Especially since 2020, affected by the overseas pandemic, the price of sea transport has risen and air transport routes have been sharply reduced, but the number of China-Europe train rides has increased against the trend, which has efficiently promoted the anti-pandemic cooperation between China and Europe and the BRI countries, and become a "life alleyway" and "fate bond" for countries to fight the pandemic together. During this period, the China-European train rides carried a heavy responsibility. It has become an important mode of transport through Central Europe and Central Asia supply chain, which is a constant flow of important materials for Central Europe and Central Asia. It acts as a bridge to protect and save lives.

(3) Bringing new business opportunities to cities in the west

For cities and major economies in the western part of the country, the China-European train transportation has effectively created new business opportunities for them. The China-Europe Class Train drives the outward development of inland cities. Chengdu, Chongqing, Zhengzhou, and other cities as the representative of the inland hinterland center cities. It is through the opening of the China-European Liner, supplemented by air freight, land and sea transport system, that it quickly built a logistics service system for global distribution and delivery. It also creates a new comprehensive comparative advantage in logistics efficiency and costs in the inland region, achieving the expansion and development of the city's export-oriented industries, cultivating the short-term of 100 billion incremental export-oriented industries, and enhancing the city's hub status, and driving the city's economic and social development.

2. Impact on the BRI countries

(1) Normalization has led to enhanced multi-level and multi-modal coordination and cooperation

The regular operation of the China-Europe Class Train is continuously strengthening multi-level and multi-modal coordination and cooperation. As a new type of international railway logistics transport organization, the China-Europe Classical Train involves many BRI countries and many market players in the international supply chain. Only by discussing together and giving full play to the role of dialogue, consultation and coordination can the interests and concerns of all parties be taken into consideration, and a smoother trade and transport system can be formed. In the process of development, the countries along the China-European Liner have formed the China-European Liner Coordination Committee on their own initiative, gradually improved the industry self-regulation mechanism through the China-European Liner Coordination Committee, and effectively mobilized the relevant market players to participate in the operation of China-European Liner through the establishment of joint venture platform companies with capital as the link, etc. The countries along the China-Europe Class Train actively discuss and coordinate to solve problems together, laying the cornerstone for the better development of the future China-Europe Class Train.

(2) Deepening economic, trade and humanistic cooperation between China and BRI countries

The China-Europe Class Train is a railway international intermodal train that runs between China and Europe, and the BRI countries. It relies on the Asia-Europe Continental Bridge, in accordance with the fixed number of trains, lines, schedules and the whole operation time, which is an important carrier to deepen the economic and trade cooperation between China and the BRI countries and an important grip to promote the BRI construction. The China-Europe Class Train has changed the land transport pattern at both ends of Europe. Previously, Europe's land transport based on foreign trade was a one-way exchange between inland and coastal ports, but now, the opening of the China-Europe Class Train has changed the pattern of Europe's inland transport into a two-way interaction, which helps Europe to open its logistics network. It is one of the reasons why the BRI countries welcome the China-Europe Class Train. Relying on the Asia-Europe Continental Bridge to connect the coastal areas and the vast market between the inland hinterland, the China-Europe Liner is not only an important sea departure alleyway for the inland areas along the route, but also the best transportation medium for both supply and demand along the route, which are deep in the inland hinterland, or a transit transport solution for Japan, South Korea and other neighboring countries to connect to the inland markets of Asia and Europe. The opening of the China-Europe shuttle train has prompted a significant increase in the logistics network utilization factor in the landlocked countries of Europe, opening up new transport routes for European producers and traders, especially providing an economic and fast way of transport for some of these small and medium-sized enterprises, enabling wider outlets for goods such as timber, grain and livestock secondary product from Russia, Poland, Germany, the Netherlands and other European countries. For BRI countries, the China-European train not only brings them the opportunity to trade and select transactions, but also brings the people of the BRI countries the benefit of selecting goods and the opportunity to consume the exotic.

The Sino-European train project also carries the mission of promoting cultural exchanges between the East and West. Where the train passes and stops, people begin to gather, and cultural exchanges become indispensable. It is hoped that one day a "culture train" will be allowed to travel between the East and West. Compared with the initial stage of its birth, Sino-European train project has attracted more and more active attention and participation along the route crown corporation, who see the opportunities brought by their participation in the Sino-European train project and have expanded such opportunities, further enhancing the influence of the Sino-European train project.

The opening of the China-Europe Class train has not only promoted the trade of goods and commodities, but also intensive exchanges of each other's culture. On the one hand, the China-Europe Class Train has promoted policy communication and collaboration between Chinese railways and national railways,

Chinese customs, and national customs. It lays the foundation for smooth economic and trade flows between China and BRI countries. On the other hand, it has also promoted cultural and livelihood exchanges in the process. From the perspective of sustainable and high-quality development, the infusion of cultural and tourism elements gives the Sino-European train project a humanistic touch and help further build the Sino-European train project brand.

2. Qassim Coal-Fired Power Station Project

(i) History of development

The Qasim coal-fired power plant is the starting point for the construction of the China-Pakistan Economic Corridor (CPEC). Located at the Qasim seaport near Karachi, the Qasim coal-fired power plant is the first power cooperation project after Chinese Premier Li Keqiang proposed the strategic concept of building the CPEC during his visit to Pakistan in May 2013. The Qasim coal-fired power station, a thermal power site in southern Pakistan, provides about 9.5 billion kilowatt hours of electricity to Pakistan annually. In April 2015, President Xi paid an important state visit to Pakistan, elevating China-Pakistan relations to an all-weather strategic partnership. On April 20, 2015, the *Project Implementation Agreement* and the *Power Purchase Agreement* were signed in the presence of the leaders of China and Pakistan (Xinhua website, 2019, November 25). On December 23, 2015, the Chairman of the Pak Private Power Infrastructure Board (PPIB) and the President of Port Qasim Power signed financing closure documents in Islamabad, Pakistan, marking the official completion of the financing closure of the Qasim coal-fired power plant project. On March 16, 2016, Port Qasim Power Company (which is held by CEC oversea investment Limited) awarded the Port Qasim Coal Unloading Terminal Project to China Hydropower Construction Group Port and Maritime Construction Company Limited (Hydropower Port and Maritime) on an EPC basis, which is an ancillary facility to the coal-fired power plant, the main body of which includes the construction of the coal unloading terminal and navigation channel, etc., for approximately US$240 million. On December 15, 2016, in the presence of the Land Registrar of Port Qasim District, Sindh, Pakistan, the Permanent Secretaries of DECC-HIT Port Qasim Project responsible official and Qasim port office signed the *Land Lease Agreement* of Port Qasim Coal-fired Power Station on behalf of both parties (CEC website, 2021).

After intense construction and active efforts by all parties in 2016 and 2017, in the afternoon of April 24, 2018, Pakistan Port Qasim Power Generation Company Limited received an approval letter signed by the Pakistan Central Power Purchase Authority agreeing to start commercial operation of the Port Qasim 2×660 MW coal-fired power station from 00:00 hours on April 25, 2018. Qasim Power Station entered commercial operation 67 days ahead of schedule, marking the completion of all the development and construction tasks of the project and achieving complete success, opening a new chapter in CEC's overseas thermal power business investment, operation, and maintenance.

(ii) Status of the project

Port Qasim coal-fired power plant project, located in the Port Qasim Industrial Park, about 37 km southeast of Karachi, Sindh Province, Pakistan, is a mixed ownership system project jointly invested by CEC and AMC, a Qatari royal family company, in accordance with the ratio of 51%; 49%, with a total investment of US$2.085 billion for the construction of two 660 MW supercritical units and a construction period of 36 months. The Qasim project is the first major energy project in the "corridor" to sign an investment agreement in the presence of the leaders of China and Pakistan. As the first "crab" enterprise, there is no precedent or experience to follow. Pakistan Port Qasim coal-fired power plant project is one of the priority projects of "China-Pakistan Economic Corridor", and the first large-scale energy project invested by a Sino-foreign cooperation (CECC and Al-Mirqab of Qatar).

Gansu Energy Pakistan Branch Office is responsible for the operation and maintenance of the Port Qasim coal-fired power station in 2020 to complete the generation of 8.893 billion kilowatt hours, setting a record high. Meanwhile, at 14:00 on May 22, 2020, the cumulative power generation capacity of Pakistan's Port Qasim coal-fired power station, which is operated and maintained by CEC Gansu Energy, has exceeded the 20 billion kWh mark since it was put into operation, creating a new milestone for Qasim power station (China Electric Power Website, 2020, May 27).

(iii) Significance of the project

1. Significance in project construction

The Qasim project is the first major energy project in the "corridor" to sign an investment agreement in the presence of the leaders of China and Pakistan. As the first pioneering enterprise, there is no precedent or experience to follow.

(1) Construction mode adopts industry chain integration mode and strict quality control

In terms of the construction model, Qassim Coal-fired Power Station has innovatively adopted the industry chain integration model, which combines the four major aspects of project investment and development, design and supervision, construction, and operation and maintenance into one. This model greatly optimizes the enterprise's management of the entire project life cycle by integrating the entire process of investment and financing, construction management, and asset management.

In terms of quality control, the construction of the Qasim coal-fired power station was carried out with strict quality control particularly by adopting a "benchmarking" approach and setting up a special exhibition area to showcase and promote standardized models in operations such as pipe welding and cement pouring. This approach has not only improved the quality of construction and eliminated safety risks on the project itself, but also has also greatly helped create a sense of standardization among the local Pakistani laborers employed on the project.

(2) High priority to environmental protection for achieving sustainable development

The Qasim coal-fired power station attaches great importance to environmental protection to achieve sustainable development, not only in the design of the power station to follow a set of advanced standards in China, Pakistan itself and international organizations such as the World Bank, but also set up an environmental administration committee, which is specifically responsible for ensuring that the environmental standards are met throughout the project's operation. In addition, the project has established numerous internal management rules and operational procedures to ensure that there are no surprises in environmental protection. The combination of these measures has resulted in the Qasim coal-fired power station winning the Pakistan Environmental Excellence Award 2018.

2. Significance for CMB

(1) Teach people to fish: localization strategy and differentiated management

The excellent conditions created by Qasim Power Station are due to the "localization strategy" and "differentiated management" that CEC has always insisted on and implemented in the international market, including increasing the proportion of foreign employees and managers, vigorously training local employees, respecting local customs and religious beliefs, etc. The Qasim project provides over 2,000 local jobs during the construction period and 500 local training and employment jobs per year during the operational period. At present, there are about 200 Pakistani managers and 1,600 Pakistani construction workers at the Qasim project site. The construction of the power station has created a large quantity of local jobs and improved the livelihood of local people.

(Chinese government procurement website, 2017, June 21).

(3) Chinese forces take responsibility for environmental protection and facilitate local economic and social development

After the commercial operation of the two units at the end of June 2018, Qasim Power Station provides Pakistan with about 9 billion kilowatt hours of electricity per year, which greatly alleviates the current power shortage in Pakistan and provides a continuous power guarantee for Pakistan's economic take-off. The project has been designed and worked on throughout with appropriate environmental measures to safeguard the good state of the local environment. The Qasim Power Station has adopted environmental technologies such as desalination and stack gas desulfurization, which are environmentally friendly to international standards. Its construction has been in compliance with environmental protection laws and regulations, and properly disposed of construction and domestic waste. Focusing on the protection of mangroves, the area of mangroves transplanted and planted is equivalent to five times the area cut down. The regular operation of the project has played a prominent role in alleviating the shortage of electricity and energy, promoting local economic development, and improving people's lives in Pakistan.

3. Mombasa-Nairobi Standard Gauge Railway in Kenya

(i) History of development

The initial phase of the project was born out of a friendly exchange between the two countries. In August 2009, the Government of Kenya and China Road and Bridge signed an understanding and cooperation memorandum entry for the Monterey Railway project. After non-stop negotiations in the early stages, finally in July 2012, the Mombasa-Nairobi Standard Gauge Railway design procurement construction turnkey contract was signed. On May 11, 2014, then Chinese Premier Li Keqiang signed a cooperation agreement with the Kenyan President Kenyatta and the prime ministers or foreign ministers of the five East African countries to build the Mombasa-Nairobi Standard Gauge Railway. On December

12, the construction of the Mombasa-Nairobi Standard Gauge Railway began. In November 2015, the Monterey Railway began laying the entire line. On May 31, 2017, the Mombasa-Nairobi Standard Gauge Railway, which is built and operated by Chinese companies, was opened to traffic (International Business Daily, 2019, April 25). On 16 December of the same year, the Nairobi container inland port was officially handed over and started operations. September 11, 2018, Mombasa-Nairobi Standard Gauge Railway Port Spur Project had its final acceptance of construction. On July 29, 2019, the Mombasa-Nairobi Standard Gauge Railway was interconnected with the Nema Railway. On July 19, 2020, the platform at Nairobi South Station in Kenya was back to life as crew members in full protective clothing checked passengers' tickets and passengers boarded in an orderly manner wearing masks. Passenger services on the Mombasa-Nairobi Standard Gauge Railway in Kenya, which were suspended due to the Covid-19 Pandemic outbreak, have officially resumed operations.

(ii) Status of the project

The full name of the Mombasa-Nairobi Railway in Kenya, connecting the Kenyan capital Nairobi and the port of Mombasa, the first port in East Africa, is 480 km long, with a total of 33 stations opened along the line, with a design capacity of 25 million tons, for internal combustion engine system, designed for 120 km/h for passengers and 80 km/h for freight, and is the first new railway built in Kenya in a century, a modern railway built with Chinese standards, Chinese technology and Chinese equipment. Nairobi South Station in Kenya's capital is one of the starting stations of the Mombasa-Nairobi Standard Gauge Railway, from where the first train departs after the railway's official opening, heading for Mombasa, East Africa's largest port. The port of Mombasa has an annual capacity of 25 million tons.

On February 18, 2020, the Mombasa-Nairobi Standard Gauge Railway has been operating safely for more than 1,300 days since it was opened to traffic on May 31, 2017. Its extension, the Nema Railway (Nairobi Malaba section of the standard gauge railway), is being extended further inland. May 31, 2020, marks the third anniversary of the opening of Kenya's Mombasa-Nairobi Standard Gauge Railway. In his 57th Autonomy Day speech on 1 June, Kenyatta referred to the Standard Track Railway in his review of infrastructure achievements since taking office and spoke highly of it, saying it would "contribute significantly to the development of cities along the railway and open up new markets in the East African region." The standard gauge railways referred to in Kenyatta's speech are the standard gauge railways that include the Mombasa-Nairobi Standard Gauge Railway, and its westward extension, the Nema Railway (Li Yiran, 2020).

(iii) Significance of the project

1. Domestic significance

(1) Realizing the BRI purpose and connecting the regional economic cooperation path of work between China and Africa

The Mombasa-Nairobi Standard Gauge Railway is a major project for China and Africa to jointly build three major networks of high-speed railways, highways, and regional aviation in Africa, starting from sub-regional cooperation. The achievement of connectivity in the East African region and Africa in general plays an important role in supporting the economic development of African countries. The friendship between China and Africa has been tested by time, history, and the people. It has endured for a long time, becoming more vibrant and thriving in the new situation. It perfectly illustrates the purpose of the BRI strategic vision. Since 2013, when President Xi launched the BRI, numerous cooperation projects have been steadily progressing under the principle of "common consultation, construction and sharing". The BRI is the road of common construction and sharing, in no way the road of seeking hegemony as maliciously portrayed by some Western media. Take the Mombasa-Nairobi Standard Gauge Railway as an example. The latest data presentation, the construction of the Mombasa-Nairobi Standard Gauge Railway has provided a cumulative total of more than 38,000 jobs for the Kenyan people, with local employees accounting for about 90% of the total workforce. It has been irrefutably proven that the Kenyan people are the biggest beneficiary and owner of this railway, and that the Mombasa-Nairobi Standard Gauge Railway exerts extremely strong demonstration effect.

(2) Become the benchmark building for China's railway output

It accelerates the pace of "Chinese standards" going abroad. The standard output is the mark of the core competitive power. In recent years, due to various technical barriers and trade protection, the pace of export of "Chinese standards" has lagged behind that of "Made in China". Many countries and regions have always been suspicious of Chinese standards, despite the recognition of Made in China. As the first complete "Chinese standard" on the African continent, the Mombasa-Nairobi Standard Gauge Railway will be of great interest to the world from today onwards. Its safety case, business, and management and even its political and cultural impact attract a lot of attention from all over the world. Relying on the strong technical strength of Chinese railways to support, the

Monterey Railway delivers a successful answer, and it is reasonable to accelerate the "Chinese standard" going out. It exemplifies the method and strategy of "going out" of domestic material and equipment driven by the Mombasa-Nairobi Standard Gauge Railway, which has important guiding significance for the follow-up international projects after the East African railway network construction project and even other regions to promote the overall industry chain of domestic railways "going out".

2. Significance for Kenya

(1) Constituting a step forward in Kenya's national development

"With a new railway for the people of Kenya, we will turn a new page and write the history of Kenya for the next 100 years." On May 31, 2017, the Mombasa-Nairobi Standard Gauge Railway, constructed and operated by China Communications Construction, was opened to traffic, as Kenyan President Kenyatta said at the opening ceremony. Time is always the best proof. Over the past year since its opening, the Monterey Railway has brought great changes to Kenya's economy, social development, and people's travel. The railway carried 1.58 million passengers and 112,000 standard cargo containers in one year, with an average passenger occupancy rate of 95.2%, and a freight capacity of 214,000 tons per month, up from the initial 22,000 tons. It has been proved that the opening of the Monel Railway has not only brought the distance between different regions and tribes in Kenya closer, but more crucially, brought the people of Kenya a convenient road, a road of affluence, a road of happiness and a road of prosperity (Lv Qiang, 2020).

(2) Kenya's Heartwarming Environmental Road

The Mombasa-Nairobi Standard Gauge Railway is an "artery" that speeds up Kenya's economic development and it is also a "warm road" that embodies green environment and improves people's livelihood. From the big data of the first anniversary of operation, the Mombasa-Nairobi Standard Gauge Railway has effectively helped improve people's livelihood and the rapid development of the local economy. Chinese railway builders can be said to have done a lot of ecological environmental protection work in the local design and construction process to maintain the harmony between animals and nature. For example, in addition to the bridged wildlife alleyway, the Mombasa-Nairobi Standard Gauge Railway has over 100 culverts at ditches to facilitate both watering for animals such as zebras and small wildlife crossing the railway, allowing for a sustainable and healthy local ecology.

4. The Russia-China Yamal Project

(i) History of development

On May 21, 2014, Chinese President Xi Jinping and Russian President Vladimir Putin witnessed in Shanghai the signing of the Sino-Russian Government's *Sino-Russian Eastern Line Rock Gas Cooperation Project Memorandum Entry* and the *Sino-Russian Eastern Line Gas Supply and Sales Contract* between China National Petroleum Corporation rock gas and Russia gas industry joint stock company (Science and Technology Daily, 2019, December 5). According to the agreement between the two sides, starting from 2018, Russia began supplying gas to China via the east line of the Sino-Russian rock gas pipeline, with the volume of gas delivered increasing year by year, eventually reaching 38 billion cubic meters per year for a total of 30 years. China and Russia are advocates and promoters of the "Silk Road on Ice". In May 2015, the heads of state of China and Russia signed a *Joint Satement on Docking Cooperation in the Construction of the Silk Road Economic Belt and the Construction of the Eurasian Economic Union* (*Ministry of Foreign Affairs,* 2017, March 16), which started the process of docking the BRI with the Eurasian Economic Union. Yamal project is the first overseas mega project implemented after the BRI was launched. It is also a model of Russia-China energy cooperation.

In June 2015, construction began on the Chinese side of the Sino-Russian East rock gas pipeline project. On January 1, 2018, the China-Russia crude oil pipeline second line was officially put into operation, and the pipeline's oil transmission capacity reached 30 million tons per year. On January 26, 2018, the Chinese government released its first Arctic policy document, *China's Arctic Policy* (*Xinhua Net*, 2018, January 26). The white paper stats that China is willing to build the "Silk Road on Ice" with all parties by relying on the development and utilization of the Arctic waterway. The co-construction of the "Silk Road on Ice" means the cooperative development of the Arctic shipping route. On July 19, the first ship of liquefied rock gas supplied to China from the Yamal liquefaction rock gas project, a mega-energy cooperation project between China and Russia, arrived onshore. In September 2018, upon invitation, President Xi Jinping attended the Fourth Eastern Economic Forum in Russia. The attendance of the two leaders at the important international conferences hosted by each other's countries has become a Russian-Chinese diplomatic tradition, providing continued momentum for pragmatic cooperation between China and Russia. On November 21, 2018, this largest regional economic cooperation work project between China and Russia at present, was fully completed and put into operation. On October 16, 2019, the entire section of the Sino-Russian East Line rock gas collar pecker course Heihe to Changling was connected, with a total line length of 1,067 km, realizing the connection with the in-service rock gas pipeline network of Hashan and Qin Shen. The third production line at the Yamal liquefaction rock gas (LNG) project in the Arctic Circle goes on stream on December 11, 2019, a year ahead of schedule. On December 20, 2019, Russia started supplying gas to China via the Sino-Russian Eastern Route rock gas pipeline (North Star Power News).

(ii) Status of the project

The Yamal project, located in the Arctic Circle, is the world's largest Arctic liquefaction rock gas project. It is located at the Yamal-Nenets Autonomous Okrug of Russia, about 2,250 km from Moscow. The Yamal LNG project is a joint cooperative development between Russia's Novatek, China National Petroleum rock gas Group Corporation, France's Total and China's Silk Road Fund. The rock gas workable reserves reached 1.3 trillion cubic meters and gas distillate workable reserves 60 million tons. The project builds three lines with an annual production capacity of 5.5 million tons LNG chain and will be able to produce liquefied rock gas 16.5 million tons and gas distillate 1 million tons per year when all of them are completed (Xinhua News Agency, 2018).

The Yamal project is an important step forward in the co-construction of the "Silk Road on Ice" between Russia and China, a model of win-win cooperation between the two sides and of great significance for the sustainable development of Russia-China relations. After the Yamal project is completed, China can get 4 million tons of liquefied gas per year from this project, providing the basis for China's clean energy supply. The successful implementation of the Yamal LNG project not only enriches China's clean energy supply, but also acts as a symbol – the linkage effect brought about by the BRI is providing positive energy for international rock gas cooperation, energy and financial cooperation (Dong Zhanwei, Li Yuanying, Gao Zhenbo, 2018)

(iii) Significance of the project

1. Domestic significance

(1) Enhancing the technological innovation capacity of China's oil rock gas project

The construction scale of the world's first Sino-Russian East Line rock gas collar pecker process, since the start of construction in June 2015 and the full construction on December 13, 2017, 48 innovative results have been achieved, 22 technical standards and management norms have been developed, a series of new pipes, new technologies, new processes, and new equipments have been verified. Operating technique there has been a qualitative leap forward, close to and catching up with the international advanced level, which has positively promoted the development of China's basic industries such as steel smelting, pipe making and equipment manufacturing. It has strongly promoted the synergistic development of the upstream, midstream, and downstream industry chains such as domestic gas fields, pipelines, gas storage and gas utilization projects.

(3) Broadening the channels of our energy imports
The Arctic is rich in oil, rock gas, minerals, and fisheries. Russia's Ministry of Natural Resources and Ecology estimates that about 100 billion tons hydrocarbon resources are contained in the Russia-owned Arctic Ocean continental platform alone (Dong Zhanwei, Li Yuanying, Gao Zhenbo, 2018). China could get 4 million tons of liquefied gas per year from this project, providing an important foundation for our country's clean energy supply. Yamal LNG project is a pearl of Sino-Russian energy industry cooperation. The project is located at the Yamal peninsula of Yamal-Nenets Autonomous Okrug in the northern part of Western Siberia within the Arctic Circle, which is an important strategic location for Russia, not only 80% of rock gas is stored here, but also an important energy transportation route for radiation worldwide (Anonymous, 2020, June 11).

(4) Accelerate the optimization of our energy structure
"Since the 12th Five-Year Plan, a period of major economic and social transformation in China, the country's energy consumption has had three irreversible aspects: the first is the increasing momentum of high-carbon energy consumption aspects such as coal and oil. The second is that the coal-based energy consumption structure is gradually shifting. The third is the growing trend in total energy. In this case, the optimal adjustment of China's energy structure. First of all, we must strictly control the excessive expansion of the total amount of energy, especially the consumption base of coal and oil, maximize the reduction of the proportion of high-carbon energy consumption, promote the development of energy diversification, accelerate the replacement of conventional fossil energy by renewable energy and new energy. The development of the Yamal project provides a new long-term and stable idea, and new channel for China's future long-term clean energy imports, which greatly helps to optimize our energy structure.

(5) Beneficial for China to open a new international shipping alleyway and deepen foreign trade and economic development

(6) The Arctic Northeast Passage is the most direct and convenient transport link between Asia and Europe alleyway. Making full use of this Arctic shipping lane, the future economic and trade cooperation between China and Europe also would reduce a lot of logistics costs and improve the regional economic cooperation between China and Europe. During the construction of the project, a total of 45 Chinese manufacturers have provided more than 100 kinds of products for the project, which has driven and promoted the technological innovation and transformation and upgrading of many domestic industries, such as steel and cable. The export value of domestic products has exceeded 10 billion US dollars. (Anonymous, 2017, December 9)

2. Significance for Russia

(1) Beneficial to the international standing of both countries
Sino-Russian Arctic energy cooperation is an important component part of the Sino-Russian "Silk Road on Ice", of which the Yamal LNG project is the flagship project. Both sides benefit from the project. First, Russia wants to stabilize its international position through cooperation with China in the current international political game. Yamal LNG project to China's exports of rock gas can ease the pressure of the growth of domestic market demand, for the continued growth of China's economy as a guarantee. Second, Russian Arctic energy development requires Chinese financial and technological input. China's energy consumption increasing foreign dependence is a major driver of Sino-Russian energy cooperation. In the face of our rapidly growing rock gas demand, our import dependence is also growing in a positive correlation. Finally, cooperation between Russia and China on the Yamal LNG project is in line with the requirements of Russia's Arctic strategy. According to the *Russian Energy Strategy Before 2030*, Russia makes the most of its rock gas energy potential and consolidates its position on the international energy market.

(2) Bringing about long-term stable development of local logistics and shipping
The Yamal project has led to the development of the Sabelta region, where the project is located. The local airport has been in operation and opened several domestic and international trade routes in February 2015, flying 7,523 flights, carrying 70,914 passengers and 9,200 tons of cargo as of November 26, 2017. The Sabetha maritime port, of which it is a member, has six berths for unloading materials and has carried a total of more than 500,000 tons of materials. The construction of the Yamal project has not only added a new northern maritime

transport to rock gas for China, but also opened the Arctic shipping lane to successfully achieve the Arctic Ocean transport through the Arctic Circle, connecting the three major economic centers of North America, East Asia and Western Europe by sea. More than 60% of the modules and components for the Yamal project were transported through the Bering Strait and via the Northeast Arctic shipping lanes, taking an average of about 16 days, which is nearly 20 days less than going through the Suez Canal (Xie Duo, 2019).

(4) Helping to upgrade and transform the Russian energy industry
The Yamal LNG project is equally important to the Russian government and Novatek, as the LNG helps Russia achieve a number of objectives: increase energy exports; give Russia access to energy consumption markets that were previously difficult to access; facilitate the introduction and digestion of new technologies; spur the development of related industries; support the economic development of key regions such as the Arctic and Far East; develop strategically important Arctic shipping lanes; and strengthen Russia's geopolitical influence in the Asia-Pacific region.

5 China-Singapore (Chongqing) Strategic Connectivity Demonstration Project

(i) History of development

The China-Singapore (Chongqing) Strategic Connectivity Demonstration Project began when Chinese Vice Premier Zhang Gaoli proposed to Singapore's Deputy Prime Minister Teo Chee Hean, during the 10th Joint Commission on Bilateral Cooperation between China and Singapore in October 2013, to undertake the third China-Singapore inter-governmental cooperation project in western China, following the Suzhou Industrial Park and the Tianjin Eco-city. Subsequently, the parties reached a consensus in this regard. In August 2014, when President Xi Jinping met with Singapore's President Tony Tan Keng Yam in Nanjing to attend the opening ceremony of the Youth Olympic Games, he formally addressed the issue of establishing a third national-level cooperation project and made it clear that the third project would be built in western China. On November 7, 2015, the People's Republic of China and the Republic of Singapore issued in Singapore the *Joint Declaration of The People's Republic of China and the Republic of Singapore on the Establishment of a Progressive All-Round Partnership* (2015) On the same day, the Chinese and Singaporean governments signed the *Framework Agreement on the Construction of the China-Singapore (Chongqing) Strategic Connectivity Demonstration Project* and its Supplementary Agreement in Singapore, officially launching the third inter-governmental cooperation project with Chongqing as the operation centre (Anonymous, 2015, November 11).

The development of the China-Singapore (Chongqing) Strategic Connectivity Project is also adding more sister cities to the list. In August 2017, Chongqing Manicipality, Guangxi Zhuang Automonous Province, Guizhou Province and Gansu Province signed the *Framework Agreement on Cooperation in Building the Southbound Alleyway of the Sino-Singapore Interconnection Project*, (2017) formulated the *Synergistic Approach on Cooperation in Building the Southbound Alleyway of the Sino-Singapore Interconnection Project* (Xinhua News Agency, 2018, September 7), and established the working mechanism of discussing, building and sharing the " the Southbound Alleyway". In November 2018, China and Singapore officially signed the *Memorandum of Understanding on Cooperation in the Construction of the New International Land and Sea Trade Alleyway of the China-Singapore (Chongqing) Strategic Connectivity Demonstration Project* (2018), which officially renamed the Southbound Alleyway as the "New Land-Sea Alleyway", further expanding the connotation, significance and key areas of the alleyway construction. Chongqing, Guangxi, Guizhou, Gansu, Qinghai, Xinjiang, Yunnan, Ningxia 8 other western provinces signed a framework agreement in Chongqing on the 7th to cooperate in building a new international land and sea trade alleyway (referred to as "land-sea new alleyway") for the Sino-Singapore Interconnection Project, which would cooperate in promoting the construction of the "land-sea new alleyway" and help China accelerate the formation of "land-sea internal and external linkages, east and west two-way mutual benefit" pattern of opening up to the outside world.

On October 20, 2020, Minister of Commerce Zhong Shan and Singapore's Minister for Manpower and Second Minister for Home Affairs Yang Liming co-chaired by video the second meeting of the Joint Working Committee on the China-Singapore (Chongqing) Strategic Connectivity Demonstration Project. On December 8, 2020, Vice Premier Han Zheng held a video meeting with Singapore's Deputy Prime Minister Ong Swee Keat in Beijing and co-chaired the 16th meeting of the Sino-Singapore Bilateral Cooperation Joint Commission, the 21st meeting of the Suzhou Industrial Park Joint Coordinating Council, the 12th meeting of the Tianjin Eco-city Joint Coordinating Council, and the 4th meeting of the Sino-Singapore (Chongqing) Strategic Connectivity Demonstration Project Joint Coordinating Council. The two sides comprehensively reviewed the implementation of the high-level consensus and the progress of practical cooperation, focusing on the BRI cooperation, economic and trade cooperation and innovation development, financial cooperation, public health cooperation, humanities exchanges and sustainable development, and bilateral cooperation projects at the national level, and fully exchanged views and jointly discussed new ideas and initiatives for cooperation (Xinhua News Agency, 2018, November 8).

(ii) Status of the project

The Sino-Singapore (Chongqing) Strategic Connectivity Demonstration Project is the third Sino-Singapore inter-governmental cooperation project between China and Singapore in the western region of China, with Chongqing, a municipality under the Central Government of China, as the project operation centre. This is the third Sino-Singapore government cooperation project with the theme of "modern connectivity and modern service economy", with Chongqing as the operation centre and financial services, aviation industry, transportation and logistics, and information and communication as the four key areas of cooperation.

The five years since 2020, during the development of the China-Singapore (Chongqing) Strategic Connectivity Project, have seen further strengthening of the Chongqing-Singapore connectivity. On September 25, 2020, the construction of UP Data Chongqing Data Center started in Chongqing Liangjiang New Area, helping Chongqing and Singapore to start comprehensive cooperation in the fields of IDC (Internet Data Center), cloud computing and big data (BRI web link, 2019, August 14).

By the end of October 2020, the China-Singapore (Chongqing) Strategic Connectivity Project has led to the participation of about 300 government departments, business associations and enterprises from both sides in the project cooperation, promoting the signing of 68 agreements and 230 cooperation projects of various types, total value of more than 32.2 billion U.S. dollars, gradually becoming a new business card for international cooperation in building the BRI together. In terms of capital flows, the Sino-Singapore cross-border financing model was established and replicated to other provinces and regions in western China (China News Network, 2020, November 19).

The project has also played an indispensable role as a bridge in cultural exchanges and public communication. By the end of 2020, in addition to the four key areas of cooperation, including financial services, aviation industry, transportation and logistics, and information and communications, the Sino-Singapore connectivity project has also yielded fruitful results in expanding cooperation between the two sides in areas such as tourism, education and research, medical and health care, and trade in agricultural products. The two sides promoted the landing of the National University of Singapore (Chongqing) Research Institute project, the official opening of Raffles Hospital Chongqing and the settlement of Sino-Singapore (Chongqing) International Cancer Hospital (China News Network, 2019, March 15).

(iii) Significance of the project

1. Domestic significance

(1) Enhancing the depth of Sino-Singaporean cooperation and expanding the benefits of cooperation between the two countries

The Sino-Singapore (Chongqing) connectivity seamlessly fits into the vision of China's BRI strategy, which aims to encourage Singapore's participation in China's BRI strategy by building an all-round interconnection between Chongqing and Singapore, leading to deeper political, economic, and cultural exchanges and cooperation between the two sides. It is a national-level strategic connectivity demonstration project jointly promoted by China and Singapore. This project is not about Chongqing bringing in a group of Singaporean companies to work on a project in Chongqing, but a collaboration between China and Singapore on the scope of interconnected services trade. It is just that its operations are centered in Chongqing, but these projects can be not only specific to the Chongqing area, but also to the southwest China, or even to the southeast Asia region's 21 Century Maritime Silk Road or the Silk Road Economic Belt.

(2) Rebirth of Chongqing's economic life for the Western Region of China to thrive

The successful implementation of the project has undoubtedly constituted the rebirth of economic life for Chongqing as a municipality directly under the Central Government, bringing significant opportunities for Chongqing to take off again. Chongqing is in the western inland region of China; the region's harsh geographical environment has led to its long-term disadvantage in opening to the outside world. The "Chongqing-Xinjiang-Eurasian Continental Bridge" effectively breaks this bottleneck, so that the Chongqing region as a mover and shaper of the Silk Road projects, to undertake the strategic mission of opening up to the west. With the deepening of cooperation, Chongqing has a role in the BRI to bear the east and enlighten the west, connecting the north and the south, which is manifested in the vast Asian and European hinterland and the Atlantic Ocean to the west, the Pacific Ocean to the east, and ASEAN and the Indian Ocean to the south, and is the liaison node of the Silk Road and the Yangtze River Economic Belt. It has a very superior location advantage in the BRI strategy.

Chongqing has strong momentum and vitality in the future, relying on the "Chongqing-Xinjiang-Europe" international logistics alleyway and the YUKUN Southeast Asia international trade alleyway. The project continues to promote the country's deepening of Sino-Singaporean cooperation, dovetailing with the BRI construction, western development and the development of the Yangtze River Economic Belt and other national strategies to enhance the level of inland opening. It is also a major historical opportunity for the development of Chongqing. The rapid development of the project is forming a modern nextwork of interconnections with Chongqing as the hub, which on the one hand accelerates the transformation and upgrading of Chongqing's industry and enhances its internationalization, and on the other hand accelerates the leading of the western part of the country to enhance the level of openness in all aspects.

(3) Building a common platform for exchange and collaboration in the southwest to promote the civic exchange between the two countries

The depth of the project's areas of cooperation and its content makes it a common platform for exchange and collaboration between regions in the process of interconnection. The Sino-Singapore interconnection project has played a radiative and driving role. It drives the western region to form a new pattern of development and opening up. Logistics channels and costs, financing channels and costs are the main arteries of economic development in the western region. It strives to achieve smooth access, multiple financing and lower costs.

The project promotes economic and trade development in the process of expanding cooperation in multiple fields, while also benefiting the people of both countries and enhancing public communication. In medical services, the establishment of Raffles Hospital Singapore in Chongqing is promoted to provide international medical and health services to the public. In education and training, cooperation has been strengthened between Chongqing and Singapore's Nanyang Technological University and the Singapore Craft Education Board to develop and exchange talents and vocational skills in key areas. Especially in the cultural tourism, the project has not only contributed to the Singapore China Coast Spring Group investment, but also to the construction of the Sino-Singapore-Jinfo Mountain Tourism and Culture Industrial Park.

2. Impact on the BRI countries

(1) Facilitate shared opening to the outside world and sharing of resources and markets between China and Singapore

Since its inception, the BRI strategy has received positive responses from countries along the routes. The development and implementation of the China-Singapore

(Chongqing) Connectivity Project is not only conducive to sharing the opportunities of opening to the outside world, but also to sharing the resources and markets of China and Singapore. China and Singapore can break through the limitations of space and time, seek policy support, break down barriers, take the cooperation between Chongqing Jiangbei International Airport and Singapore Changi International Airport as the basis and the starting point to truly realize China-Singapore (Chongqing) connectivity. It makes joint efforts to build the pivot of the Air Silk Road, and eventually realize the interconnection with the BRI countries and even the whole world.

(2) Provides a convenient transportation channel for the exchange of people along the road

From Singapore's perspective, the 21st Century Maritime Silk Road and the Silk Road Economic Belt should not operate independently, but in synergy and complementarity with each other (the Information Office of the State Council, 2015, July 28). Keeping this in mind, Singapore proposed the China-Singapore Connectivity Project Southbound alleyway (referred to as Southbound alleyway) in 2017, enabling this alleyway to connect Chongqing, a key node of the Silk Road Economic Belt, to the port of Qinzhou in the Guangxi Zhuang Autonomous Region by rail, and to Southeast Asian countries and the wider region by sea through the port of Qinzhou. It plays the role of Chongqing and Singapore as a "double hub", expanding the radiation and drive effect. Its north connects the Silk Road Economic Belt, south through Guangxi and Singapore and other ASEAN countries by sea to connect the 21st Century Maritime Silk Road. The use of railways, highways, water transport, aviation and other modes of transport creates a new land-sea alleyway. The formation of the BRI through the western region of China completes the loop.

6. China and Belarus Industrial Park

(i) History of development

In October 2010, during President Lukashenko's visit to China, the Ministry of Economy of Belarus and CIGI signed the *Cooperation Agreement on the Establishment of China-Belarus Industrial Park in the Republic of Belarus* (2015). On September 18, 2011, *China and Belarus Signed the Agreement between the People's Republic of China and the Republic of Belarus on China-Belarus Industrial Park* (2014), which marked the official development of the China-Belarus Industrial

Park (Great store) and started the planning and construction. In July 2013, China and Belarus established a comprehensive strategic partnership, ushering in a new era in the development of relations between the two countries. This industrial park is a major strategic project jointly identified by President Xi Jinping and President Lukashenka and jointly built by the two countries, to which the leaders of the two countries attach great importance and are committed to making it a pearl among the BRI projects. In September 2014, Vice Premier Zhang Gaoli visited Belarus and reached a consensus with Belarusian leaders on accelerating the construction of the China-Belarus Industrial Park. The two sides agreed that the Ministry of Commerce of China and the Ministry of Economy of Belarus jointly establish the Industrial Park Coordination task group to promote the construction of the Industrial Park. On December 22, 2014, Zhong Shan, International Trade Negotiator and Vice-Minister of the Ministry of Commerce of China, and Snopkov, Minister of Economy of Belarus, signed the *Protocol on Cooperation between the Ministry of Commerce of China and the Ministry of Economy of Belarus on Building the Silk Road Economic Belt* in Beijing (2014). The two sides have been jointly promoting the construction of the "Silk Road Economic Belt" within the framework of the Economic and Trade Sub-Committee of the Intergovernmental Cooperation Committee of the two countries, and have been comprehensively enhancing trade, investment, economic and technological, industrial park cooperation and infrastructure connectivity. On May 10, 2015, the two sides signed the *Treaty on Friendship and Cooperation between the People's Republic of China and the Republic of Belarus* (2017), laying a solid legal foundation for the further development of relations between the two countries.

In 2018, the two sides signed a contract to develop friendly provincial-state relations letter of intent and cooperation between the Sixth Academy of Aerospace Science and Technology, Shaanxi Baoguang holding company and other enterprises and the construction of China-Belarus Industrial Park has steadily advanced. At the beginning of 2019, the China-Belarus Industrial Park ushered in good news – Belarusian President Lukashenko signed a presidential decree granting China-Belarus industrial park territorial special economic zone status, which allows enterprises and residents of the park to enjoy the customs facilitation policy of the Eurasian Economic Union to the maximum extent in terms of logistics and transport and production activities. This is a recognition of the fact that the China-Belarus Industrial Park promotes the economic development of Belarus. On December 14, 2020, during the fourth meeting of the Intergovernmental Cooperation Committee between China and Belarus, China's Vice Minister of Commerce and Deputy Representative for International Trade

Negotiations Yu Jianhua and Belarusian Minister of Economy Chervyakov officially signed by video the *Joint Declaration of the Ministry of Commerce of the People's Republic of China and the Ministry of Economy of the Republic of Belarus on the Launching of Negotiations on the "China-Belarus Agreement on Trade in Services and Investment*" (2014). The Ministry of Foreign Affairs of Belarus said on December 31, 2020 that during its presidency of the CIS in 2021, Belarus would promote dovetailing with China's BRI in logistics and jointly promote the development of China-Belarus industrial park (Jiang Pingping, Chang Xuemei, 2016).

(ii) Status of the project

China-Belarus Industrial Park, known as China-Belarus Industrial Park, is in Minsk Oblast of Belarus, an important hub of the Silk Road Economic Belt linking Europe and Asia. With a planned area of 91.5 km^2, the China-Belarus Industrial Park is a landmark project of China-Belarus cooperation in building the Silk Road Economic Belt. Compared with other overseas parks, an important advantage of the China-Belarus Industrial Park is the establishment of the China-Belarus Industrial Investment Fund. This fund is a national fund initiated by SASAC, established by China Merchants Group, and managed by China Merchants Capital, which mainly invests in the enterprises and high-quality projects in the China-Belarus Industrial Park. In 2020 Belarus saw a significant increase in exports to China of chicken, beef, dairy products, rapeseed oil, soybean oil and confectionery, and began exporting canned meat, starch, and sugar to China. To expand dairy exports to China, 14 special dairy container trains were sent to China from 2018 to 2020 from Belarus (Liu Chang, 2021). Based on the international political and economic situation and domestic development strategy, the China-Belarus Industrial Park was jointly built by China and Belarus in the context of the BRI. The China-Belarus Industrial Park is a landmark project of cooperation between the two countries and a link in the construction of the community of mankind for a shared future, a product of bilateral or multilateral regional economic cooperation for the development of regional economic integration in the new era. The development of the park has also been recognized by the international community and has received awards from the World Free Special Economic Zone Alliance (FEMOZA) for the fastest growing special economic zone in the world, the 2019 Global Free Zone of the Year Award from *Foreign Direct Investment* magazine, and the Best Project in Central and Eastern Europe Award from *Asia Money* magazine for the BRI (Liu Liliang, 2020, May 12).

(iii) Significance of the project

1. Domestic significance

(1) Contribute to China's strategy of deepening cooperation with Eastern Europe
Belarus is an important node of the BRI in the Eurasian region and is a natural hub for connecting the Eurasian continent. There are many favorable conditions for the development of China-Belarus Industrial Park. The more obvious ones are the comprehensive strategic partnership and good economic and trade relations between China and Belarus. Other conditions include the stabilization of the macroeconomic situation in the country in recent years, the completion of the construction of public facilities, the advantageous geographical location of Belarus, and the experience of Belarus in the construction of free economic zones. In recent years, China and Central and Eastern European countries have made solid progress in economic and trade cooperation, including Belarus with the China and Belarus industrial park. The China-Belarus Industrial Park dovetails the development strategy, project planning and standard system of both sides, and comprehensively enhances the in-depth cooperation in the whole field of trade, investment, and infrastructure construction, which is of great significance to smoothly promote the construction of the BRI.

(2) Beneficial to the overall development of Chinese companies in the EU
EU member countries are technologically advanced, with good product quality, high standards and mature markets. Industrial Park enterprises are still dominated by Chinese and Belarus enterprises. Theoir products do not meet the EU standards. It is relatively weak in the EU market in terms of competitive power, but the experience in the China-Belarus Industrial Park makes these enterprises grow in the competition and enhance the vitality and innovation of these enterprises. The work of the park achieved better results in 2018. The number of enterprises and residents in the park has increased. The number of industries involved is also increasing. Many products have been put into production in the park. The construction of the Chinese-aided residential building and the scientific technical achievement transformation cooperation center project have started one after another.

(3) Deepening cultural exchanges between the peoples of China and Belarus
China-Belarus relations are based on a high level of mutual trust and equal treatment. China respects the internal and external policies and the development path chosen by Belarus in the light of its national realities and supports the efforts of Belarus to safeguard national sovereignty and national dignity, maintain social

stability and promote economic and social development. In turn, Belarus supports China's independent choices in national development and international affairs. China is particularly grateful to Belarus for the valuable support it has given to China on issues related to China's sovereignty, such as Taiwan and Tibet. The Chinese Cultural Centre in Minsk was established in December 2016. Zhang Hongwei, responsible official of the Cultural Office of the Chinese Embassy, said that in the past two years, the Chinese Cultural Center had held more than 270 cultural and tourism events of various kinds, involving more than 20,000 people (Wu Yan, Qu Song, 2017) .

2. Significance for Belarus

(1) Strengthening the exchange of connectivity facilities between Belarus and China

Unlike parks in Southeast Asia, Africa and other regions, the influence of the China-Belarus Industrial Park is far enhanced through the China-European train. At present, 12 round-trip China-Europe container train rides transit Belarus every day (Xinhua News Agency, 2018, October 16). The China-Belarus industrial zone has been able to exert the effect from point to point, through the China-European train. In 2018, the two sides signed a contract to develop friendly provincial and state relations letter of intent. The cooperation between the Sixth Academy of Aerospace Science and Technology, Shaanxi Baoguang Holding Company and other enterprises and the China-Belarus Industrial Park has been steadily promoted. The Belarusian Center was officially inaugurated at Xi'an Foreign Studies University, Shaanxi. The direct flight project between Xi'an Xianyang International Airport and Minsk Airport has been actively promoted. The China-European liner "Chang'an" has been opened to Minsk, the capital of Belarus, and achieved regular operation.

(2) Building a cultural platform for communication between China and Belarus

Both the Chinese and the Belarus peoples have excellent qualities of hard work, bravery, kindness, and simplicity, as well as the glorious tradition of defying violence and resisting foreign invasion. The people of Belarus appreciate the splendid culture and long history of China and admire the achievements of the Chinese people in the cause of nation-building. The Chinese people are also pleased to see the stable development of Belarus and value its practices and accumulated experience in the cause of nation-building. The Sino-Belarusian friendship and cooperation relations have stood the test of time and international changes. The establishment and development of the Sino-Belarusian industrial park has further promoted friendship between the Chinese and Belarusian peoples. As of December 2018, Belarus established four Confucius Institutes, two Confucius Classrooms, and 20 primary and secondary schools offer Chinese language instructions. In 2018, a four-day exhibition of Chinese films also opened at the Central Cinema in Minsk (Wu Yan, Qu Song, 2017)).

(3) To promote the development of foreign trade in the local industry of Belarus
Pragmatic cooperation and mutual benefit are the strong driving forces behind the development of China-Belarus relations. In the economic and trade fields, China and Belarus are actively exploring the potential for cooperation, exchanging benefits, and seeking to achieve complementary advantages. China imports potassium fertilizer from Belarus for agricultural production. Belarus imports some daily consumer goods needed by people from China, etc. The two sides are now actively promoting cooperation in the fields of machinery manufacturing, telecommunications, and thermal electric plants. The Belarusian presidency of the CIS in 2021 promoted a logistical interface with the Chinese BRI. Exports to China in the first 11 months of 2020 alone were already close to US$239 million, almost twice as much as in all of 2019. Exports to China now account for 4.5% of all agricultural and food exports, making China the third largest export market for Belarus' agricultural and food products.

(4) Contributing to the social development of Belarus in favour of future growth industry development
In 2020, the park was classified by the Belarus government as the first 5G pilot zone and the first unmanned vehicle pilot zone. The park has attracted numerous high-quality enterprises with great strength in science and technology innovation. The high-tech temperature measurement and pandemic prevention intelligent robots produced by Chinese enterprises in the park are also used in Minsk airport and other public places, playing a role in the prevention and control of the Covid-19 Pandemic in Belarus. After 2020, the second five-year plan being planned for the China-Belarus Industrial Park began to focus on the development of growth industry, making the China-Belarus Industrial Park an industrial gathering platform for expanding regional cooperation and economic and trade exchanges between the two countries. The plan focuses on the direction of "industrialization, internationalization, digitalization and ecology" (Economic Daily, 2020, May 12). It also focuses on the development of growth industry, moving from the construction period to high-quality development. It makes the China-Belarus Industrial Park to become an industrial gathering platform for expanding regional cooperation and promotes economic and trade exchanges between the two countries.

7. Xiamen University Malaysia

(i) History of development

On January 21, 2013, the Malaysian government officially invited Xiamen University to establish a branch campus in Malaysia. On October 4 of the same year, in the presence of Chinese leaders, an agreement was signed for China Development Bank to fully support the construction of the Xiamen University Malaysia campus. On July 3, 2014, a groundbreaking ceremony for the branch campus was held in the presence of Malaysia's Prime Minister Najib. On October 17, 2014, China Electric Power Construction Corporation broke ground on the Malaysian campus of Xiamen University, marking the official birth of China's first overseas university. In February 2016, the branch welcomed its first 200 students and officially began operating. In September 2016, the first group of 440 Chinese students arrived in Malaysia to begin their study abroad journey at the branch. On September 9, 2017, the second phase of the Xiamen University Malaysia campus project began construction, with the first phase to include four new high-rise student residences, a student cafeteria, an international academic exchange center and a water music plaza. In 2017, the school administration decided that it was not conducive to teaching and learning for teachers to live in student housing for extended periods of time, and for that reason leased several houses off-campus as teacher apartments. On April 25, 2018, Xiamen University Malaysia and the Australian National University signed an understanding memorandum entry on academic cooperation. This aims to promote exchange and cooperation between the two universities in the field of reciprocity, especially in student exchanges. On May 3, 2018, Xiamen University Malaysia and Dublin Business School, Ireland signed an understanding memorandum entry on academic collaboration. In June 2018, Xiamen University Malaysia signed an understanding memorandum entry and other cooperation agreements with the University of Cincinnati regarding academic cooperation. In September 2019, Xiamen University Malaysia welcomed its first batch of undergraduate graduates, with 371 students, including 242 Malaysian students, receiving their degree certificates from Xiamen University President Zhang Rong.

In November 2015, during his visit to Malaysia, Premier Li Keqiang said that China and Malaysia are good neighbors, good partners, and good friends. The then Malaysian Prime Minister Najib strongly agreed with President Xi Jinping and Premier Li Keqiang's positioning of the relationship between the two countries (Xinhua News Agency, 2016). The opening of the Xiamen University

Malaysia campus is an "excellent example of the strength of the relationship between Malaysia and China". In 2018, Xiamen University Malaysia is open to students from twenty provinces and regions, including Hebei, Jiangsu, Zhejiang, and Liaoning. All majors are taught in English, except for Chinese Language and Literature and Chinese Medicine, which are bilingual. From 2019 onwards, the school requires candidates to score 80% of the full marks in English for admission (Qi Wei, 2016)

(ii) Status of the project

Xiamen University Malaysia (XUMM), officially opened in 2016, is the first branch of a leading Chinese university established overseas. It is a direct overseas campus of Xiamen University, wholly owned by XUMM. Located 45 km southwest of Kuala Lumpur, the capital of Malaysia, it is conveniently situated between Putrajaya, the federal administrative centre of Malaysia, and Kuala Lumpur International Airport. The campus covers an area of 900 acres, floor area of building is planned to be 475,500 square meters. The total scale of enrollment is planned to be 10,000 students, which covers three levels of education: undergraduate, master's and doctoral (http://www.xmu.edu.my/main.htm).

December 13, 2020, northern hemisphere entered the cold of winter. But the XU campus on the outskirts of Kuala Lumpur, the capital of Malaysia, was still a lush, tropical landscape. More than 1,300 Malaysian and Chinese students are currently enrolled at the school. XAU Malaysia currently offers 12 majors, all of which are taught in English according to international standards, except for Chinese and Chinese Medicine. It is expected that by 2022, XMSU Malaysia have more than 5,000 students enrolled, with an eventual student population of 10,000. In the future, the university will train a large number of international talents needed by the BRI countries. Wang Ruifang, president of XU Malaysia, said that. The school aims to better serve Malaysia, Southeast Asia, and Chinese companies on the BRI. The talents trained by the university should serve the local economy and social development.

(iii) Significance of the project

1. Domestic significance

(1) Providing more opportunities for domestic students to pursue higher education
XMSU Malaysia can provide more pathways for Chinese students from independent Chinese secondary schools in Malaysia as well as from neighboring countries to pursue higher education. For example, the principal of Han Kong Secondary

School in Penang, Malaysia, a Chinese-language independent secondary school, Chou Mei Li, said that the Chinese-language independent secondary school takes the general examination administered by the General Board of Education of Malaysia every year, but the results of this examination are not recognized by the Malaysian Ministry of Education, resulting in the pathway for high school students of Han Kong Secondary School being limited to foreign universities and domestic private universities in Malaysia. The establishment of a branch in Malaysia provides more pathways for students from similar Chinese independent schools to pursue higher education.

(2) Contribute to the development of international talent
The interest and donations of people in both countries are a strong boost to the construction of the branch campuses. The branch strives to be an international university with first-class teaching and research, and a multicultural mix, dedicated to nurturing young talents with dignity and wisdom to become good citizens of the region and contribute to the well-being and social progress of Malaysia, China, and other BRI countries. These people also continue to shine as part of the cultural exchange on the future platform of Sino-Malaysian diplomatic relations.

2. Significance for Malaysia

(1) Enriching the local higher education structure
Malaysia's higher education system was born out of the British higher education system in the former colony. By 2012, more than a dozen universities from the UK, Australia and the Netherlands were operating in Malaysia, gradually forming a combination of European, American, and local higher education structures. As the first leading Asian university in Malaysia, Xiamen University has a strong reputation in Southeast Asia. The presence of Xiamen University in Malaysia enriches the local higher education culture and it is making a beneficial impact on its higher education system. The establishment of XAU in Malaysia creates a healthy competition with local private and national universities, enhancing the quality of teaching and learning, making the development of higher education in Malaysia more diversified, and attracting more international students to study in Malaysia, as well as providing more choices for Malaysian students.

(2) Beneficial to boost the local economy in Malaysia
The creation of XAU Malaysia obviously pulls in a significant number of transient people to Malaysia each year and provides employment and increased social stability during the time the school is being established and maintained. During the

school's education period, the school provides more opportunities for Malaysians to study and experience a different culture, as well as train more intellectuals and useful people for the Malaysian nation. The school was held to both drive the local economy and contribute to the supply of talent for future economic development.

References

Anonymous. (2015). China and New Zealand signed the third intergovernmental cooperation project agreement and exchanged letters on launching negotiations on the upgrading of the China-New Zealand free trade agreement (2015, November 11). http://sg.mofcom.gov.cn/article/dtxx/201511/20151101161046.shtml.

Anonymous. (2017a). Sino-Russian cooperation "Yamal Project" officially put into production to open up Arctic waterways to boost the "Silk Road on Ice" (2017, December 9). http://m.news.cctv.com/2017/12/09/ARTIZqJ3xHCEZOI55TmkEJ66171209.shtml.

Anonymous. (2020). Yes! China has won 1.6 trillion natural gas orders and reached a cooperation project with Russia! (2020, June 11).

BRI web link. (2019, August 14). Set up a demonstration project of strategic connectivity between China and Singapore (Chongqing). https://www.yidaiyilu.gov.cn/xwzx/dfdt/99993.htm

CEC website link. (2021). http://pr.powerchina.cn

China is Malaysia's true friend and strategic partner -- An interview with Malaysian Prime Minister Najib, Xinhua News Agency (2016, October 30). http://www.gov.cn/xinwen/2016-10/30/content_5126227.htm

China's Arctic Policy (2018, January 26). Xinhua News Agency. https://www.yidaiyilu.gov.cn/zchj/zcfg/45994.htm.

China's first ship Yamal LNG arrived in China, Xinhua News Agency (2018, July 19). http://www.scio.gov.cn/m/31773/35507/35513/35521/Document/1634183/1634183.htm.

Dong Zhanwei, Li Yuanying, Gao Zhenbo (2018). North star power news (2018, December 20). http://news.bjx.com.cn/html/20181220/950545.shtml.

Dong Zhanwei, Li Yuanying, Gao Zhenbo (2018). North star power news (2018, December 20). http://news.bjx.com.cn/html/20181220/950545.shtml.

Hanlin. (2021). 'The current development of China-European Liner', refer to Hanlin Logistics, Sohu.com.,https://www.sohu.com/a/463766515_787509?scm=1102.xchannel:325:100002.0.6.0&spm=smpc.channel_248.block3_308_NDdFbm_1_fd.1.1678707436123q0ScVRQ_324

Henan Provincial People's Government. (2014, December 17). The second domestic transport coordination meeting of the China-Europe Class Train was held in Zheng yesterday, Henan Daily. http://www.henan.gov.cn/2014/12-17/346965.html.

Jiang Pinpin, Chang Xuemei. (2016). Joint statement of the people's Republic of China and the Republic of Belarus on the establishment of a comprehensive strategic partnership of mutual trust and win-win cooperation (2016, September 30). http://cpc.people.com.cn/n1/2016/0930/c64387-28751583.html.

Joint Statement on Docking Cooperation in the Construction of the Silk Road Economic Belt and the Construction of the Eurasian Economic Union (2017, March 16). Ministry of Foreign Affairs. https://www.yidaiyilu.gov.cn/zchj/sbwj/2427.htm.

Li Yiran. (2020). Xinhua Review: China-Africa relations have been in the same boat for 20 years, and Sino-African relations continue to move forward (2020, October 13). https://www.yidaiyilu.gov.cn/xwzx/gnxw/151226.htm.

Liu Chang. (2021). Exclusive interview: Belarus is optimistic about the prospect of cooperation in agricultural products and food export to China -- Interview with Belarusian Minister of Agriculture and Food Kropko (2021, January 21).

Liu Liliang. (2020). China-Belarus Industrial Park has entered a new stage of high-quality development. China Securities Network. (May 12). https://finance.eastmoney.com/a2/202005121482897052.html

Liu Xian. (2014). China's multi-party coordination and joint efforts to improve the China-European train competitive power, China News Network. August 14. https://www.chinanews.com/gn/2014/08-14/6494702.shtml.

Lv Qiang. (2020). Kenya-Inner Mongolia Railway resumes passenger service (2020, July 19). https://www.yidaiyilu.gov.cn/xwzx/hwxw/138256.htm.

Mombasa-Nairobi Standard Gauge Railway Operation Project: A Classic Case of Belt and Road Initiative's Foreign Investment and Cooperation (2019, April 25), International Business Daily. http://shanghaibiz.sh-itc.net/article/dwtz/dwtzhwtz/201904/1472517_1.html.

Nearly half of the signed cooperation projects in the China-Singapore connectivity project are in the financial sector (2019, March 15). China News Network. https://www.yidaiyilu.gov.cn/xwzx/gnxw/82699.htm

Pakistan's Qasim Power Station has generated more than 20 billion kWh of electricity (2020, May 27). China Electric Power Website. http://mm.chinapower.com.cn/xw/gjxw/20200527/20146.html

PPP project for coal-fired power station in Qasim Port, Pakistan (2017, June 21). Chinese government procurement website. http://www.ccgp.gov.cn/ppp/gj/201706/t20170621_8410859.htm

Qi Wei. (2016). Xiamen University, Malaysia: The Pearl embedded in "Belt and Road Initiative". (2016, September 27). http://www.scio.gov.cn/ztk/wh/slxy/31209/document/1492497/1492497.htm

Singapore looks forward to new opportunities brought by the construction of the Maritime Silk Road, the Information Office of the State Council (2015, July 28). http://www.scio.gov.cn/ztk/wh/slxy/31200/Document/1442474/1442474.htm

The China-Singapore connectivity project has gradually become a new business card for international cooperation of "Belt and Road Initiative" (2020, November 19). China News Network. https://www.yidaiyilu.gov.cn/xwzx/dfdt/154344.htm

The "circle of friends" of the southbound channel of the China-Singapore interconnection project has been further expanded. (2018, September 7). Xinhua News Agency. http://m.xinhuanet.com/cq/2018-09/07/c_1123392868.htm

The opening of China-Russia East Route Operation to promote the intellectualization of Oil and Gas Pipeline (2019, December 5). Science and Technology Daily. https://www.yidaiyilu.gov.cn/xwzx/gnxw/111796.htm.

The total amount of US US$21.9 billion of the 137projects signed on the 3rd anniversary of the China-Singapore connectivity project (2018, November 8). Xinhua News Agency. https://www.yidaiyilu.gov.cn/xwzx/gnxw/71067.htm

Wu Yan, Qu Song. (2017). Chinese investment helps Belarus to develop national industrial culture and integrate into Chinese cultural landscape (2017, August 17). https://www.yidaiyilu.gov.cn/xwzx/hwxw/23046.htm

Xie Duo. (2019). Chairman of the Silk Road Fund: Serving "Belt and Road Initiative" to promote financing (2019, May 6). https://www.yidaiyilu.gov.cn/ghsl/gnzjgd/89041.htm.

Xinhua News Agency. (2018, October 16). In the first 9 months, the volume of trains passing through Belarus and Central Europe increased by 25%.Xinhuanet.http://m.xinhuanet.com/2018-10/06/c_1123522921.htm.

Xinhua website. (2019, November 25). http://www.xinhuanet.com/power/2019-11/25/c_1210368166.htm

V

Analysis of the Stabilization Mechanism of the Co-construction of the BRI for China and the World Economy in the Year of the Pandemic

Introduction

The outbreak of Covid-19 Pandemic in early 2020 successively caused a serious impact on the Chinese and world economies. The rapid propagation velocity of the pandemic and the wide range of transmission have plunged the global economy into recessionary difficulties, making it an unprecedented global catastrophe faced by human society. Since the outbreak, China has been playing a leading role in the global recovery process by taking strong measures to effectively contain the spread of the pandemic and taking the lead in recovering from the pandemic and achieving economic growth. In the years of the pandemic, the co-construction of the BRI has been fruitful in bucking the trend. China has become a bridge between the domestic and foreign economic cycles in the double cycle by linking its own stability to the global recovery through the co-construction of the BRI.

Facts prove that the common construction of the BRI during the pandemic has been not only playing a positive role in stabilizing the Chinese economy, but also has been demonstrating a strong vitality in helping the global economic recovery. It has made an important contribution to stabilizing the BRI countries and the global economy. It is a powerful grasp of China's participation in global governance, opening to the outside world and building a community of a shared future for mankind. Looking ahead, China will continue to contribute Chinese wisdom, Chinese solutions, and Chinese power to the world economy through the BRI, helping the world economy to get out of the predicament and accelerate its recovery as soon as possible.

1. Overview of the impact of the covid-19 pandemic on the global economy

The outbreak of the Covid-19 Pandemic had two successive waves of impact. The first mainly hit the Chinese economy and, after the global pandemic, created a second wave of impact on the major developed economies in Europe and the United States. The pandemic not only hit China and the world economy from both the demand and supply sides, but also brought the global production system to a halt, factor flows were blocked, industrial chains and supply chain were broken, economic activities went into shock, and the global economy fell into recession.

(i) Impact on the Chinese economy

As the first country hit by the outbreak of Covid-19 Pandemic, China has effectively curbed the spread of the pandemic through strong measures, but the pandemic has nevertheless had a significant negative impact on the economy. The first is to hit the tertiary sector from the demand side, where the worst hit are mainly the transport and storage, wholesale and retail, tourism, accommodation, catering, offline entertainment, agriculture, forestry, fishery, and animal husbandry sectors. According to the National Bureau of Statistics, the cumulative growth rates of consumption, investment, and imports and exports in China all experienced significant declines in February 2020 following the outbreak. According to the National Bureau of Statistics, the cumulative growth of China's total social consumer goods retail sales, permanent assets investment volume, and total import and export value all experienced a precipitous decline in February 2020, with consumption and investment falling by more than 20%, and import and export falling by more than 10% until the end of 2020, continuing negative growth. It is thus clear that the Covid-19 pandemic has not only had a huge impact on the Chinese economy in the short term, it also has had a negative impact in the long term. (National Bureau of Statistics, 2021)

Furthermore, according to the National Bureau of Statistics, hitting the manufacturing sector from the supply side, China's manufacturing purchasing managers' index, non-manufacturing business activity index and composite PMI output index all experienced significant declines in February 2020. Among them, the 2020 China Manufacturing Purchasing Managers' Index was 35.7, down 27% year-on-year; the Non-Manufacturing Business Activity Index was 29.6, down 45% year-on-year; and the Integrated PMI Industry Index was 28.9, down 45% year-on-year. As effective progress is made in the prevention and control of the pandemic, the indicators return to normal levels in March 2020 (National Bureau of Statistics, 2021).As can be seen, the Covid-19 pandemic outbreak hit the Chinese economy in the short term, but timely and efficient pandemic prevention allowed the Chinese economy to recover quickly from the outbreak.

Finally, the pandemic has also had a serious impact on other areas. First, the pressure on employment has increased significantly. The employment situation has become even more severe as the pressure on enterprises to reduce staff has increased in the face of a significant slowdown in economic growth, coupled with the rising risk of shutdown or closure of some small and micro enterprises. The national urban survey unemployment rate rose to 6.2% in February 2020. The urban survey unemployment rate in 31 major cities stood at 5.7%, up 1 and 0.5% average points, respectively, from December 2019 (National Bureau of Statistics, 2021). Second, the gap between local government revenues and expenditures widens rapidly. It is difficult to close it quickly with the end of the pandemic, as business production and operation activities contract and local government revenues plummet, while expenditures to deal with the pandemic increase significantly. Third, the financial system risk rose, debt defaults increased, the banking system non-performing rate rose, the risk of small and medium-sized banks increased, the stock market decline may again trigger the risk of equity pledge financing, fight of capital pressure increased.

(ii) Shocks to the global economy

1. Impact on global capital market

With the global pandemic of the Covid-19, the capital market shock brought about by the Covid-19 pandemic swept through almost all the major economies of the world, increasing the domestic financial risks in countries around the world. As the U.S. increasingly becomes a region hit hard by the Covid-19 Pandemic, U.S. financial markets are in violent turmoil, with the value of changes in several market indicators setting all-time peaks.

In the stock market, there were four primary meltdowns in the U.S. stock market in the eight trading days between March 9 and 18, 2020. This is the second time a meltdown has occurred since the U.S. stock market established a meltdown mechanism in 1987, and four meltdowns in a row is unprecedented. The mid-March week saw stock market declines at record levels since the international financial crisis of 2008, and saw the Nasdaq, Dow Jones Industrials, and S&P 500 fall back to levels seen at the end of 2016. At bond market, the 1-year and 3-month U.S. Treasury bonds rates of return were once close to zero, and the 10-year Treasury bond rate of return fell below 1% for the first time since 1953. In the foreign exchange market, the dollar index also experienced a violent shock, falling from late February until mid-March when it began to rise again. And in the bulk commodity market, the bull market in gold ended, with London gold falling 8.59% in a week in March, and oil prices plummeting as Organization of Petroleum Exporting Countries international oil prices fluctuated sharply due to the breakdown of negotiations with Western countries over oil production and prices.

Since the United States is at the heart of global financial markets, its financial volatility is directly transmitted and passed on to global financial markets. The dramatic volatility in the U.S. financial markets was quickly transmitted to financial markets in other parts of the world and produced chain reaction in the global financial markets, causing a severe contraction in the world financial asset and a substantial devaluation of international wealth.

U.S. pandemic prevention and control policies exacerbate the huge risk of dollar devaluation borne by countries around the world. To counteract the negative impact of the Covid-19 pandemic on the real economy and financial markets of the United States, the United States has adopted a Mombasa-Nairobi Standard Gaugeary financial policy of massive liquidity inflows into the financial markets, which has not only increased the instability of the domestic financial markets of the United States, but also exacerbated the huge risk of countries around the world bearing the devaluation of the dollar. In response to the external shock of the Covid-19 pandemic, the United States federal reserve bank continued to inject liquidity into the financial markets, launching financial support policies such as interest rate cuts, restarting the mercantile paper financing facility, big scale quantitative easing programs, and "unlimited money printing" in the form of purchases of Treasury bonds and agency loan on security backed securities to mitigate the release rate risk of financial asset impairment. This pandemic prevention and control type of financial policy of the United States, which has allowed the dollar to depreciate, not only fails to address the risks facing the real economy of the United States, but also exacerbates the depreciation of the dollar and pushes up the debt level of the United States federal government. The new "dollar flood" in the United States induced by the Covid-19 pandemic is flooding the world, with countries around the world bearing the burden of the new crown financial crisis in the United States and competing to trigger potential crises in domestic financial markets around the world.

In fact, the stock market crash and meltdown not only happened in the United States, but major economies such as Europe, Japan and Korea also experienced violent turmoil in the financial markets. US, Japanese, British, French, German and Korean stock indices fell by more than 20% in half a month and entered technical bear markets one after another.

2. Impact on global trade

The prevention and control of the Covid-19 pandemic has not only directly affected the pace of cross-border trade development, but also changed the regional and industrial structure of world trade and reshaped the map of its future development. The Global Investment Trends Monitor, published by the United Nations Conference on Trade and Development, shows that the total global FDI was about US$859 billion in 2020, down 42% year-on-year. (United Nations, 2021, January). Global trade volume according to UNCTAD also shrank by about 9% in 2020 (United Nations Conference on Trade and Development, 2020, October 29). The "embargo" measures taken by countries during the pandemic have reduced trade flows and slowed down the process of investment projects. At the same time, multi-national enterprises have been affected by uncertainty about the global economic outlook and is reassessing investments in new projects.

First, the prevention and control of the Covid-19 pandemic has led directly to a sharp decline in the rate of development of world trade. The current severe global shortage of medical supplies, such as personal protective equipment and ventilators for patients suffer from the severity of the new pneumonia pandemic. The varying degrees of trade restrictions adopted by many countries to "help themselves", have led directly to the emergence of tariff and non-tariff barriers that impede the development of cross-border transactions. On the one hand, in order to increase their own material security capacity to prevent and control the Covid-19 pandemic, some countries have resorted to differential import tariff or export tariff measures and implemented the policy of "rewarding entry and restricting exit" for pandemic prevention and control products, distorting the allocation effect of the international market; on the other hand, in order to reduce the risk of pandemic transmission brought about by world trade in goods, some countries have directly adopted non-tariff barriers to restrict cross-border trade in non-Covid-19 pandemic prevention and control products. These policies, which restrict cross-border transactions, not only reduce the efficiency of transactions in international markets, but also lead to a significant decline in the volume of global trade and greatly hinder the development of global trade. In short, whenever there is an upward trend of Covid-19 pandemics in their countries, countries preferentially adopt various non-tariff barriers to control exports of medical supplies, equipment, and drugs, and reduce Covid-19 imported risks by reducing the frequency of cross-border trading activities.

Second, the Covid-19 pandemic prevention and control has changed the commodity structure of world trade. In the process of preventing and controlling the new pandemic in countries around the world, an important measure to prevent "medical crowding out" and relieve the shortage of national pandemic prevention and control products is to expand the supply capacity of national pandemic prevention and control equipment and medical supplies. Therefore, many countries increase the supply of pandemic prevention and control equipment and medical supplies in two ways. On the one hand, it encourages a number of domestic manufacturing industries to switch to the production of medical equipment or medical supplies, or increase investment in new production facility for pandemic prevention and control equipment or medical supplies, so as to guarantee the supply of pandemic prevention and control products; on the other hand, it increases the import capacity of pandemic prevention and control equipment or medical supplies in the international market, and strictly control the outflow of domestic pandemic prevention and control products. As a result, the intense competition among countries for products for the prevention and control of the Covid-19 pandemic has changed the commodity structure of world trade.

Finally, the Covid-19 pandemic prevention and control has changed the flow of world trade. Due to the variability in the timing of the global spread of the Covid-19 pandemic, and the differences in national strategies for the prevention and control of the Covid-19 pandemic and its prevention and control effectiveness, countries around the world have different attitudes towards the flow of imports and exports of medical equipment and medical supplies. After achieving early positive results in the prevention and control of the Covid-19 pandemic, China increased its exports of pandemic prevention and control products to the world market and became a major net supplier of certain pandemic prevention and control products globally, while Europe and the United States, where the Covid-19 pandemic was severe, became the main suppliers of pandemic prevention and control products and net importer. Thus, against the background of the still global spread of the Covid-19 pandemic, the structure of trade flows for certain commodities undergoes sustained changes in the short term and depends on the development of the Covid-19 pandemic on a global scale and the actual effectiveness of the pandemic prevention and control in each country.

2. Impact on global value chain

The global spread of the Covid-19 pandemic and the sharp rise in the number of infected people, countries have taken measures to "seal off cities" and "seal off countries", which has a negative impact on the production, transportation, and employment of the global manufacturing industry. The global production value chain is seriously hit, intensifying the trend of global supply chain "decoupling".

First, the Covid-19 pandemic prevention and control directly interrupt a link in the global production chain system. From the perspective of the global allocation of production, supply chain is an important link in the global production chain and value chain. To control the global spread of the Covid-19 pandemic, many countries have adopted interdiction measures that cut off global economic activity. On the one hand, restricting the movement of people and goods across borders, which is a common practice in many countries to prevent the global spread of new Covid-19 pandemic, can lead to sudden disruptions in the supply chain of global production chains and to a "wedge effect" between countries. On the other hand, countries with severe new crown outbreaks have generally taken isolation measures such as furloughs, production shutdowns, and work stoppages to reduce clustering or maintain some social distance within the country, with serious inhibiting effect on the global production chain in the domestic production chain. Thus, the continued development of the Covid-19 pandemic is bound to amplify the economic risks of global production chain interruption.

Second, the prevention and control of the Covid-19 pandemic has forced countries to reorganize their production chain processes. The global outbreak of the Covid-19 pandemic has caused governments to rethink the significance of the life and health industry and to weigh the economic efficiency and social benefits of industrial development. An important reflection is that an over-reliance on global production chains in the medical and health industry, which is a matter of life safety and physical health, seriously undermines a country's ability to provide comprehensive management of major public health events. Therefore, the country's medical and health care industry cannot be configured to produce entirely abroad simply to pursue globalization economic benefit, and the direction of the country's industrial development must be adjusted to meet the real needs of the country's response to public health emergencies. In fact, in the wake of the

Covid-19 pandemic, many countries around the world not only converted traditional manufacturing companies to produce medical equipment and protective gear, but also increased the intensity of investment in biomedical and medical equipment. They initiated a process re-engineering of their production chains in order to increase their capacity to cope with the impact of public health security events.

Finally, the impact of the Covid-19 pandemic may have directly damaged some links in the global production chain. A core protective measure against interpersonal transmission of the Covid-19 pandemic is to reduce opportunities for interpersonal contact and big scale social aggregation, and to also reduce intensive contact-based economic activities. A "neighbourhood avoidance effect", may change people's consumption behaviour and habits in the long term and may be difficult to reverse even after the Covid-19 pandemic is over. Changes in people's lifestyle and consumption habits does not only directly affect the development of sectors with strong social aggregation, such as tourism, culture and sports, aviation, and transport, but also generates a "bullwhip effect" from changes on the demand side, which in turn leads to permanent changes on the supply side of the global production chain system, and some parts of the global production chain may suffer substantial damage. In short, the Covid-19 pandemic leads to fundamental changes in consumption and production behaviour that can profoundly affect the process of re-engineering global production chains.

2. Annual growth against the trend in the co-construction of the BRI in the year of the pandemic

(i) Fruitful policy communication

In 2020, the policy communication on building the BRI was fruitful. China signed the BRI cooperation documents with many countries and international organizations. For the first time, China has signed a cooperation document on the BRI with the African Union, a regional international organization. She has signed relevant cooperation documents with Kiribati, Ukraine, the International Seabed Authority and other countries and international organizations. The contents of the policy cooperation documents are gradually being refined. At the beginning of 2021, in less than two months, China signed the BRI cooperation agreements with two more countries, the Democratic Republic of Congo (DRC) and Botswana.

Table 5.1 Cooperation documents signed by China with countries and international organizations along the BRI route, 2020–2021

TIME	Contents
January 6, 2020	Signing of various cooperation documents with Kiribati, including the joint promotion of the BRI construction
November 11, 2020	Signing of the *Understanding Memorandum Entry with the International Seabed Authority on Promoting the Construction of the Silk Road Economic Belt and the 21st Century Maritime Silk Road.*
Mid-December 2020	Signing of the *Cooperation Plan Between the Government of the People's Republic of China and the African Union on jointly promoting the construction of the BRI with the African Union.*
December 23, 2020	Signing with Ukraine of the *Cooperation Plan between the Government of the People's Republic of China and the Government of Ukraine on Building the Silk Road Economic Belt and the 21st Century Maritime Silk Road.*
January 6, 2021	Signed an understanding with the Government of the Democratic Republic of the Congo to promote the co-construction of the BRI memorandum entry.
January 7, 2021	Signed cooperation documents with Botswana, including the understanding between the two Governments on promoting the co-construction of BRI memorandum entry.

Data source: compiled by the author from BRI.com

During the pandemic, the BRI member countries made full use of modern network technologies in multilateral cooperation meetings within the framework of the BRI. Due to the requirements of the pandemic control, countries strictly control their foreign activities, which has led to a significant reduction in "face-to-face" meetings, but the communication between the BRI member countries has not stopped because of the pandemic control. By the end of 2020, the BRI member countries had organized more than 30 multilateral meetings in the form of videos on topics such as health care, the BRI quality development, and the United Nations 2030 Agenda for Sustainable Development.

Table 5.2 Selected important multilateral BRI meetings initiated by China in 2020

TIME	Contents
June 18, 2020	China held a video conference on the theme of "Strengthening 'BRI' international cooperation and joining hands to combat the Covid-19 pandemic", with the participation of foreign ministers or ministerial officials from 25 countries and the United Nations and WHO responsible official to discuss the role of BRI international cooperation in policy communication, science and technology, supply chain stability and other aspects for the prevention and control of the pandemic.
December 14, 2020	International video seminar "China's New Development Pattern and High-Quality Building BRI" was successfully held, in which more than 30 foreign dignitaries, experts and scholars discussed the way of BRI high quality development.
December 18, 2020	The BRI International Cooperation Summit Advisory Committee was formed and its first meeting held, which emphasized the need to make the BRI a road to cooperation, health, recovery and growth.

Data source: compiled by the author from BRI.com

(ii) Steady progress in facility connectivity

During the pandemic, the co-construction of the BRI facilities did not stop because of the pandemic, and many landmark projects were successfully completed during this period, promoting connectivity and thus economic recovery on the BRI regions.

First, the China-Laos Railway and the China-Laos Expressway have made a breakthrough. Both the China-Laos Railway and the China-Laos Expressway are landmark projects in the docking of the BRI and Laos' strategy of "turning a land-locked country into a land-linked country". The breakthrough of the two projects marks the successful "going out" of our enterprises and plays an immeasurable role in the internal and external circulation of our economy and local socioeconomic development after the opening of the railway. On the one hand, the pandemic did not postpone the construction schedule of the China-Laos

Railway, and the 17.5-kilometer-long An'ding Tunnel, the first longest tunnel of the China-Laos-Kunming-Vientiane Railway, was successfully completed, laying the foundation for the completion and opening of the entire line. By the end of 2020, with the completion of the An'ding Tunnel, 84 of the 93 tunnels have been completed, and 98.8% of the tunnel process has been completed, which can substantially increase the capacity of the areas along the railway after operation and vigorously promote the economic and social development of the areas radiated by the railway. On the other hand, the opening of key sections of the China-Laos Expressway has laid the foundation for local economic recovery. On December 20, 2020, Laos' first expressway, the Vientiane to Vanlung section of the China-Laos Expressway, was opened to traffic 13 months ahead of schedule. The China-Laos Expressway is jointly invested by China and Laos, and the first phase connects the Laos capital Vientiane with the tourist city of Vanrong, with a total length of about 110 km, accounting for a quarter of the total length. The opening of the road shortens the drive from Vientiane to Vanrong to one hour from the previous 3.5 hours, laying the foundation for local tourism and economic development and boosting economic recovery. After the completion of the whole high-speed journey, it can strongly strengthen the exchange and cooperation between China and Laos in the fields of economy, trade, culture, education, and health, as well as promote the economic development of the north-central region of Laos.

Second, sundry items in the BRI areas are progressing well. In 2020, the construction in the BRI regions was well advanced in many areas such as railways, roads and bridges, ports, aviation, energy, digital technology, health, and medical facilities connectivity, etc., enhancing global connectivity. During the pandemic, a numerous projects constructed by and invested in by Chinese companies met high standards. It can be seen, in the BRI regions, the formation of an all-round, multi-level and complex infrastructure network with railways, highways, shipping, aviation, pipelines and integrated space information networks as its core is being accelerated.

Table 5.3 Progress of projects along the BRI route constructed by and invested in by Chinese enterprises in 2020.

Infrastructure category	**TIME**	**Item designation**	**Project progress**
railways	20 March	Direct inbound fruit train from Vietnam to Pingxiang Railway Port in Guangxi, China	Achieve regular operation
	17 August	Vietnam's interconnected international alleyway anti-dong railway laid	Completion of the last major 720-ton box girder of the double-lane bridge at Shima 2
	25 November	Hungarian-Serbian Railway, Zemun to Batajnica, Serbia, left line	Successfully completed and opened for operation
roadside bridge	Feb. 26	Cambodia Route 58	officially open to traffic
	June 22	The most difficult shield interval tunnel in KL MRT Phase 2, Malaysia: shield interval from RMN Station to South Cave	Successful penetration
	25 October	China-Pakistan Economic Corridor's first major rail project and Pakistan's first metro: Lahore Rail Orange Line, Pakistan	Officially opened for operation
	10 December	Padma Bridge Project, Bangladesh	Main bridge steel girders successfully joined together
port	18 March	China Postal Airlines Limited trade route	First implementation of Zhengzhou-Seoul freight trade route
Energy	2 March	World's largest combined cycle power plant: Saudi Aramco Jizan 3850 MW power project 12 combustion engine	Successful grid connection with load

Table 5.3 Continued

Infrastructure category	TIME	Item designation	Project progress
	18 May	The first "Made in China" clean coal-fired power plant in the Middle East: Unit 1 of the 4x600 MW clean coal-fired power plant project in Hasjan, Dubai, UAE	One successful grid connection
	December 3	The middle section of the Sino-Russian Eastern Route rock gas pipeline	Officially put into operation
digital technology	May 14	Widest coverage on the African continent, connecting the Middle East and the African continent ocean bottom cable 2 Africa	Start of construction
	August 3	Brazil's 5G construction bid	Brazil's vice president welcomes Huawei's participation
Healthcare	April 7	First project of cooperation between China and Egypt to produce face masks	Officially commissioned in an industrial park in the Egyptian capital Cairo
	9 May	"FireEye" virus detection laboratory using Chinese technology and built by a Chinese company	Inaugurated in the Philippines

Data source: compiled by the author from BRI.com

(iii) Trade flows grow against the trend

While global investment has plummeted and trade has shrunk under the impact of the pandemic, investment in the BRI regions has shown significant headwinds. *The Global Investment Trends Monitor Report*, published by the United Nations Conference on Trade and Development, shows that the total global FDI is about US$859 billion in 2020, down 42% year-on-year. Global trade volume according to UNCTAD also shrank by about 9% in 2020. The "embargo" measures taken by countries during the pandemic have reduced trade flows and slowed down the process of investment projects. At the same time, multi-national enterprise has been affected by uncertainty about the global economic outlook and is reassessing investments in new projects.

However, in 2020, China's non-financial direct investment in BRI countries showed a reverse growth trend, totaling US$17.79 billion, up 18.3% year-on-year, accounting for 16.2% of the total in the same period, up 2.6% average points from the previous year. China's total imports and exports with BRI-related countries totaled CNY 9.37 trillion, up 1% year-on-year (People's Daily, 2021, February 19). The BRI constructions has shown strong resilience.

(iv) Deepening of financial integration

During the pandemic, 29 countries have approved the BRI financing governing principles. The Ministry of Finance of China and the Ministries of Finance of 26 countries approved the BRI financing governing principles in 2017. The principles include recognizing that a good financing system and financing environment cannot be achieved without the strong support of governments of the BRI countrues, encouraging these countries to establish common platforms, and supporting financial resources to serve the real economic development in these countries, among other 15 financing governing principles. As of the end of November 2020, 29 countries, including China, Singapore, and Saudi Arabia, had approved the above governing principle (Wang Shidan, 2017. http://www.gov.cn/xinwen/2017-05/15/content_5194223.htm). Driven by international market demand, the internationalization of the CNY has made positive progress. The internationalization progression of CNY, 2020 presents the following two new features that deserve attention. One is the very rapid growth in the amount of CNY cross-border receipts and payments in 2020. The amount for the first three quarters of 2020 surpassed all of 2019, already reaching more than CNY 20 trillion. Second, the influence of CNY denomination is expanding. Two more futures, low-sulfur fuel oil futures and international copper futures, were added to the range of CNY-denominated bulk commodity futures in 2020, bringing the number of internationalized futures to six. In the global economic recession situation, the CNY internationalization can still make positive progress is enough to show that the internationalization of the CNY is an inherent demand of the international market.

(v) Reaching consensus on interpersonal contacts

In the year of the pandemic, BRI countries reached a broad consensus on the joint creation of a digital Silk Road. At the BRI High-level Video Conference on

International Cooperation held on June 18, 2020, China and foreign ministers or ministerial officials of 25 countries, as well as the heads of the United Nations and WHO, jointly issued the *BRI Joint Statement of the High-level Video Conference on International Cooperation* (Xinhua News Agency, 2020, June 19), which clearly stated that "against the backdrop of the pandemic impacting global economic and social development, countries need to strengthen cooperation in the digital economy, medical industry and food safety, and foster new economic growth points in areas such as e-commerce, smart cities, AI and the application of big data technologies, draw on international good practices and narrow the digital divide". The statement clarifies that the construction of the Digital Silk Road is a new highlight of cooperation in the BRI and is one of the key agreements on the BRI to be reached in 2020 (Ministry of Foreign Affairs, 2020, June 20).

In addition, the construction of the Health Silk Road is escorting lives in the year of the pandemic. Both before and after the outbreak, China would always have upheld the concept of a human community of a shared future in its international dealings and has actively built a human health community in the BRI regions. The international cooperation in the fight against the pandemic is further proof of the fact that China has always upheld the concept of a human health community in the BRI regions.

3. The stabilization mechanism of the co-construction of the BRI for the Chinese economy

(i) Import and export trade with the BRI countries stabilizes China's foreign trade

Under the impact of the pandemic, in order to stabilize the fundamentals of foreign trade, the General Administration of Customs of China has issued documents such as Ten Measures on Coping with the Impact of the Pandemic and Promoting Stable Growth of Foreign Trade, Measures on Customs Support for the Development of China-Europe Class Trains, Measures on Customs Support for the Development of Comprehensive Free Trade Zones and List of Measures on Coordinating the Prevention and Control of Pandemics and Customs Clearance Facilitation at Ports to cope with the downward pressure of foreign trade.

(http://www.customs.gov.cn/tianjin_customs/yqfk83/zzcs/3550846/index.html

http://www.gov.cn/xinwen/2020-02/25/content_5482880.htm
http://www.ithc.cn/m/marticle/362767.html
http://www.customs.gov.cn/customs/xwfb34/302425/2892073/index.html)

The main measures include five aspects: (1) simplifying the customs clearance operation process, optimizing the process of enterprise registration, approval, customs clearance and inspection for the special period of pandemic prevention and control; (2) streamlining documents and supporting materials, further deepening the reform of "management and administration", exempting enterprises from submitting contracts, packing lists and other accompanying documents in the import and export declaration process; (3) reducing customs clearance costs for enterprises, reducing or waiving late filing fees for imported goods not declared in time due to the pandemic, and extending the time limit for electronic payment of aggregated tax declarations; (4) supporting the cultivation of new foreign trade growth points, promoting the full implementation of global repair and re-manufacturing business in the comprehensive free trade zone, gathering new momentum for foreign trade and creating a new platform for opening up to the outside world; (5) actively responding to restrictive measures abroad, stepping up efforts to negotiate with foreign countries, and urging countries that have taken restrictive measures on China's trade in goods due to the pandemic to cancel or adjust their restrictive import measures on my goods.

According to the National Bureau of Statistics, in 2020, China's imports and exports of foreign trade totaled CNY 179.326 billion, up 4.1% year-on-year, while imports totaled CNY 142.231 billion, down slightly by 0.7% year-on-year, and net exports totaled CNY 370.96 billion, up 27.1% year-on-year. Among them, the trade situation with the BRI countries and the overall situation is generally consistent, exports to BRI countries CNY 5,426.3 billion, up 3.2%, imports CNY 3,943.3 billion, down 1.7%, net exports CNY 1483 billion, up 18.8%, accounting for 40% of the total for the same period.

In the face of a relatively serious break in global supply chain, trade growth between China and countries along the BRI has remained positive, demonstrating the strong vitality of contributing to the BRI in promoting trade. China's net exports to the BRI countries accounted for 40% of the total in the same period, which shows that the co-construction of the BRI occupies an important position in China's foreign trade and is of great positive significance in stabilizing mutual economic development and safeguarding people's livelihood.

Table 5.4 China's total imports and exports and exports with the BRI countries in 2019 and 2020 (unit: CNY billion)

Import and Export	**2019**	**2020**	**Growth in 2020**
Total export volume	172342	179326	4.1%
Total imports	143162	142231	-0.7%
Total net exports	**29180**	**37096**	**27.1%**
Exports to BRI	52585	54263	3.2%
Imports of BRI	40105	39433	-1.7%
Net exports to BRI	**12480**	**14830**	**18.8%**

Data source: National Statistical Office

Foreign trade is an important guarantee for China's healthy and sustainable economic development and a fundamental way to integrate into the global economy. As one of the most popular international public goods in the new era, the BRI is a great practice of China to promote global governance, and it is of great significance in promoting global infrastructure construction and building a global trade network. The global spread of the pandemic, while increasing uncertainty and downside risks to global growth, presents opportunities for optimizing our trade structure. The smooth flow of trade in the BRI reflects the resilience of China's foreign trade and the vast space for the development of foreign trade, bringing new opportunities for global trade. In the post-pandemic era, the construction of BRI should be used as an opportunity to promote high-quality trade development through the promotion of multilateral cooperation mechanisms and free trade.

(ii) Investment in the BRI countries by China stabilizes China's attraction of foreign investment

Under the impact of the pandemic, in order to stabilize the fundamentals of foreign investment, China's Ministry of Commerce, the Ministry of Industry and Information Technology, the Central Bank, customs and many others have introduced measures to support enterprises to resume work and production to tide over the difficulties from various aspects such as taxation, credit insurance and financing. For example, in late March 2020, the Ministry of Commerce and the Ministry of Finance jointly issued the *Notice on the Use of Special Funds for Domestic and Foreign Trade to Support the Work of Stabilizing Foreign Trade,*

Stabilizing Foreign Investment and Promoting Consumption. It proposes full support for stabilizing the basic plate of foreign trade and foreign investment, and special funds for foreign trade and economic development should be tilted towards foreign trade and economic areas that are more affected by the pandemic, etc.; (Ministry of Commerce & Ministry of Finance, 2020, March 23) on April 1, 2020, the Ministry of Commerce issued the Notice on Furthe*r Reform and Opening Up to Achieve Excellent Work Performance of Stabilizing Foreign Investmen*t. In response to the pandemic, it put forward 24 specific initiatives and work requirements in five aspects (Ministry of Commerce, 2020, April 6)

Due to the impact of the pandemic, global cross-border investment has declined, but foreign investment in China has increased slightly, and the investment cooperation between China and the BRI countries has improved steadily. According to the Ministry of Commerce, in 2020, direct investment in China from countries along the BRI is US$8.27 billion, basically the same as in 2019, 4,294 new enterprises were established in China by BRI countries, creating valuable jobs for China. In the first quarter of 2021, 1,241 new enterprises were established in China by the BRI countries, up 44% year-on-year, with actual investment of US$3.25 billion, up 64.6% year-on-year.

Table 5.5 Foreign direct investment in China, 2019 and 2020 (in CNY billion)

Overseas Foreign Direct Investment (OFDI)	**2019**	**2020**	**Growth in 2020**
Total Amount	9415	10000	6.2%
Direct investment in China from BRI countries	576	574	-0.3%

Data source: Ministry of Commerce

As a major trading nation, China's foreign trade stabilization is not only a necessity to help the national economy run smoothly and grow continuously, but also an important measure to release the pressure of overcapacity and exchange time and space for structural adjustment and industrial upgrading. It is a major responsibility and mission to stabilize global economic growth expectations and promote world economic recovery. Foreign investment and foreign trade go hand in hand. Foreign investment can not only inject momentum into national economic growth and industrial upgrading, but also provide capital, technology, management experience, advanced equipment and so on, which in turn can promote higher quality development of China's enterprises. In terms of stabilizing foreign trade and foreign investment for China to deal with the impact of the pandemic, to promote economic operation back on track, to complete the annual economic development goals and tasks are of great significance. The joint construction of the BRI as an important component part of China's foreign trade and foreign investment, since the pandemic, has been playing a positive role in stabilizing the Chinese economy.

4. Stabilization mechanisms for the economies along the BRI route

(i) The growth of China's direct investment against the trend in the BRI countries

In 2020, China invested US$110.2 billion in outbound non-financial investment, down 0.4% year-on-year, but invested US$17.79 billion in direct investment in BRI countries, up 18.3% year-on-year and accounting for 16.2% of the total for the same period, up 2.6% from the same period last year, mainly in Singapore, Indonesia, Vietnam, Laos, Malaysia, Cambodia, Thailand, the United Arab Emirates, Kazakhstan and Israel. This shows that even under the very serious pandemic situation, China's investment in the BRI major related countries not only did not decrease, but also increased significantly, and the counter-trend data fully demonstrates the strong resilience and vitality of the BRI, injecting a warm current for the economic "winter" of the BRI countries.

Table 5.6 China's outward foreign direct investment and outward contracted works, 2019 and 2020 (in USD billion)

Foreign Economic Relations	Subdivision	2019	2020	Growth in 2020
OFDI (non-financial)	Total Amount	1106	1102	-0.4%
	Direct investment in the BRI countries	**150**	**178**	**18.7%**
External contracting work completed revenue from operation	Total Amount	11928	10756	-9.8%
	Completion of revenue from operation for the BRI countries	980	911	-7.0%

Data source: Ministry of Commerce

In terms of foreign contracting projects, Chinese enterprises signed 5,611 new contracts for foreign contracting projects in 61 BRI countries, with a new contract value of US$141.46 billion, down 8.7% year-on-year, accounting for 55.4% of the new contracts signed for foreign contracting projects in China during the same period; China's foreign contracting projects completed a turnover of US$156 billion, down 9.8% year-on-year, of which US$91.12 billion was completed in BRI countries, down 7% year-on-year, accounting for 58.4% of the total for the same period. It has been a strong support for the recovery and development of the economies of the BRI countries.

In the first four months of 2021, China's outward non-financial direct investment was US$34.25 billion, down 5.2% year-on-year, but it invested US$5.96 billion in BRI countries, up 1.8% age points from the previous year. In terms of foreign contracting projects, Chinese enterprises signed 1,465 new contracts for foreign contracting projects in 59 BRI countries, with a new contract value of 41.56 billion U.S. dollars, up 25.2% year-on-year, accounting for 58.2% of the new contracts signed for foreign contracting projects in China during the same period; and completed revenue from operation 2.46 billion U.S. dollars, up 12.9% year-on-year, accounting for 55.1% of the total amount for the same period.

The 2021 BRI Survey Report published by the International Finance Forum states that the 2020 Covid-19 pandemic has a clear impact on GDP growth in countries around the world. According to the April 2021 IMF Petroleum Fund organization (IMF), the total global economy shrank by 3.3% in 2020, the worst economic recession since the Great Depression of the 1930s. More than two-thirds of the central banks interviewed said that the outbreak had negatively impacted the implementation of the BRI projects. Nevertheless, 75% of the central banks surveyed still believe that the BRI contributed to economic recovery in their jurisdictions after the pandemic (International Financial Forum, 2021, May, 30).

(ii) China-European Liner to help smooth trade flows in the BRI countries

During the Covid-19 pandemic, flights were grounded, and shipping was blocked, and the China-Europe train showed unique advantages in the situation of pandemic prevention and control because it was transported in sections and did not involve personnel quarantine, and it became a strategic alleyway to support trade between Asia and Europe, playing the role of an international transport artery against the pandemic. The safe and efficient operation of the China-Europe Class Train has continuously transported pandemic prevention materials and production raw materials to Europe, which has become an important logistics alleyway for the smooth international transportation of pandemic prevention materials and the protection of trade between China and Europe. It has become a "life alleyway" and "fate link" for countries to join hands to fight the pandemic. It has played a good demonstration role for countries to resume work and production and boost their economy.

Data shows 2020 China-Europe train operation exceeded 10,000 train rides. The first 10 months of 2020 traffic exceeded the whole year of 2019, becoming a "steel transport camel caravan" to help countries fight pandemics. It has connected 92 cities in 21 countries in Europe, has strongly guaranteed the security and stability of the international supply chain.

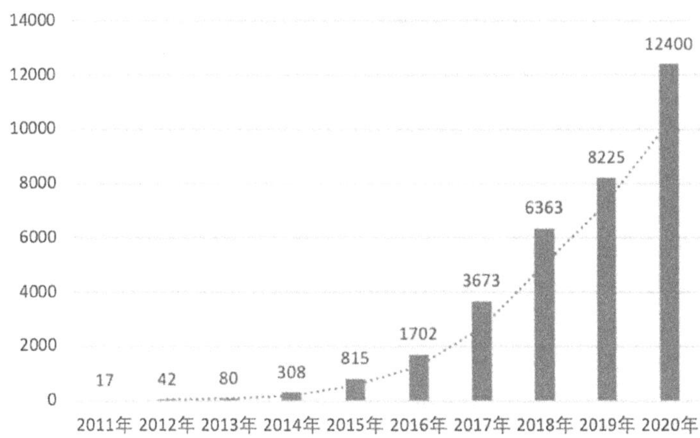

Figure 5.1 Number of Central European Classes in 2011–2020
Data source: based on open data (https://www.yidaiyilu.gov.cn/xwzx/gnxw/223960.htm)

In addition to the China-European Liner, the TIR (Transport International Routier) cross-border road transport has emerged during the pandemic, playing a flexible and convenient "door-to-door" transport advantage, transporting all kinds of goods urgently needed for countries and regions along the way, helping countries to resume work and production, and becoming a new highlight of the BRI interconnection. Up to now, China has opened up the entire territory for TIR transport, extending to 27 countries outside its borders. Three of the six BRI economic corridors have already realized TIR transport. According to the TIR international convention, the mutual recognition of customs control and non-opening of containers by countries along the route, which ensures the safety of cargo transportation and effectively reduces close human contact. It has become a new way of cross-border transportation in the context of the pandemic. TIR transport is mainly for goods with high value and strong timeliness, such as cold chain fresh food, lithium batteries and other special transport products, which complement the goods transported by CEB. The trans-Asian-European TIR transport route passes through Kazakhstan, Russia, Belarus and finally arrives in Poland, with a total length of more than 7,000 km and a freight time of only about two weeks, which saves 5 days compared to railway transport and has obvious benefits.

5. Co-constructing the BRI stabilization mechanism for the world economy

The global economy has been hit hard by the Covid-19 pandemic. Countries around the world are facing unprecedented hardships and challenges. Thanks to the joint efforts of the Chinese government and people, the pandemic in China has been stability control and social production has gradually returned to order against the backdrop of the normalization of pandemic prevention and control, and the Chinese economy is recovering extremely well and growing against the trend. Nowadays, China's steady economic growth has brought confidence and hope to the world, and the co-construction of the BRI has benefited the development of many countries and become an important channel for global economic recovery.

(i) Contributing to the global fight against the pandemic and helping the global economy to "come alive"

At the beginning of the outbreak, the demand for testing kits, protective clothing, medical masks and other pandemic prevention materials increased dramatically in various countries. As a global producer, China quickly resumed work and production after effectively controlling the pandemic, and worked hard to provide a large amount of much-needed pandemic prevention materials for the global fight against the pandemic, while the co-construction of the BRI played a bridging role in helping the global fight against the pandemic, and set up a "life alleyway" to fight the pandemic. In 2020, the BRI cooperating countries supported each other in the pandemic, united against the pandemic and developed together, and China actively played its role as the largest supplier of anti-pandemic materials worldwide, providing more than 280 shipments of emergency anti-pandemic materials to more than 150 countries and international organizations. For example, the Republic of Kenya, a country along the BRI route, was seriously affected by the pandemic. In June 2020, 18 Chinese enterprises donated 3 million worth of pandemic prevention materials to Kenya, including KN95 masks, disposable medical masks, disinfectant, respirators, protective clothing, etc. In addition, the co-construction of the BRI is a "health road" that ensures people's health. China has also sent medical expert teams to the BRI countries. Chinese enterprises have built temporary hospitals and organized the production of medical protection materials urgently needed locally according to the actual needs of the countries where they are located.

As the fight against the pandemic gradually evolves into vaccination, China has strengthened cooperation in vaccine research and development, production and distribution, making the vaccine a truly accessible and affordable public product for people in all countries. BRI countries are beginning to see a ray of hope for opening their doors and resuming pre-pandemic contacts. By the end of May 2021, China had provided or was providing vaccine assistance to more than 80 countries, exported vaccines to more than 40 countries. China had provided more than 350 million doses of vaccines to the international community.

The BRI countries became the first recipients of assistance and beneficial owner. In the sense that universal access to vaccines means early unblocking, smooth international communication and economic revitalization. The BRI countries are the first to experience and reap the real benefits of public goods derived from vaccines. China has taken practical action to promote the equitable distribution of vaccines, promote international cooperation in combating the pandemic and do its utmost to provide assistance to other countries, especially developing countries, in obtaining vaccines, reflecting the role of a responsible power and winning positive comments from the international community.

(ii) Cross-border e-commerce to drive global trade

At the beginning of the global pandemic of the Covid-19 pandemic, WTO had made a forecast in early April 2020 that global trade in goods would fall by 12.9–31.9% in 2020, but the WTO updated its forecast in early October that global trade in goods would only fall by 9.2% in 2020. Among other things, e-commerce has played an important role in facilitating the development of cross-border trade in the context of the pandemic. In China, for example, national e-tailing sales grew by 10.9% year-on-year in 2020. Under the framework of the international cooperation on the BRI, China has been actively building "silk road e-commerce" with countries and regions along the BRI in the past few years. By the end of 2019, China's cross-border e-commerce trade has achieved full coverage of countries and regions along the BRI, which has had a positive impact on promoting e-commerce development in both China and the relevant countries.

In the context of the pandemic, cross-border e-commerce, with the unique advantages of online marketing and online transactions, avoids physical contact between people and reduces many intermediate links, becoming an important channel for international trade in goods and providing new opportunities for developing international markets. The scale of online trading of goods in various countries climbed rapidly during the pandemic, and cross-border e-commerce business models continued to innovate, with B2B, B2B2C, cross-border social e-commerce, cross-border live streaming and short video greatly driving the development of cross-border e-commerce trade.

According to preliminary customs statistics, in 2020 China's cross-border e-commerce import and export CNY 1.69 trillion, an increase of 31.1%, of which exports CNY 1.12 trillion; an increase of 40.1%, imports CNY 0.57 trillion; an increase of 16.5%, cross-border e-commerce has become an important force in China's "stable foreign trade". With regard to policies on cross-border e-commerce, as of November 12, 2020, China has proposed a "Chinese solution" for cross-border e-commerce in WTO, G20, APEC and other multilateral occasions, and signed e-commerce cooperation memorandum entry with 22 countries. Therefore, during the pandemic, cross-border e-commerce not only did not slump, but the trade volume increased significantly in 2020, even in the first three quarters of 2020 has exceeded 2019, and cross-border e-commerce "China solution" is also recognized by countries around the world.

On May 15, 2020, driven by UN Under-Secretary-General Vera Songwe, some ambassadors in China, including Rwandan Ambassador to China James Kimonyo, Belgian Consulate General in Shanghai Commercial Consul Lin Bersen, Chilean Commercial Consul in Shanghai Wu De, etc., came into the Tmall international live broadcast to promote their "national goods", opening a new mode of e-commerce diplomacy, with eye-popping benefits. Rwanda Gorilla Coffee Beans sold 1.5 tons in just one second, equivalent to last year's entire sales. One winery in Chile even sold its entire inventory for the year, replenishing it immediately. Belgium's dietary fiber powder sold in 5 minutes what it had sold in the previous 10 days.

On June 11, a group of diplomats from 10 Central and Eastern European countries, including ambassadors, consuls and counsellors from Serbia, Latvia, Montenegro, Slovenia, Poland, the Czech Republic, Slovakia, Bosnia and Herzegovina, Hungary, Bulgaria, etc., promoted their special products in the Pindo live broadcast rooms in Beijing, Shanghai and Ningbo as "global good recommendation officers", and the sales were exceptionally hot. At present, 10,000 Russian companies are registered on Sotheby's to sell Russian products to the Chinese market. 50 companies from Kazakhstan registered on Taobao to promote cross-border e-commerce between China and Kazakhstan.

On June 20, the General Administration of Customs (GAC) issued the *Announcement on the Launching of the Pilot Project of Business-to-Business Export Supervision for Cross-border E-Commerce*, adding "9710" and "9810" as two new customs supervision methods, i.e., B2B direct export for cross-border e-commerce and overseas warehouse for cross-border e-commerce export, and launching the pilot project of business-to-business (B2B) export supervision for cross-border e-commerce in 22 cities across China, which provides important policy protection to promote the healthy development of cross-border e-commerce and help countries resume work and production (General Office of the General Administration of Customs, 2020, June 14).

On October 13, China and Uzbekistan held an exchange and matchmaking meeting on the theme of "Tashkent-Shanghai: The Prospects of E-Commerce", in which Chinese and Uzbek enterprises signed a deal worth US$3 million for candy supply contract through Alibaba's e-commerce platform. On October 23, Yangling City, Shaanxi Province, held the 27th Agricultural High-tech Conference Cross-border E-Commerce Industry Forum, planning to use the "Yangling Agricultural Science" agricultural products cross-border e-commerce base and trading platform to build a national and global e-commerce network of agricultural products, and promote a large number of Chinese high-quality agricultural products out of the country and to the world. On November 4, China announced the establishment of 10 import trade promotion innovation demonstration zones, aiming to promote innovation in the import trade regulatory system, more complete service functions and more flexible transaction models, ultimately expanding the scale of imports, optimizing the structure of goods, promoting the quality and efficiency of industries, raising consumption levels and helping to build a new development pattern, which provides a new opportunity to expand trade between China and BRI countries.

(iii) Providing global financial security

Before the pandemic, there were 26 international BRI special investment funds with a combined investment size of over 1 trillion CNY. Among the special investment funds, the most popular one is the Silk Road Fund. The special investment fund is financed by China, with a focus on medium- and long-term equity investments, to provide financial support for the co-construction of the BRI. In 2020, the Silk Road Fund signed 10 new projects with an additional committed investment amount of about USD 800 million and CNY 11.4 billion. By the end of 2020, the Silk Road Fund had signed 49 projects of various types through equity-based financing, with a committed investment of about US$11.7 billion and CNY 43.8 billion, covering many BRI countries. 70% of the signed investment funds are used in large international cooperation projects such as power plant development, infrastructure construction, port shipping, and high-end manufacturing.

The "Silk Road Fund" is similar to payback period long private equity investments, and has similarities to International Finance Corporation and the African Development Bank's Common Development Fund. These funds do not have too short-term return needs and have relatively long investment cycles, so they are heavily weighted towards equity investments. Moreover, these types of investments tend not to favor equity in the stock market public company, but rather favor more start-up and unlisted projects. "The Silk Road Fund", with its focus on equity investments, is more flexible and simpler in structure and decision-making than lending-based multilateral development institutions. It is worth noting that equity investments do not require the same financial guarantees and regular principal and interest recovery as debt investments, which can facilitate long-term start-up investments in the BRI investment and financing environment.

Since the outbreak of the Covid-19 pandemic, AIIB has announced the establishment of the "Covid-19 pandemic Crisis Recovery Fund" to provide financial support for members' urgent public health needs. The emergency fund is used to support member countries to overcome the emergency economic, financial and public health stress caused by the pandemic, benefiting 19 countries, including Vietnam, Georgia, Pakistan, Turkey and Kazakhstan. The initial amount of the contingency fund was US$10 billion, which was increased to US$13 billion in July 2020. 27 projects have been approved and total charge has exceeded US$7 billion. It can be seen that the cooperation in the field of BRI medical care, emergency response and other interpersonal exchanges, led by the capital financing, has not only expanded its "circle of friends", but also paved the way for the internationalization of the CNY. The AIIB also issued CNY 3 billion of "panda bonds" between Bank of China and bond market to provide financial support for global pandemic prevention and control.

In addition, China is an active participant in the G20 debt relief initiative and has the largest amount of debt relief among the G20 members, some of which are partner countries with which China has built the BRI. These have directly contributed positively to safeguarding the basic livelihood of the people of China and the countries concerned.

(iv) Promoting the building of a human community of a shared future

The Covid-19 pandemic has sounded an alarm to all mankind, reminding them to start from the height of the "human community of a shared future". It transcends economic rationality and responds to global issues at a higher rational level, so that mankind can break through the "prisoner's dilemma" and achieve a win-win cooperation. In the face of the pandemic, the future and destiny of all countries in the world have never been more closely intertwined. After the baptism of the struggle against the pandemic, the relationship between China and regional countries is deepened and sublimated.

Throughout the history of human development, mankind has revived in the game of catastrophes and disasters. They developed and progressed in the struggle against fate. Globalization is the inevitable development of history. The future development of human society cannot retreat to the state of "isolation" of individual countries. Globalization is also a manifestation of human progress and a way out of the problem of the new pneumonia pandemic. Global problems must be solved through global cooperation.

Over the past seven years, the BRI has become a fundamental platform for China to build a human community of a shared future together with the rest of the world. It has been proved that the BRI has played an important and positive role in stabilizing China's economy, helping the economic development of countries along the route and promoting global economic recovery in the year of the pandemic. Looking ahead, China continues to contribute Chinese wisdom, Chinese solutions and Chinese power to the world economy through the BRI, helping the world economy to emerge from its predicament and accelerate its recovery as soon as possible, while building a community of a shared future for mankind has also become a widely accepted global consensus.

References

National Bureau of Statistics (2021) .https://data.stats.gov.cn/easyquery.htm? cn=A01

Global Investment Trend Monitor, No. 38 (2021, January), United Nations, https://unctad.org/webflyer/global-investment-trend-monitor-no-38.

The United Nations Conference on Trade and Development predicts that global trade will decline by 7% in 2020 (2020, October 29). http://www.mofcom.gov.cn/article/tongjiziliao/fuwzn/oymytj/202010/20201003011530.shtml.

In 2020, Chinese enterprises invested 17.79 billion US dollars in non-financial direct investment along the "Belt and Road Initiative" route (2021, February 19). People's Daily. http://fec.mofcom.gov.cn/article/fwydyl/zgzx/202102/20210203039372.shtml.

Joint Statement of the High-level Video Conference on "Belt and Road" International Cooperation (2020, June 19). Xinhua News Agency. http://www.gov.cn/xinwen/2020-06/19/content_5520682.htm.

Joint Statement of "Belt and Road Initiative" High-level Video Conference on International Cooperation (2020, June 20). Ministry of Foreign Affairs. https://baijiahao.baidu.com/s?id=1669997823082733342&wfr=spider&for=pc.

The General Office of the Ministry of Commerce and the General Office of the Ministry of Finance issued a circular on making good use of special funds for domestic and foreign trade to support stable foreign trade, stabilize foreign investment and promote consumption (2020, March 23). http://sydney.mofcom.gov.cn/article/ddfg/tzzhch/202003/20200302947400.shtml.

The Ministry of Commerce issued a circular to deal with the epidemic situation and further reform and opening up to do a good job in stabilizing foreign investment (2020, April 6). http://www.gov.cn/xinwen/2020-04/06/content_5499510.htm.

The 2021 China report of the International Financial Forum shows that the "Belt and Road Initiative" initiative contributes to green recovery in the post-epidemic era (2021, May, 30). http://www.gov.cn/xinwen/2021-05/30/content_5614026.htm.

Announcement on carrying out the pilot project of export supervision by cross-border e-commerce enterprises (2020, June 14). General Office of the General Administration of Customs. https://baijiahao.baidu.com/s?id=1669406738665157808&wfr=spider&for=pc.

VI

Basic Ideas and Requirements for the Co-construction of the BRI Quality Development

Introduction

On March 11, 2021, the Fourth Session of the 13th National People's Congress voted to adopt a resolution on the outline of the 14th Five-Year Plan and the Vision 2035 for National Economic and Social Development. In Chapter 12 entitled "Achieving a high level of opening up to the outside world and seek for new opportunities of win-win cooperation" of *The Outline of the 14th Five-Year Plan and Vision 2035 of the National Economic and Social Development of the People's Republic of China*, the content of Chapter 41 "Promoting the high-quality development of building the Belt and Road" specifies clear ideas and requirements for the high-quality development of co-constructing the Belt and Road (Xinhua News Agency, 2020, November 3).

By drawing on the past experience of the high-quality development of the BRI, clarifying the results of the current high-quality development, and referring to the direction of the future high-quality development, its basic ideas and requirements can be categorized and summarized. The basic requirements for the high-quality development of the BRI can be sorted out in accordance with the five aspects of "Five Links" development: policy communication, infrastructure connectivity, smooth trade, financial integration, and civic exchanges.

1. The basic idea of co-constructing the BRI quality development

The 14th Five-Year Plan clearly states that the basic idea of promoting the co-construction of BRI for high-quality development is to "adhere to the principle of co-business, co-construction and sharing, uphold the concept of green-open-integrity, deepen pragmatic cooperation, strengthen safety and security, and promote common development" (Qian Zhongbin, 2021, March 13).

Among them, adherence to the principle of co-business, co-construction and sharing means advocating multilateralism, with all parties negotiating on an equal footing, sharing responsibilities and benefiting together. Co-business means to draw on collective wisdom, so that the construction of the BRI can take into account the interests and concerns of all parties as far as possible and reflect the wisdom and creativity of all parties. Co-construction means to give full play to the strengths and potential of each party, to gather sand into a tower, to accumulate water into an abyss, and to promote it consistently. Sharing is to provide more positive results fairly available to all parties, and strive to build a human community of interests and shared future along the BRI.

We uphold the concept of "green-open-integrity", i.e., "green" as the base color of development. "Open" as the premise of development. It insists that all cooperation operate under the supervision and justice. "Green" means actively promoting green infrastructure construction, green investment, green finance, and properly addressing the challenges of environmental protection and climate change with the ultimate goal of protecting the common home on which humanity depends. "Open" encompasses two aspects: on the one hand, it means supporting an open economy and an inclusive and non-discriminatory global market, and on the other hand, it means welcoming all interested countries to participate in cooperation and flagrantly opposing protectionism. Integrity means that all parties should build a culture of integrity and fight corruption, so as to build a strong guarantee for the BRI to go deeper and more stable, and to build a clean Silk Road.

Deepen practical cooperation relies on project-driven. Countries along the route, in terms of infrastructure connectivity, promote the construction of railways, highways and other land-based large alleyway. It accelerates the construction of sea ports, improve oil and gas pipelines, power transmission and communication network; in terms of physical regional economic cooperation work, it vigorously promote the construction of economic corridors, perform good economic and trade, manage industrial cooperation parks, as well as further promote investment, aggregation of industries, drive employment, and take the path of innovative development; in terms of trade and investment liberalization and facilitation, it promotes the construction of free trade zones, strengthen the rules and standards system compatible with each other. It provides a better business environment and institutional safeguards, fully release the positive effects of connectivity. In terms of financial cooperation, it expands financing channels, prompts innovative financing methods, reduces financing costs, and opens up financing as a key link in the advancement of the project. In terms of interpersonal exchanges, it carries out in-depth exchanges and cooperation in the field of humanities, so that cooperation is more inclusive, the basis for cooperation is more solid, the people of all countries become the main force and beneficial owner of the BRI construction.

Strengthening security means firmly safeguarding the interests of sovereignty, security and development, and creating a favourable external environment for the co-construction of the BRI high-quality development. Resolutely safeguard sovereign interests, starts from the height of safeguarding the security of overseas interests, takes the national security concept as the leader, adheres to the new security concept of common, comprehensive, cooperative and sustainable. It explores with countries along the route to jointly build the BRI system structure of overseas security risk identification, prevention and control, and emergency response. We resolutely safeguard security interests, strengthen bottom-line thinking, enhance pre-judgement and research, fully consider the multiple risks that may be encountered in cooperation, improve risk warning and prevention mechanisms, effectively respond to challenges posed by changes in the external environment, and achieve progress in a stable manner. We resolutely safeguard the interests of development, focus on the construction of major projects such as infrastructure and cooperation in production capacity, resolve various key issues, strengthen cultural exchanges and cooperation in talent, and ensure that more cooperation achievements are put into practice.

In order to promote common development means to focus on the fundamental issue of development. It leads and promotes balanced development of international relations by proposing new and viable plans and programmes. The BRI strives to maintain cooperation and dialogue with existing regional cooperation mechanisms in major countries and regions. It exchanges and communicates with established development-oriented international organizations, in an effort to create a balanced and sustainable international cooperation system conducive to the common development of all countries. It promotes the construction of a new type of international relations, fully unleash the development potential of all countries, and to achieve a win-win situation of greater economic integration, greater development linkage and greater sharing of results.

2. Basic requirements for the co-construction of the BRI quality development

(i) Strengthening development strategies and policy alignment

The first and foremost requirement for the high-quality development of the BRI is to strengthen the effective convergence of strategies and policies among countries and regions along the route. Specifically, we should actively promote the alignment of strategies, plans and mechanisms, and strengthen the linkage of policies, rules and standards. Innovative docking methods is used to promote the implementation of signed documents, promote the negotiation of investment protection agreements and double taxation avoidance agreements with more countries. It strengthens cooperation in customs, taxation and supervision, and promotes the implementation of a higher level of customs clearance integration. It expands the areas of rule docking and strengthens cooperation in rule docking in financing, trade, energy, numerical information, agriculture and other fields. It promotes effective dovetailing and synergy between the BRI and regional and international development agendas.

Development strategies, plans and mechanisms are major choices and top-level designs that affect the future of a country. They are global, long-term and leading in nature (He Lifeng, 2017, May 14). Development strategy docking is the highest level of policy communication and coordination between countries, which is conducive to enhancing political mutual trust and consensus on cooperation, seeking cooperation greatest common measure from a macroscopic perspective, finding the direction of common action and achieving common development. Policies, rules and standards are the refinement and vehicle for major choices and top-level design. The alignment of development rules between countries is to clarify the vision set by development strategies into concrete timetables and road maps on the basis of consensus, while making use of the comparative advantages of each country to achieve development goals in a step-by-step manner at lower cost and with greater efficiency. The alignment of rules and standards aims to effectively connect the relevant implementing agencies in each country, builds smooth channels and mechanisms for exchange, communication and consultation, and aligns relevant resources more effectively to solve problems and difficulties faced in planning and implementation and project implementation in a timely manner.

The requirements for strengthening development strategies and policy alignment should be clarified in the following three key areas. First, innovative docking methods. Through practice, countries and regions along the BRI route have jointly explored and formed an effective working method of strategic docking, planning docking, mechanism docking and project docking. In the future, there is also a need to strengthen the interface in a more flexible manner, taking into account the national circumstances of each party. For example, after the signing of the intergovernmental cooperation understanding memorandum entry, it is possible to start with top-level design and adopt the approach of preparing cooperation plans, establishing docking platforms and implementing project lists. It is also possible to cut directly from practical cooperation and take the lead in launching a number of mature projects on which both sides have consensus and conditions are available, so that all parties can share the early harvest results as early as possible.

Second, it expands the area of docking. The establishment of an analytical framework for the sustainability of project assets and debts should be studied to jointly improve asset utilization and debt management. Consider launching the Clean Silk Road Initiative to improve the anti-corruption legal system and mechanisms. The countries along the route should form a mechanism to share their experiences in improving the business environment and strengthening corporate compliance, while trying to carry out pilot projects on the "one-sheet system" of financial settlement and financing rules for intermodal railway transport, improve the quality and efficiency of the operation of China-Europe trains, and continue to promote the strengthening of all-round docking cooperation in the fields of trade, energy, numerical information, agriculture, ecology and poverty reduction.

Third, it improves the efficiency of docking. It is necessary to improve the efficiency of docking as a prerequisite, actively promote the docking of norms and standards in key areas or processes, reduce unnecessary losses caused by different systems and standards with maximum effort, and facilitate the enhancement of intercountry exchanges in various areas such as science and technology, trade and commerce, education, tourism, culture, health and sports. In the process of strategy and policy alignment, the focus on efficiency is conducive to the pooling of resources from all parties, the export of industrial clusters, the transformation of short-term benefits to drive long-term industrial cluster upgrading drive, and the transformation from the former unilateral and bilateral interest relationship-based subject to a multilateral community of interest.

(ii) Deepening the interconnection of facilities

The BRI quality development should focus on infrastructure connectivity. Facility connectivity is the basis for cooperative development and a priority area for the construction of the BRI. Since the introduction of the BRI, the countries along the route have taken infrastructure connectivity planning and technical standards docking as an entry point. Based on infrastructure construction project, they have pragmatically promoted in-depth cooperation among countries in the fields of railways, highways, water transport, civil aviation and postal services, and promoted continuous new progress in regional transport connectivity. In the future, we also continue to further promote the interconnection of facilities.

Specifically, we should promote the 4-in-1 connectivity of land-sea-sky-networks, taking the "Six Corridors, Six roads, Multiple Countries and Ports" as the basic framework, building an interconnection network led by economic corridors such as the New Asia-Europe Continental Bridge, with China-Europe Classes, the New Land-Sea alleyway and other major alleyway and information highways as framework. With railways, ports and pipeline networks as the backbone, it creates a new international land-sea trade alleyway. It focuses on key alleyway and key cities, promoting the construction of major cooperation projects in an orderly manner. It integrates the goals of high quality, sustainability, risk resistance, reasonable price and inclusiveness into the entire process of project construction. It improves the quality of China-European train operations and promotes the development of rules for international land transport trade. It expands the impact of the "Silk Road Shipping" brand and promotes the construction of the BRI core areas in Fujian and Xinjiang. It also promotes the construction of the BRI spatial information corridor and the construction of the "Air Silk Road".

Among them, by accelerating the 4-in-1 connectivity of land-sea-sky-networks, we further prosper multifaceted synergistic cooperation. In order to achieve this, it takes the six Corridors and six Roads as the main direction and content of infrastructure interconnection, and the multiple countries and ports as the anchor point, as well as by building a number of important ports and node cities with BRI countries. At the same time, the interconnection network formed so far has taken shape. It is led by economic corridors such as the New Asia-Europe Continental Bridge, with major alleyway and information highways such as the China-European Train and the New Land-Sea alleyway, and the information highway as framework. It relies on railways, ports and pipeline networks. The new international land-sea trade alleyway created has begun to bear fruit. The work of building an efficient and economical trade alleyway will continue to be promoted in depth in the future. The focus on key alleyway, key cities and critical item purposes helps to coordinate resource allocation and prioritize the opening of missing links, unblocking bottlenecks and improving access. The direction and extent of focus should be scientifically planned in the future to promote the efficient development of key links in an orderly manner.

The requirements for deepening the interconnection of facilities should be clarified with the following three key tasks. First, deepen integration and development. Infrastructure connectivity, industrial park construction and industrial investment should be better integrated. The training of local employees should be done from the height of the talent strategy to accelerate the efficiency of labour of the partner and effectively help the long-term sustainable development of the partner's economy.

Second, it strengthens financial security. The foundation of facility connectivity is finance. The difficulties in financing facility connectivity projects are effectively addressed to provide financial support and security for facility connectivity. Consideration could be given to making full use of existing platforms to actively explore models for investment and financing of major infrastructure projects. The role of equity investment funds and regional development finance institutions could also be targeted to make portfolio investments in key facility linkage projects.

Third, it focuses on the development of cooperative demonstration projects. It considers uniting countries along the route to actively carry out a number of third party cooperation demonstration projects through the joint third party cooperation strategic planning, co-construction of project banks, information banks and data pool, and the joint third party project feasibility studies. It introduces advanced technology, management model advantages, construction capacity advantages and other advanced experiences in many aspects. It promotes the realization of a win-win situation in which infrastructure construction is organically combined with the advantages of each host country factor endowment.

(iii) Sustained expansion of two-way trade and investment

The high-quality development of the BRI should continue to expand two-way trade and investment, and achieve sustainable trade development among countries along the route. Specifically, it promotes the optimization and upgrading of trade and investment cooperation with the co-establishing BRI countries, and actively develop silk road e-commerce. It deepens international production capacity cooperation, expands third party market cooperation, constructs a mutually beneficial and win-win industrial chain, supplies chain cooperation system, and expands two-way trade and investment.

Among them, the optimization and upgrading of trade and investment cooperation includes vigorously optimizing the trade structure and exploring new trade platforms such as various types of cross-border e-commerce, including Silk Road e-commerce. Deepening international cooperation in production capacity covers the promotion of imports and exports, trade in goods and services, trade and two-way investment, and the coordinated development of trade and industry. It focuses on promoting the orderly and free flow of international and domestic factors, the efficient allocation of resources and the deep integration of markets, promoting the basic balance of payments, achieving high-quality trade development, and creating a new situation of international trade that is open, cooperative, inclusive and mutually beneficial.

The sustained expansion of two-way trade and investment requires the following four key tasks to be identified. First, to improve the quality of trade. Efforts should be made to expand two-way trade with co-construction countries, especially to increase imports quality products and let more countries share the dividends of China's super big scale market. It focuses on the development of "silk road e-commerce", promoting cross-border e-commerce and other new business models and the BRI integrated development, for China-European train, land-sea new alleyway and other trade big alleyway construction to add new dynamic energy. It should also promote the construction of a high-standard free trade zone and create a free and convenient international environment for the co-construction of the BRI.

Second, innovative investment cooperation. It tries to promote the construction of a number of high-quality cooperation projects, implements a number of foreign aid projects that benefit the people's livelihood. It encourages powerful domestic enterprises to go out by innovating foreign investment and cooperation methods, promoting the upgrading of foreign contracting projects, and expanding cooperation in the third party market. It is necessary to attract more co-construction crown corporation to invest in China and promote the construction of mutually beneficial and win-win industrial chains supply chain and value chain.

Third, it creates a platform vehicle. Continuing to run every import expo, China International Trade in Services Fair and a series of other major exhibitions.

It plays the role of Canton Fair, China ASEAN Expo, China-Africa Economic and Trade Fair, China-Russia Expo, China-Central and Eastern European Countries Expo and a series of exhibition platforms, in order to deepen the BRI economic and trade cooperation to create more strong support platform.

Fourth, it strengthens mechanisms. It should actively promote the docking of strategies, plans and mechanisms with the co-construction countries, build more trade smoothness task group and investment cooperation task group under the framework of bilateral economic and trade joint committees and mixed committees. It improves the working mechanisms of trade in services and e-commerce, and provides institutional guarantee for the co-construction of the BRI.

(iv) Adherence to market-oriented financing

The co-constructed BRI high-quality development should always adhere to the market-oriented financing. Specifically, this includes adhering to an enterprise-oriented and market-oriented approach, following international practice and the principle of debt sustainability, and improving the diversified investment and financing system. Innovative financing cooperation framework, play the role of the BRI special loan, Silk Road Fund, etc. It establishes and improves the BRI financial cooperation network, promotes the interconnection of financial infrastructures, and supports the joint participation of multilateral and national financial institutions in investment and financing. Also, it improves the BRI risk prevention, control and security system, strengthen legal services, and effectively prevent and resolve various risks.

Among them, a diversified investment and financing system helps to strengthen financial cooperation and promote capital financing, creating a stable financing environment for the construction of the BRI and actively guiding various types of capital to participate in the development of the real economy and the creation of industrial chains. At the same time, we should adhere to the market-oriented approach, following the laws of the market, adhering to the market-led approach, and responding to the needs of the market-oriented operation of the BRI construction. Financial support should mainly be provided in the form of commercial funds, with the provision of concessionary funds being minimized, and government funds should mainly be used to leverage private sector investment. In addition, it is important to focus on the importance of improving risk prevention, control and security systems while establishing a sound BRI financial cooperation network. Enterprises are the main body of the BRI. Chinese enterprises going abroad should practice the concept of high-quality development. It strengthens the awareness of risk prevention, making efforts to improve the ability of risk prevention and control, operating legally and compliantly. It fulfills their social responsibility and becomes the ambassador of the BRI.

It adheres to the requirements of market-oriented financing that can be combined with the new development pattern of China's international and domestic dual cycle. The following two key tasks should be clarified: (1) increase investment support for the weaknesses of industrial chain and supply chain, and allow a smooth domestic circulation. Some areas of key core technologies of the industry chain and supply chain exist a "stuck neck" problem, which is an important factor impeding the development of China's industry. Efforts must be made to make up for the shortcomings and weaknesses of the industrial Chain and supply chain, giving play to the key role of investment in optimizing the supply structure, and smooth out the large domestic circulation. It should increase investment support for innovative subjects, guide them to increase investment in research and development, strive to break through key core technologies, and promote independent and controllable industrial development. It should also improve the institutional policy system that encourages independent and original innovation, stimulate the enthusiasm of innovative subjects, and form a long-term mechanism to stimulate investment in innovation. It vigorously supports the construction of technology platforms, innovation platforms and industrial infrastructure, enhances industrial infrastructure capabilities, promotes the integration and development of basic research and industrial technology innovation. It promotes the digital transformation, intelligent upgrading and integration and innovation of industries. It explores the mechanism of diversified investment in the construction and operation of major scientific and technological infrastructure. It provides financing support for important products and key core technology research and development. It further clarifies the functional positioning of government-funded industrial investment funds and promotes the organic unification of public policy objectives, and economic benefit objectives. It plays an important role in investment policies such as investment subsidies to enhance the level of autonomy of key core components and software.

Secondly, we should strengthen investment in domestic and international connectivity facilities, and promote the smooth flow of the domestic and international double cycle. In accordance with the need for coordinated development of introduction of foreign investment and outward investment, imports and exports, it focuses on interconnection projects, accelerates the improvement of transportation infrastructure and logistics facility networks. It strengthens the construction of a large open alleyway, makes full use of the two domestic and international markets and resources, and encourages the domestic and international double cycle. It improves open infrastructure such as ports, cross-border transport and information alleyway, accelerates the construction of open logistics networks and cross-border mail delivery systems, strengthens the construction of air ports and hubs. It further improves the layout of national logistics hubs and improves the efficiency of logistics operations. It focuses on key alleyway, key cities and critical item projects, building high-quality, sustainable, safe and reliable, reasonably priced, inclusive and accessible infrastructure with all parties concerned. It promotes connectivity on land, at sea, in the sky and online. It accelerates the construction of international logistics alleyway, hubs and network systems. It makes up for shortcomings such as overseas feeder logistics networks. It strengthens maritime infrastructure links with countries along the Maritime Silk Road, promotes the high-quality development of the BRI by linking infrastructure networks and improves the logistics supply chain system. It attracts more high-quality resources and elements worldwide, and encourages the domestic and international double cycle.

(v) Deepening exchanges and cooperation in key areas

The co-construction of the BRI high-quality development should deepen exchanges and cooperation in various key areas. Specifically, this includes deepening humanities cooperation in the fields of public health, digital economy, green development, science and technology, education, culture and the arts. It strengthens exchanges between parliaments, political parties and civil society organizations. It is closely communicating with women, youth, persons with disabilities and other groups, so as to form a pattern of multifaceted and interactive humanities exchanges. It promotes the implementation of the BRI Science and Technology Innovation Action Plan, the construction of the Digital Silk Road and the Innovation Silk Road. It strengthens exchanges and cooperation in addressing climate change, maritime cooperation, wildlife protection, desertification control, etc. It promotes the construction of the Green Silk Road. We are actively cooperating with the BRI countries on health care, and infectious disease prevention and control, and building a healthy Silk Road.

Among them, strengthening humanistic cooperation and diversified exchanges and communications in multiple fields is of far-reaching significance in promoting the construction of the BRI, fostering common development of countries along the route and promoting the building of a human community of a shared future. On the one hand, deepening cooperation and exchanges in key areas inherits the spirit of the Silk Road. By promoting exchanges and mingling among the people of various countries, it helps the people's hearts and minds to reach out to the people's sentiments, achieving the effect of enhancing trust, promoting friendship, deepening cooperation and common development. It promotes a stable development model of conceptual identity, convergence of interests and friendship. On the other hand, deepening cooperation and exchanges in key areas is of great significance in promoting "hard cooperation". The interpersonal exchanges achieved through rich forms of cooperation and exchanges have a positive role in promoting the construction of the BRI, promoting common development of countries along the route and promoting the building of a human community of a shared future.

The requirements for deepening cooperation and exchange in key areas should be clarified in the following five key tasks. First, it should have a close communication in science and technology and deepen humanistic exchanges. Deepening science and technology humanities exchanges and enhancing mutual trust and understanding among the science and technology community is the basis for promoting the BRI science and technology innovation cooperation. It is also the core of ongoing humanities exchange activities with countries along the route. Cooperating with countries along the route to jointly train scientific and technological talents, it expands the scale of the programme for outstanding young scientists to work in China. It builds a number of training centres and training bases of different types, and extensively conduct training in advanced appropriate technology, science and technology management and policy, scientific and technological assessment, and scientific and technological entrepreneurship. It implements the International Science and Technology Missionary Program, carries out science and technology volunteerism, addresses technical matters and meeting technology needs. It cooperates in science activities and promote youth science exchange. We should continue to strengthen science, technology and innovation policy communication and support policy capacity building in countries along the route. It is actively working with countries along the route to jointly carry out science and technology innovation planning, science and technology innovation policy formulation and national innovation system construction. It promotes joint assessment of major science and technology activities and forms a collaborative network on science and technology innovation policies. It builds a multi-level platform for scientific and humanistic exchanges. It makes full use of the platforms of Boao Forum for Asia, China-ASEAN Expo, China-Asia-Europe Expo, China-South Asia Expo, China-Arab States Expo, China-Russia Expo,

West China International Expo, etc., and continue to build science and technology innovation cooperation platforms such as the International Technology Transfer and Innovation Cooperation Conference with ASEAN, South Asia and Arab States, China-Central and Eastern European Countries Innovation and Technology Cooperation and International Technology Transfer Seminar.

Second, it strengthens platforms and promotes technology transfer. It is proposed to build a number of national joint laboratories (joint research centers), combining the major scientific and technological needs of countries along the route, encouraging cooperation between China's scientific research institutions, higher education institutions and enterprises and relevant institutions of countries along the route. It builds joint laboratories (joint research centers) around key areas, jointly promotes high-level scientific research, carries out exchanges and training of scientific and technological talents. It promotes appropriate technology transfer and transformation of results, builds long-term and stable cooperative relationships, and enhances the scientific and technological capacity of countries along the route. At the same time, consideration should also be given to jointly building a number of technology transfer centres, giving full play to the role of China's technology transfer centres with ASEAN, Central Asia, South Asia and Arab countries, etc. It further improves the construction of technology transfer collaboration networks and information matching platforms, encouraging technology transfer centres to build international technology transfer service alliances. It jointly promotes the transfer of advanced appropriate technology, strengthening the combination of China's science and technology, talent, information and other resources with the needs of countries along the routes. It also deepens cooperation between industry, academia and research. In addition, consideration should be given to jointly building a number of advanced appropriate technology demonstration and promotion bases, combining the technological needs of countries along the route, and encouraging China's scientific research institutions, higher education institutions and enterprises to actively promote advanced appropriate technology in key areas.

Third, it supports the construction of major projects and promotes the interconnection of scientific and technological resources. It strengthens science and technology to support rail and road intermodal transport links. It attempts to make breakthroughs in scientific research and development of special environmental condition for the railway and highway construction, technical equipment and other adaptable key techniques, as well as strengthen the interface of technical standards. It takes important coastal ports as nodes, builds environmental, hydro-meteorological, marine and other relevant data monitoring networks, breakthrough ports, water alleyway construction, shipping support and security systems, etc. The construction and upgrading of power grids in countries along the route should be supported. The interconnection of power grid construction should be accelerated, relying on extra-high voltage and smart grid technologies. It supports the interconnection of information communication network, carries out research on the construction of cross-border land cables, intercontinental submarine optical cables and communication network networks. It accelerates the construction of data sharing platforms and information service facilities. It realizes the high-speed transmission, correlation and integration of scientific data resources and service sharing. It actively promotes the interconnection of scientific research instruments and facilities, scientific research data, scientific and technological literature, biological germplasm and other scientific and technological resources. It promotes the sharing of data, literature and other scientific and technological resources on the basis of large-scale research infrastructures that have been opened and shared on a pilot basis for international use in the first instance. It establishes an integrated earth recording geometry and scientific data sharing service platform to interconnect major earth observation data centres in Asia and the Pacific. It establishes a biotechnology information network to realize the common sharing of biological resources and technological achievement data pool.

Fourth, we build a special park together to encourage enterprises to innovate and start a business. In the future, a number of distinctive science and technology parks should be built together. China's high-tech zones, independent innovation demonstration zones, agricultural science and technology parks, marine science and technology industrial parks, environmental protection industrial parks and green building materials industrial parks should be guided and encouraged to take the initiative to dock with countries along the route. It encourages competent enterprises to build science and technology parks with countries along the route and explore diversified construction models. It actively encourages science and technology-based enterprises to innovate and start up in countries along the route. It cultivates a number of multinational innovative enterprises with international competitive power and promote mutually beneficial cooperation in new industries and new business models. It encourages enterprises in a position to do so to set up R&D centers in countries along the route with strong scientific and technological strength. It strengthens the protection and utilization of intellectual property rights, and promotes the industry to climb up to the middle and high end of value chain.

Fifth, it focuses on common technologies and strengthen cooperative study. It focuses on key commonalities technical matters faced by countries along the route in economic and social development, and strengthen cooperative study. In terms of fundamental research, it conducts cooperative study on major scientific problems in high-energy physics, biophysics, ecological climate, astronomical sight, extreme weather climate, focusing on the relationship between ice belts and climate change, and integrated Earth observation. In addressing common challenges, the focus is on joint research and development in areas as ecology and the environment, energy security, population health, food security, natural disasters and the protection and preservation of cultural heritage. International scientific and technological organizations and relevant regional organizations are encouraged to participate actively.

References

He Lifeng. (2017). Chairman of the National Development and Reform Commission, at the BRI International Cooperation Summit on May 14, 2017, "Strengthening Policy Communication and Making the Four Docking Points".

Liu Chang. (2020). The proposal of the Central Committee of the Communist Party of China on the formulation of the 14th Five-Year Plan for National Economic and Social Development and the visionary goals for 2035 (2020, November 3). Xinhua News Agency. http://www.gov.cn/zhengce/2020-11/03/cont ent_5556991.htm.

Qian Zhongbin. (2021). The outline of the 14th Five-Year Plan and Vision 2035 for the National Economic and Social Development of the People's Republic of China (2021, March 13). http://www.xinhua net.com/2021-03/13/c_1127205564.htm.

VII

Institutional and Mechanism Innovations for Co-constructing the BRI

Introduction

The BRI is in line with the trend of globalization and is a product of globalization. It has gradually gained wide support from the international community in the eight years since it was proposed. The United Nations General Assembly at its 71st session adopted a resolution welcoming the BRI and urging the international community to provide "the necessary safety environment" to ensure the "full implementation" of development initiatives such as the BRI, which was unanimously endorsed by 193 MemberStates (Zheng Chengqiong, 2016, December 8). It must not be overlooked that the BRI covers a wide geographical area and has complex and diverse cross-border economic and trade relations, and that the existing international system cannot meet its current demands. Therefore, innovation is an important expulsive force quantity for the high-quality development of the BRI. Through institutional and mechanism innovation, the construction of the BRI continues to deepen, and to move steadily and farther.

Over the past eight years, the BRI high-quality development has accumulated some advanced experience in institutional and mechanism innovation, while also clarifying the work of institutional and mechanism innovation ideas and directions for the future. Specifically, the construction of a new system of higher-level open economy is a major breakthrough in institutional innovation for the high-quality development of the BRI, laying a solid foundation for the BRI to continue to play a positive role. At the same time, the framework, content and indicators of the Five Links based on policy communication, infrastructure connectivity, smooth trade, financial integration, and civic exchanges also provide systematic support and ideas for the innovation of mechanisms to build the BRI high-quality development.

1. Building a new system for a higher level of open economy

The *Proposal* of the 5th Plenary Session of the 19th CPC Central Committee put forward the main objectives of economic and social development for the 14th Five-Year Plan period, emphasizing that "a new system of a higher level of open economy will basically take shape". The 14th Five-Year Plan clearly proposes to build a new system of higher-level open economy. It specifically points out the construction paths in four areas, such as accelerating system-based opening, enhancing the functions of the opening platform, optimizing the layout of regional opening and improving the security guarantee system for opening up, which points out the way forward and provides a basis for China to implement a higher level of opening up to the outside world (Liu Chang, 2020, November 3).

Since the reform and opening up, China has adhered to the basic State policy of opening up to the outside world. It opens its doors for construction and achieves a great historical turnaround from being closed and semi-closed to being open on all fronts. At present, China's development environment is facing profound and complex changes, posing new requirements and challenges to the expansion of external opening. Building a new system of a higher level of open economy is not only a strategic choice made in response to changes in the development environment, but also an inherent requirement for improving the socialist market economy system and promoting deep market-oriented reforms with a high level of openness. During the "14th Five-Year Plan" period, in order to shape China's international regional economic cooperation work and new competitive advantages, China accelerates the construction of a new development pattern in which the domestic cycle is the mainstay, and the domestic and international cycles promote each other. The new development pattern is not a closed country internal recycle, but a more open domestic and international double cycle. There is a greater need to unswervingly expand openness, promote the transformation from commodity and factor flow-based openness to rules and other system-based openness. It builds a new system of higher-level open economy.

Building a new system of a higher level of open economy requires further strengthening of reform and innovation and the formation of new advantages in international cooperation and competition. (1) It is to accelerate institutional openness. Through such mechanisms and initiatives as improving the management system of pre-entry national treatment plus negative list for foreign investment, it establishes a management system of negative list for cross-border trade in services, steadily promotes the opening up of the financial sector, prudently promotes the internationalization of the CNY, and improves the management services of the external interface. We achieve a good opening-up situation such as promoting orderly competition between foreign capital affiliate and domestic enterprises in a fair market environment, domestic and foreign capital market interconnection. A new mutually beneficial cooperative relationship is based on the free use of the CNY, and high-quality and efficient external interface. (2) the function of the platform for opening up to the outside world is enhanced. It coordinates and promotes the construction of various open platforms, improves the layout of free trade pilot zone and free trade ports, innovates and upgrades key development and opening pilot zone and border regional economic cooperation work zones, It speeds up the promotion of trade innovation, promotes investment upgrading and builds new highlands for opening up to the outside world. (3) the regional open layout should be optimized. All regions should expand their openness based on their own advantages. It strengthens inter-regional open linkage, builds an open pattern of linkage between the land and sea, and encourages mutual assistance between east and west. (4) The upgrading of an open security system is necessary. Along with a higher level of openness to the outside world, the construction of a corresponding regulatory and risk prevention and control system help build a new high-level, sustainable and stable open economy. It is necessary to improve the systems of national safety review for foreign investment, anti-monopoly review and national technical safety list management, the list of unreliable entities. It should also establish an early warning system for the risk of global supply of important resources and products, strengthen the detection of balance of payments and foreign assets and liabilities, build a system for the protection of overseas interests and risk prevention, and safeguard the safety and legitimate rights and interests of the Chinese citizens and institutions overseas while achieving a win-win situation for all parties.

Over the past eight years, in building a new system of a higher level of open economy, China has taken a series of concrete actions within the framework of the systemic project of opening up to the outside world, including promoting the construction of Hainan Free Trade Port, joining the RCEP (Regional Comprehensive Economic Partnership Agreement), joining the CPTPP (Comprehensive and Progressive Trans-Pacific Partnership Agreement), opening the Shanghai Import Expo, and expanding the free trade zone to 21 provinces (Xu Huiyun, Lin Xiaozhao, Zhu Yanran, 2020, September 22). The overall goal of constructing the pilot point for experimentation of the new system of open economy: through a pilot experiment within a period of 2 years, it formulates a number of replicated and promotional experiences and patterns that are grounded in the creation of new mechanisms for resource allocation in the market, new modes of economic operation, new patterns of all-round opening and new advantages in international cooperation, etc. By doing so, it further expands the breadth and depth of China's opening to the outside world and lays a solid foundation for the basic formation of a new system of open economy and the creation of a new situation of all-round opening in the Thirteenth Five-Year Plan period. The comprehensive pilot experiment on building a new system of open economy focuses on six areas of exploration: a new mode of open economy operation and management; a new mechanism of coordinated opening of various parks; a new way of promoting international investment cooperation; a new system of quality and efficiency-oriented foreign trade promotion; a new pattern of all-round opening; and a new initiative of financial services for open economy.

Building a new system of open economy is not only "open", but also "reform". In the process of conceiving the pilot programme and soliciting opinions, some coastal, inland and border provinces and cities have reflected that the current construction of a new system of open economy still faces many institutional obstacles and needs to gnaw hard bones. It selects some cities and regions for early and pilot implementation, starting with the easy ones before the difficult ones, moving from point to point. It advances gradually, and then replicates and expands after the experience is gained. It helps truly break down institutional barriers and accelerates the formation of a new system for opening up to the outside world. In recent years, relevant departments and pilot regions have been brave enough to explore new mechanisms for allocating resources in the market. It explores new modes of economic operation and management, building a new pattern of all-round openness and fostering new advantages in international cooperation and competition, and in turn, the pilot work has achieved positive results.

Building a new system of open economy is the institutional guarantee to overcome the unprecedented changes of the past century. It implements a high level of opening up to the outside world, with emphasis on promoting trade and investment liberalization and facilitation. It deepens the opening up of commodity and factor flows, and expands the opening up of rules, regulations, management, standards and other systems. A comprehensive understanding and accurate grasp of the scientific connotation of building a new system of open economy is of great significance for us to understand globalization. It adapts to globalization and leads globalization in the new development stage and starts a new journey of building a comprehensive socialist modern country.

At present, the construction of a new system of open economy also requires the establishment of a system of open security. In order to implement a high level of openness to the outside world and accelerate the construction of a new development pattern, we must handle the relationship between openness and security, development and safety, and construct a regulatory and risk prevention and control system that matches the higher level of openness in order to make our cause go steady and far. The current phase of work is being carried out in five main areas.

First, we should improve the early warning system for industrial damage, enrich trade adjustment assistance, trade remedies and other policy tools, and properly respond to economic and trade frictions. We should play the role of industry associations and chambers of commerce, improve risk monitoring, analysis and early warning mechanisms, strengthen risk coordination and consultation within industries and industries, strengthen training on trade remedy rules and practices, enhance the comprehensive capacity and level of application of trade remedy rules, and guide enterprises to take the initiative to prevent risks. It improves the foreign trade investigation system, strengthen the sharing of information and data among government departments, customs, business associations, think tanks and enterprises, conduct regular tracking and research on key countries, key institutions, key enterprises and key products, and establish a good trade adjustment assistance system in a targeted manner to effectively respond to economic and trade frictions.

Second, we should improve the system of national safety review for foreign investment, anti-monopoly review and national technical safety list management, and the list of unreliable entities. We should also improve national safety review system for foreign investment in sensitive areas, effective prevention and mitigation of national security risks, and escort for a higher level of opening up to the outside world. A good anti-monopoly review system avoids vicious mergers and acquisitions, safeguard consumer interests and market stability, and guarantee the healthy development of the industry. It establishes and gradually improves the national technical security list management system, and improves the system of lists of unreliable entities to effectively prevent and resolve national security risks.

Third, a global supply chain risk early warning system for important resources and products should be established to strengthen international supply chain safeguard cooperation. It establishes a market-based guarantee system for energy, ores, metals, food and other bulk commodity based on the state-owned economy and with enterprises as the mainstay. It enhances the commercial reserve capacity of important domestic resource products, improves the comprehensive bargaining power of domestic finance and term market and stabilizes global prices of important resources and products. It deepens international technical and industrial cooperation, strengthens international cooperation in the field of intellectual property protection. It enhances cooperation with key countries in key areas of the industrial chain technology chain, supply chain, and maintains international supply chain security and stability.

Fourth, we strengthen balance-of-payments monitoring and maintain a basic balance of payments and the basic stability of foreign exchange reserves. It improves the balance of payments monitoring and statistical system, tracks and studies the relationship between the internationalization of the CNY, exchanges rate fluctuations, the balance of payments and changes in foreign exchange reserves. It identifies the direction, characteristics and deep-seated causes of short-term capital flows. It should strengthen the monitoring of foreign assets and liabilities. It improves state enterprise the mechanism for binding foreign liabilities, and establishes an early warning and supervision system for the full-caliber foreign debt level. It needs to improve the classification and grading supervision system for overseas investment, establishes a rating mechanism for overseas assets of key enterprises, and improves the internal risk disclosure system for state enterprise overseas investment and overseas assets.

Fifth, it builds an early warning prevention and protection system for overseas interests risks. It plays the role of a bridge and link with overseas office. It makes full use of the power of foreign enterprises, social organizations and multilateral organizations, strengthens exchanges and interactions with local governments, social organizations, communities and political parties. It does a good job of regular risk tracking and analysis of key countries, key regions and major projects. It optimizes and enhances the capacity of overseas office infrastructure protection, improves the institutional mechanism for consular protection, enhances the comprehensive support service capacity of foreign aid for national interests overseas, and safeguards the safety and legitimate rights and interests of Chinese citizens, overseas Chinese, enterprises and institutions overseas.

2. Building strategic planning docking systems and policy-standard linkage platforms

One of the important measures to accelerate system-based opening is to build a system and regulatory model that is in line with international pass line discipline. It continues to promote the docking of strategies, plans and mechanisms, and continuously strengthens the linkage of policies, rules and standards. Innovative docking methods is used to promote the implementation of signed documents, promote the negotiation of investment protection agreements and double taxation avoidance agreements with more countries, strengthen cooperation in customs, taxation and supervision, and promote the implementation of a higher level of customs clearance integration. It expands the areas of rule docking and strengthens cooperation in rule docking in financing, trade, energy, numerical information, agriculture and other fields. It promotes effective dovetailing and synergy between the BRI and regional and international development agendas. In addition, it should actively participates in the reform of the global economic governance system, actively participates in international regional economic cooperation work, safeguards the multilateral trading system, participates in WTO reform, promotes and participates in various bilateral and multilateral investment and trade cooperation mechanisms, and build a global-oriented high-standard free trade network. It gives the grassroots greater room for innovation and exploration, and mobilizes to the greatest extent the enthusiasm, initiative and creativity of the grassroots in practicing innovation and open development.

Since the launch of the BRI, many countries have sought strategic alignment with it. Up to now, the BRI has been dovetailed with Russia's "Eurasian Economic Union", Mongolia's "Steppe Road" strategy, Kazakhstan's "Avenue of Light", Turkey's "Middle Corridor", Europe's "Juncker Investment Plan", Vietnam's "Two Corridors and One Circle", the United Kingdom's "Northern Economic Centre for England" and other countries and regions' strategic planning (BRI Portal). States or international organizations are the actors in strategic planning docking. There are many actors in docking; individuals, enterprises, States and international organizations are all partly involved in various docking activities, but not all of them have the legal status of subjects under international law. The State or international organization is the basic actor in the policy interface and has the international legal personality to participate independently in global economic and trade governance. At the same time, dovetailing is based on the fact that there are points of convergence in the economic development of each country. This is the most essential feature of docking, as it nurtures the young shoots of the elements of global governance. The prerequisite for docking is that each country has its own development strategy and that further cooperation is considered on the basis of the points of convergence between the parties. In terms of content, the docking is an economic development strategy, the scope of which is much larger than international trade, manifesting itself in the integration of trade, investment rules and industry policy, and the expansion of global economic and trade governance rules from focusing on border rules to rules within borders. In terms of behavior, the docking reached consensus on areas of common concern, reflecting the characteristic of acting within one's means, a way of behavior that is quick to reach agreement, low cost, and addresses the current dysfunctional global economic and trade governance.

From the perspective of strategic planning docking and collaboration, during the eight-year construction period, the BRI has formed strategic rules docking and collaboration with the United Nations, the World Bank and IMF Petroleum Fund organizations, and has taken important issues such as the "sustainable development agenda", "efficient investment and financing programmes" and "innovative financial guarantee systems" as points of convergence for cooperation between the BRI and the above-mentioned international organizations respectively.

In terms of policy and standards connectivity, standardization has a fundamental and supportive role in the construction of the BRI. Comprehensively deepening bilateral and multilateral practical cooperation and interconnection in standardization with countries and regions along the route and actively promoting mutual recognition of standards is conducive to the overseas promotion and application of China's standards, promoting the "going out" of Chinese standards, enhancing the level of internationalization of standards. It better supports and serves China's industries, products, technologies, projects and other "going out". China attaches great importance to deepening mutually beneficial cooperation in standardization with countries along the route, promoting policy, facility, trade and capital flows with mutual recognition and joint development of standards. It supports connectivity and production capacity cooperation. China's advantageous industries, such as high-speed rail, electric power, nuclear energy, aerospace and aviation, information technology and high-end equipment manufacturing, should also actively make use of the internationalization of Chinese standards to raise the level and sophistication of going global, drive Chinese products, technologies and services to go global with Chinese standards, and shape the advantages, reputation and image of Chinese enterprises and brands with high-quality, high Chinese standards. This is the key to promoting the transformation and upgrading of China's going out and improving quality and efficiency.

In order to promote the construction of the BRI, standards and policies and rules complement each other and advance together, providing an important mechanism for interconnection, China released the *Action Plan for the Common Construction of the BRI by Standards Link (2018–2020)* in early 2018. (Standard Unicom to jointly build "Belt and Road Initiative" Action Plan (2018–2020), 2018) The plan adheres to the principles of demand-oriented, standard-led, innovative cooperation and mutual benefit. It takes the initiative to strengthen the docking of standardization strategies and mutual compatibility of standard systems with countries along the route, vigorously promotes the internationalization of Chinese standards, strengthens the organic convergence of standards and policies and rules, creates a "hard mechanism" for cooperation with standard "soft connectivity", strives to improve the compatibility of standard systems, supports the construction of infrastructure connectivity, promotes international cooperation in production capacity and equipment manufacturing, facilitates investment and trade and deepens humanistic exchanges. It provides solid technical support and powerful mechanism to promote the construction of the BRI.

The BRI is rich in connotation and can be summarized as "Five Links", such as policy communication, infrastructure connectivity, smooth trade, financial integration, and civic exchanges, leading from point to point and from line to line, forming a new pattern of economic, trade and humanistic cooperation based on a partnership network. Building the BRI means that all countries participates together and make use of their own strengths and potential, thus promoting regional openness and integration. As the BRI is a brand-new cooperation model, with various subjects and levels of cooperation, how to carry out cooperation in specific areas in the light of the actual situation of the region and countries require countries to negotiate on the better construction of a strategic planning docking system and a platform for linking policies and standards in the future.

3. Building a 4-in-1 "land-sea-sky-internet" network system

In the process of promoting infrastructure interconnection, it builds a 4-in-1 network system of land-sea-sky-internet, taking the "Six Corridors, Six Roads, Multiple Countries and Ports" as the basic framework, and setting up an interconnection network led by economic corridors such as the New Asia-Europe

Continental Bridge. Grounded in the main structure of the China-Europe Classes, the new land-sea alleyway, other major alleyway and information highways, as well as supported by railways, ports and pipeline networks, creates a new international land-sea trade alleyway. It also continues to focus on key alleyway and key cities, orderly promotes the construction of major cooperation projects, and integrates the goals of high quality, sustainability, risk resistance, reasonable price and inclusive accessibility into the entire process of project construction.

Facility connectivity is one of the five priorities of cooperation under the BRI, of which infrastructure connectivity is a priority area for the BRI construction. The construction of infrastructure connectivity in our border areas and cross-provincial junction areas is relatively lagging behind. The shortcomings and ink bottle effect are more prominent. The focus on infrastructure connectivity as the BRI construction is not only conducive to making up for the shortcomings in infrastructure construction, but also helps promote economic and trade development in border areas and cross-provincial areas. In the process of promoting infrastructure connectivity in BRI countries, work has been carried out in three main areas.

First, it promotes cross-border infrastructure connectivity. In this regard, we should promote the construction of cross-border missing road and rail sections, bottleneck sections and port infrastructure, accelerate the construction of cross-border oil and gas pipelines and cross-border power projects and power transmission alleyway, promote the construction of ports and water transport alleyway, and steadily promote the construction of cross-border fiber optic cables, intercontinental submarine fiber optic cables and air information alleyway. At the same time, it is actively seeking to establish a national-level working mechanism for cross-border infrastructure connectivity with neighboring countries to set up sub-national joint task group with the border areas of neighboring countries to jointly study and propose cross-border infrastructure connectivity projects, and then formulate action plans for implementation.

Second, it has innovative investment and financing models. It should make full use of China Development Bank, The Import-Export Bank of China and relevant banks in neighboring countries to carry out syndicated loan or joint financing. It should work with the combined efforts of international development finance organizations such as the Asian Infrastructure investment bank, the World Bank, Asian Development Bank and regional multilateral financial institutions. At the same time, the government-social capital cooperation (PPP) model is used to support domestic infrastructure investment enterprises, large engineering contractors and commercial infrastructure-type investment funds and other financial capital to form a consortium to jointly participate in investment, construction and operation.

Third, it promotes cross-provincial infrastructure development. Due to the different levels of economic and social development in different provinces and the different directions of flow of people, logistics and other important factors of production between different provinces, the enthusiasm of neighboring provinces for investment and construction of infrastructure across provincial borders varies greatly, which to a large extent has become an important shortcoming and weakness in the field of infrastructure in China at this stage. For example, due to inconsistent progress in the construction of transport infrastructure and inconsistent scale and standards, some cross-provincial roads have even become "cut-off roads". In view of this, it is important to focus on supporting the construction of transport infrastructure such as inter-provincial railways, highways and waterways, as well as energy infrastructure such as power grids. It opens up the "last mile" of inter-provincial transport and energy and electricity interconnections. In terms of funding, the integrated and integrated development of cross-provincial infrastructure projects with industrial-type projects in tourism, logistics and resource development and utilization along the routes or in the surrounding areas should be encouraged to improve the financial efficiency of the entire project. Support the joint introduction of PPP models in neighboring provinces or in neighboring regions across provincial borders.

Infrastructure connectivity is an important foundation and priority area of cooperation for the construction of the BRI, an important support for promoting the economic development of countries along the route, benefiting the general public and enhancing the well-being of the region. It is an important basis for realizing practical cooperation and mutual benefit. In the area of energy and communications facilities connectivity and technical cooperation, breakthroughs have been made in cooperation on major energy projects and infrastructure networking. China has carried out extensive cooperation with countries along the route in the fields of oil rock gas, coal, power and refining and chemical industries. It has worked with relevant countries to maintain the safe operation of cross-border oil and gas pipeline networks and promote energy optimal allocation of resources between countries and regions.

The importance of promoting the BRI facility connectivity is conducive to breaking through development bottlenecks, accelerating economic development and improving people's livelihood in the region. It is also important to improve the environment for economic and trade cooperation in the region, promoting complementary advantages and achieving win-win development. It achieves early harvest as soon as possible to strengthen confidence in the BRI. It is conducive to breaking through development bottlenecks and accelerating the economic development and livelihood improvement of the countries in the region. Promoting the BRI facility connection is conducive to improving the economic and trade cooperation environment in the region, promoting complementary advantages and achieving win-win development. Interconnected and standard-compatible infrastructure is a basic prerequisite and important condition for trade in goods, allocation of funds, movement of people and industrial cooperation. From global experience, the deepening of regional economic cooperation cannot be separated from the improvement of the level of infrastructure connectivity. Many regional economic organizations, such as the European Union and ASEAN, have put in place plans, designed mechanisms and formulated policies to improve connectivity. SCO is also committed to promoting cooperation in transport facilitation. The BRI countries have greater potential for complementary development, with the Middle East and Central Asia rich in natural resources, Southeast Asia, South Asia and Africa with intensive labour resources and obvious cost advantages, China with a complete industrial system and a broad consumer market, and Europe with a developed economy and advanced technology. After the formation of a large alleyway and facility connection network, resources, production, markets and technology are well dovetailed. Land-sea integration entails that landlocked countries can break the "landlocked" constraint and marine countries can expand the economic hinterland.

The promotion of the BRI facility linkage is conducive to achieving early harvest as soon as possible and strengthening confidence in the BRI. First, infrastructure development and connectivity are in line with international trends and easier to achieve alignment. There is a broad global consensus, with many regions and countries identifying infrastructure development and connectivity as important areas, and putting in place corresponding implementation plans. Secondly, strengthening infrastructure development is a way to make up for shortcomings and break bottlenecks. Once completed, it has a significant effect on economic promotion and livelihood improvement. Third, China's international engineering contracting strength is strong and its construction capacity is guaranteed. "Built in China" has a high cost performance in terms of quality, cost and efficiency. Infrastructure projects are usually large, far-reaching, indicative and conducive to enhancing confidence in the BRI.

President Xi Jinping pointed out that "the Silk Road must first have a road. Only when there is a road can people travel and things flow smoothly." (Unicom leads Development Partners to focus on Cooperation: speech at the Host

Partnership Dialogue on "strengthening the Partnership for Connectivity", 2014, November 8) Facility connectivity is the basis for cooperative development and a priority area for the construction of the BRI. Since the introduction of the BRI, China has adhered to the principle of common consultation, common construction and sharing. It actively planned and taken the initiative, utilized existing bilateral cooperation mechanisms, international organizations and multilateral forums and other effective platforms. It promoted the formation of consensus, took infrastructure connectivity planning and technical standards docking as the entry point. It relied on infrastructure construction project, pragmatically promoted in-depth cooperation with countries along the route in the fields of railways, highways, water transport, civil aviation, postal services and other areas. It promotes regional transport connectivity and continued to make new progress. It continues to form a 4-in-1 network system of land-sea-sky-internet in the future.

After China proposed the BRI, it has focused on strengthening cooperation with countries along the route in regional infrastructure connectivity. Various departments and regions have actively docked and communicated with countries along the route to pragmatically promote bilateral and regional infrastructure connectivity construction and cooperation. In building the 4-in-1 network system of land-sea-sky-internet, the current concentrated trends include three aspects: (1) the continuous expansion and deepening of cooperation areas and contents. From the "hardware connection" in the construction of infrastructure such as transportation, communication, energy, pipelines and ports, to the "software connection" in the construction of customs clearance, inspection, certification, financing and other operation and management, financial support services and other systems, the areas of cooperation have been expanding from easy to difficult, and the content of cooperation has begun to deepen gradually to policy communication, system docking, mutual assistance in law enforcement, mutual recognition of standards and information service platform construction. (2) The forms and levels of cooperation are increasingly diversified. The subjects of cooperation include the central and local levels of the countries along the route. The mechanisms of cooperation include regular meetings between senior officials and working levels, and counterpart communication between enterprises and industry associations; the forms of cooperation include the promotion of key projects, the signing of sectoral cooperation memorandum entry and the signing of sub-regional or bilateral agreements with certain binding effect in special areas. (3) The results of the cooperation are beginning to emerge. The strengthening of regional infrastructure construction and connectivity has received positive responses and wide recognition from countries and international organizations along the routes. While accelerating the construction of regional and international logistics corridors such as the China-Europe Land-Sea Express and the China-Europe Liner, customs and quality inspection and other agencies have carried out coordinated cooperation on customs clearance at ports and improved customs clearance facilitation, with the average inspection rate and passage time falling by 50%. According to the World Investment Report, connectivity has emerged as an important factor in improving the Asian region investment climate, with FDI and cross-border M&A in infrastructure flowing into East and Southeast Asia bucking the trend during the global downturn in cross-border investment (Anonymous, 2016, May 4).

4. Improving the construction of domestic supply chain and international industrial chains

Trade is an important engine of economic growth. As one of the "Five Links" of the BRI, smooth trade is a key element in promoting the construction of the BRI. Improving the construction of domestic supply chain and international industrial chains is an attempt to build the BRI trade flows together. Through the improvement of the domestic supply chain and the construction of the international industrial chain; on the one hand, Chinese manufacturing, Chinese construction and Chinese services are welcomed by more countries and regions along the route. More products, services, technologies and capital from countries and regions along the route are entering China in a steady stream. We have achieved a good situation in which trade exchanges are expanding, investment cooperation continues to deepen, economic and trade cooperation zones are accelerating, major projects are being built in an orderly manner, and a network of free trade zones is being formed.

In order to improve the construction of domestic supply chain and international industrial chains, and to ensure the stable and smooth flow of trade industrial chains and supply chain, China has made efforts to integrate the development of trade industries, identified a number of foreign trade transformation and upgrading bases, cultivated a number of processing trade industrial parks. It also builds a number of national import trade promotion innovation demonstration zones to expand the scale of imports and drive related industries to improve quality and increase efficiency.

The industrial chain and supply chain is the key to the smooth flow of the economic cycle of a large country. By deeply promoting the structural reform on the supply side, the core competitive power of China's industrial chain supply chain has been continuously strengthened and its position in the global industrial chain supply chain continues to climb. At present and in the coming period, China's development remains in a period of important strategic opportunities, but both opportunities and challenges have evolved in new ways. In the face of the new mission and new tasks and new requirements to open a new journey of building a modern socialist country, the importance of upgrading the level of modernization of the industrial chain supply chain has become more prominent. Upgrading the level of modernization of the industrial Chain supply chain is an urgent need to accelerate the development of a modern industrial system. After years of rapid development, China has formed a complete industrial system of large scale and complete support. It is the only country in the world that has all the industrial categories in the United Nations Industrial Classification. However, there are still many "breakpoints" and "blockages" in the industrial chain supply chain, some core links and key technique are restricted, the industrial base capacity is insufficient, the national economic cycle is not smooth, and there is a structural imbalance. We must adhere to the main line of deepening the structural reform on the supply side, use a systematic approach to solve the structural problems of the industrial chain and supply chain, smooth the various links of production, distribution, circulation and consumption. We must realize the effective linkage and efficient operation of upstream and downstream, production, supply and marketing, promote the synergistic development of the real economy, science and technology innovation, modern finance and human resources, and accelerate the development of a modern industrial system.

Upgrading the level of modernization of the industrial chain supply chain is a critical way to promote the high-quality development of the real economy, especially the manufacturing industry. It is an important initiative to coordinate development and security. A good development environment is a prerequisite and guarantee to enhance the level of modernization of the industry chain and supply chain. Therefore, in order to optimize the development environment of the industrial chain and supply chain, it strengthens factor support, enhances international industrial security cooperation, forms an industrial chain and supply chain with stronger innovation, higher added value, more secure and reliable, and forms a new comparative advantage. The relevant work includes four main areas: (1) Deepen reforms in key areas. President Xi Jinping stressed the need to continue to use the key move of reform to its fullest extent. (Unswervingly continue to make full use of the key trick of reform, 2020, May 3) At present, upgrading the industrial chain and supply chain modernization surface level faces many deep-rooted institutional mechanisms obstacles, it is necessary to further increase reform and innovation, landing reform initiatives, cracking development problems, and stabilizing market expectations. It promotes factor market reforms, breaks down institutional barriers that hinder the market-based allocation of production factors and the flow of goods and services, promotes the balanced development of finance and real estate with the real economy, and effectively reverses the tendency of "de-realization to deficiency". It improves the institutional mechanism for promoting consumption, deepen the reform of the investment and financing system, and nurture the expansion of domestic demand. It deepens the state enterprise mixed ownership system reform, accelerates the improvement of the modern enterprise system with Chinese characteristics, and enhances the vitality and competitive power of enterprises. (2) It strengthens the support of elements. Technology, capital, talent and data are the core elements that support the industry chain supply chain. We should enhance the application of science and technology support to industries, improving the technological innovation system with enterprises as the main body, focusing on a market orientation, and establish a deep integration of industry, academia and research. It constructs an institutional mechanism for finance to effectively support the development of the real economy, promotes the construction of the capital market infrastructure system, promotes a significant reduction in comprehensive financing costs, and ensures that new financing flows to manufacturing, small and medium-sized micro enterprises in a focused manner. It strengthens the support of talents, enhancing the training of talents in science, engineering, agriculture and medicine, implementing knowledge and technology updating projects, and building an army of knowledgeable, skilled and innovative workers. It accelerates the cultivation of data element markets, promotes the open sharing of government data, enhances the value of social data resources, and strengthens the integration and security protection of data resources. (3) It creates a world class business environment. The business environment is the soil for industrial development and an important manifestation of the country's economic soft power. We must continue to optimize the market-oriented, rule of law and internationalized business environment to further strengthen our market attractive force and economy competitive power. It continues to promote the reform of "management and service", speeds up the transformation of government functions, innovates government management methods, opens up two-way communication channels between government and enterprises, and builds a friendly and clear relationship between government and business. It builds a high-standard market system, strengthens the fundamental status of competition policy, and accelerates the transformation of industry policy to a universal and functional one. It protects entrepreneurs legal interests according to law, strengthens the protection of property rights and intellectual property rights, and strengthens confidence in the development of enterprises. (4) International industrial security cooperation should be strengthened. President Xi Jinping profoundly pointed out that opening up and cooperating is a move in line with the general trend. Resolving the contradictions that have arisen in the process of economic globalization and coping with the impact of the Covid-19 pandemic have increased the need for mutual support, solidarity and cooperation among all countries of the world. In accordance with the requirements of building a new system of a higher level of open economy, market access restrictions should be further relaxed, and a large domestic cycle should be used to attract global resource elements. It promotes the docking of strategies, plans and mechanisms with BRI countries, strengthen the connection of policies, rules and standards, and deepen complementary cooperation in the industrial chain supply chain. It strengthens intergovernmental macroeconomic policy coordination, upholds the multilateral trading system, promotes the establishment of a global industrial chain and supply chain emergency coordination and management mechanism, and opens up a new situation of win-win cooperation.

5. Improve diversified standardized investment and financing system

As one of the "Five Links" of the BRI, financial integration provides a solid guarantee for the linkage of facilities and smooth trade. It promotes the optimization and upgrading of trade and investment cooperation with the co-construction of the BRI countries, and actively develop silk road e-commerce. It deepens international production capacity cooperation, expands third party market cooperation, constructs a mutually beneficial and win-win industrial chain supply chain cooperation system, and expands two-way trade and investment. It adheres to an enterprise-oriented and market-oriented approach, follows international practice and the principle of debt sustainability, and improves the diversified investment and financing system. The innovative financing cooperation framework play the role of the BRI special loan, Silk Road Fund, etc. It establishes and improves the BRI financial cooperation network, promotes the interconnection of financial infrastructures, and supports the joint participation of multilateral and national financial institutions in investment and financing. It improves the BRI risk prevention and control security system, strengthens legal services, and effectively prevents and resolves various risks.

Capital financing is the guarantee for the stability of the BRI, and an open and market-oriented investment and financing system is an inherent requirement for the high-quality development of the BRI. It is necessary to follow the laws of the market, adhere to the market-led, multi-party participation, strengthen debt and risk management, achieve sustainable development, and promote the construction of the BRI in conjunction with the opening up of China's financial sector, so as to play the role of opening up to support the construction of the BRI and practice the concept of green development. All parties should adhere to the principle of joint enterprise, market-oriented operation and mutual benefit. They need to build an open and market-oriented investment and financing system. Since the introduction of the BRI, China has focused closely on the financing needs of infrastructure connectivity and trade and investment cooperation of countries along the route, mobilized the active participation of various stakeholders, and collaborated to build the BRI investment and financing system of common consultation, common construction and sharing.

Specific construction efforts include: (1) The integration of domestic and international resources. The Government's official development assistance, development finance, policy finance, commercial finance and other domestic financial resources have been fully utilized to jointly promote the construction of the BRI financing mechanism. Over the years, China has accumulated rich practical experience in the field of policy and development finance, and the participation of the market-oriented Silk Road Fund and commercial financial institutions has been increasing. They have many advantages, such as strong and influential, and are well positioned to make a big impact in the construction of the BRI. At the same time, multilateral development agencies, foreign sovereign wealth funds and institutional investor have huge funds and rich knowledge resources, so they should be fully mobilized to jointly create key projects with controlled risks, considerable benefits and shared interests, so that international and domestic resources can complement each other and make joint efforts. (2) The integration of government and market forces. It is necessary to effectively play the role of government policy guidance and support, promote the dovetailing of development strategies of countries along the route, shape regional infrastructure planning and financing arrangements, and create a transparent, friendly and non-discriminatory financing environment. For strategic projects of great significance, government official development assistance, policy finance, development finance, sovereign wealth funds and investments of a public nature such as state enterprise should form a synergy to drive private capital to enter and promote the government and social capital cooperation model (PPP) to continuously reach a new level. It encourages innovative financing models, channels, tools and services under the premise of risk control, develop and improve the equity financing market and local currency bond market to form a full set of financial policy portfolio tools. Fundamentally, the construction of the BRI still relies on the power of the market. It is necessary to follow the laws of the market and international general line discipline, give full play to the decisive role of the market in financial resource allocation and the main role of various types of enterprises, and realize a good pattern of joint participation of funds from all parties. (3) Multilateral and bilateral cooperation should be integrated. Most of the projects under the BRI construction are bilateral cooperation, especially in infrastructure, trade and investment and capacity cooperation. We should attach great importance to bilateral financing cooperation, make good use of the two financial resources and two financial markets of our country and the country where the funds are used, and build bilateral projects in an integrated manner. At the same time, the strong convening power of multilateral development institutions should be brought into full play to explore areas of cooperation that are consistent with the respective purposes and development strategies of multilateral development institutions. With the construction of the BRI, and cooperation related to the BRI should be carried out in various forms, such as joint financing, parallel financing, tripartite cooperation and multiparty cooperation, in order to promote common development. (4) Integrate immediate and long-term tasks. In the near future, limited funds should be concentrated on key regions and countries, and critical item projects and demonstration projects should be grasped, so that more can be done with less, and efforts can be concentrated on major issues, so that the countries concerned can continue to have a tangible sense of achievement. In the medium and long term, the financing needs of the construction of the "Six Corridors, Six Roads, Multiple Countries and Ports" should be comprehensively considered. The construction of the BRI should be provided with stronger, more stable and sustainable financing support through diversified financing channels and improved fund management mechanisms by leading from point to point and from line to line.

In the process of developing a diversified and standardized investment and financing system, a series of financial institutions have been established to specifically respond to the need for financing for the construction of the BRI. Among them, Asia Infrastructure Investment Bank was officially established on January

16, 2016 as an intergovernmental Asian regional multilateral development institution focusing on supporting infrastructure development in five major areas, including energy, transport, rural development, urban development and logistics. It was established to promote the process of building connectivity and economic integration in the Asian region and to strengthen cooperation between China and other Asian countries and regions. The AIIB currently has 57 member countries and is headquartered in Beijing. The AIIB is the twin of the BRI and an important instrument for the realization of the BRI strategy, as well as an important tool for infrastructure development in the region. By its very nature, the AIIB, as an important financing platform for BRI countries, especially developing countries, can help solve the problem of large demand for infrastructure construction funds in countries along the route, but the supply of funds from their own governments is insufficient. The AIIB is an important public sector within the BRI area division as far as a multilateral international organization is concerned, with national governments as shareholders. It does not invest in infrastructure in the region on a profit-seeking primary target basis, but is more concerned with the externalities and overall strategic nature of the construction of infrastructure in the region. With the AIIB as the main investment body, countries in the region that are relatively economically backward and politically unstable are provided with relatively efficient, stable and large-scale financing channels, which help these countries establish basic infrastructure systems in the short term and help them move towards a virtuous cycle of investment and financing. The AIIB, as the main body of investment in the region, can effectively regulate the direction and scale of investment in regional infrastructure, which, both from the perspective of promoting the balanced development of infrastructure in the region in the short term and from the perspective of long-term strategy, can make the construction of infrastructure in the BRI region move towards a virtuous development path. It can make the most urgent development needs, so that those areas that most urgently need to change the state of infrastructure have good access to financing.

In order to further improve the scale of funding sources and the level of financial services, it is necessary to innovatively use a variety of financial products to support the construction of the BRI project, making full use of "two excellent" loans, commercial loans, equity investment funds, insurance and other financial tools, and take mixed loans, combined investment and loans, combined loans and guarantees, etc., to improve financing solutions. It guides the joint participation of the government, financial institutions and private capital. By adopting the public-private partnership approach, it encourages the use of the PPP model, provide flexible and diverse financing options for projects. It gives full play to financial support for project construction. It encourages resident enterprise to explore issuing bonds to host country in order to expand the scale and scope of funds raised and accelerate financial market integration. In addition, financial institutions can actively explore the use of offshore assets and equity as collateral and facilitate the formation of a collateral trading market to provide protection for investments or loans. In the future, it makes full use of the resources of the international capital market. It combines with China's 14th Five-Year Plan, and the development goals and directions of the countries along the BRI. In order to establish a diversified investment and financing system, it further builds an investment and financing platform that is in line with the international capital market.

6. Building a coordination mechanism for green humanities exchanges

The establishment and operation of the BRI working group, through specific work to coordinate the actions of the various ministries under the central leadership of the BRI component part. In this framework, the establishment and promotion of a green humanities exchange and coordination mechanism for the co-construction of the BRI provides a good basis for win-win cooperation in the construction of the soft power of the BRI. By early 2021, China had signed 205 cooperation documents with 171 countries and international organizations to build the BRI (BRI Portal). Since 2020, in the face of the complex international situation, especially the impact of the Covid-19 pandemic, China and the countries concerned stay strongly together and overcome the difficulties, as a result the promotion of the co-construction of the BRI has made new progress and yields new results. A number of major projects progressed smoothly, especially the "Health Silk Road" and "Digital Silk Road" construction results are obvious. The BRI circle of friends is expanding and real friendship is revealed in the times of difficulties. This fully demonstrates that the BRI and the concept of building a community of a shared future for mankind are deeply rooted in people's hearts.

In 2019, President Xi Jinping stressed at the second BRI international cooperation summit that "we should adhere to the concept of openness, greenness and integrity, make green a base color, promote green infrastructure construction, green investment and green finance, and protect the common home on which we depend." (Jin Qiu, 2019). Green BRI and the United Nations 2030 Agenda for Sustainable Development are highly compatible and complementary in terms of concepts, principles and goals, and are recognized by the international community as one of the solutions to promote the implementation of the sustainable development agenda.

In the process of building high-quality development in the BRI, China is strengthening its own ecological civilization, while actively working with the BRI co-construction countries to build the "Green Silk Road", strengthen cooperation on ecological environment, biodiversity conservation and climate change, promote the continuous cohesion of global environmental governance, and inject new momentum into the implementation of the 2030 sustainable development goals. China has been an active participant and contributor to the global climate governance process and the multilateral process on biodiversity. It has effectively fulfilled its obligations under climate change, biodiversity and other environment-related treaties. It has achieved the 2020 targets related to addressing climate change and establishing wilderness area ahead of schedule. President Xi Jinping declared at the general debate of the 75th UN General Assembly that "China will increase its autonomous national contribution, adopt more vigorous policies and measures, strive to peak CO2 emissions by 2030, and work towards achieving carbon neutrality by 2060." (Xi Jinping, 2020, September 22).This oath demonstrates China's firm determination to actively respond to climate change and follow a green and low-carbon development path. It reflects China's responsibility to build a community of a shared future, and represents a new effort and contribution by China to address climate change. The international community highly appreciates China's climate commitment and believes that it has a positive impact on the global response to climate change. It helps to drive BRI clean energy investment and helps developing countries embark on a low-carbon path.

Since 2013, the Chinese government has issued a series of policy documents to strengthen BRI ecological environmental protection work. *The Guidelines on Foreign Investment Cooperation Environmental Protection* issued in 2013 put forward the requirement that enterprises should carry out pollution prevention and control work in accordance with the host country environmental protection with the requirements of laws, regulations and standards, pollutant emission which should comply with host country pollutant emission standard provisions and reduce the adverse impact on local biodiversity. (*The Guidelines on Foreign Investment Cooperation Environmental Protection*, 2022, January 5) *The Vision and Action for Promoting the Construction of the Silk Road Economic Belt and the 21st Century Maritime Silk Road*, released in 2015, requires enterprises to take the initiative to assume social responsibility and strictly protect biodiversity and the ecological environment. (Vision and actions on jointly building Belt and Road,

2015, March 25). In 2017, *The Guidance on Promoting Green BRI Construction* and *The BRI Ecological and Environmental Protection Cooperation Plan* were released, clarifying the overall idea and task measures for the construction of green BRI and proposing to "promote enterprises' compliance with ecological and environmental laws and regulations, policies and standards of host countries". (*The Guidance on Promoting Green BRI Construction* and *The BRI Ecological and Environmental Protection Cooperation Plan*, 2017, May 16) By implementing President Xi Jinping's initiative, the Ministry of Ecology and Environment and the Chinese and foreign partners jointly launched the BRI International Alliance for Green Development. It promotes green technology exchange, builds BRI environmental technology exchange and transfer center (Shenzhen) and other environmental protection industry technology cooperation platform. It holds the BRI green innovation conference, and promotes the construction of green technology innovation and entrepreneurship base and international high-end environmental protection industrial park. Domestic and foreign research institutions have jointly developed the Rapid Environmental Risk Assessment Tool (ERST) for Chinese offshore investment projects, which can give recommendations for assessing the potential impacts on biodiversity and environmental resources at an early stage of project development, and identifying and managing the environmental and social risks of the project. In terms of promoting green infrastructure, China has already built a large number of solar, wind and other renewable energy projects in the co-construction countries, helping to accelerate the transformation of host country energy supply to an efficient, clean and diversified direction.

The BRI is an international cooperation initiative with deep historical and cultural roots. The people's friendship lies in the people's affinity. The people's affinity lies in the fellow-feeling. The people's fellow-feeling is the basis of social and public opinion for the construction of "One Belt and One Road" to be stable and far-reaching. The humanities exchange is the most dependable bridge link for the people's fellow-feeling. Carrying out BRI humanities exchanges can help promote the exchange of civilizations and dialogues that are harmonious, different and inclusive, which is of great significance in promoting national foreign exchanges, promoting the international dissemination of Chinese culture, maintaining world peace and development and building a community of a shared future for mankind. Since the 18th Party Congress, under the leadership of the Party Central Committee with President Xi Jinping at the core, China has taken a path of humanities exchanges with distinctive Chinese characteristics, and humanities exchanges have achieved tremendous development. The BRI humanities exchange, as an important component part of humanities exchange work with Chinese characteristics, has achieved fruitful results in the steady march, and continues to promote the co-construction of BRI in a deep and practical way.

Since the introduction of the BRI, China and the participating countries of the BRI have adhered to the principles of leadership by Head of State diplomacy, leadership by high-level visits, demonstration by high-end mechanisms and a combination of bilateral and multilateral cooperation, and have actively set up humanities exchange and cooperation mechanisms in various forms and with wide coverage. The consensus on humanities exchanges is growing, the channels are more open, the forms and contents are richer, and the BRI humanities exchange and cooperation platform is taking shape with multilateral participation, pluralism and openness, and the sharing of multiple subjects. Every major diplomatic event or international conference is gradually institutionalized and normalized, and a series of high-caliber large-scale foreign cultural activities have become successful examples of showcasing the essence of Chinese culture, becoming the organic component part of the head of state's diplomacy. Eight high-level cultural exchange mechanisms at the vice-prime ministerial level have been established with relevant countries, including China-Russia, China-US, China-UK, China-France, China-Germany, China-South Africa, China-Indonesia and China-Europe, to bring into play their radiating and leading role in regional humanities exchanges.

China and the participating countries of the BRI have carried out a rich variety of humanities exchanges and cooperation in a wide range of fields, including education, science and technology, culture, health, tourism and sports, through a variety of subjects such as political parties, parliaments, media, think tanks, youth and social organizations. The countries concerned are organizing a number of activities such as cultural years, tourism years, art festivals, film and television bridges, seminars, book fairs, sports events, think tank dialogues, etc. Innovative cooperation methods such as talent training, works creation, copyright cooperation, joint publishing, capital cooperation, project evaluation, etc. continue to promote the BRI humanities exchanges to be more professional, detailed and practical, and benefit more people's livelihood. With the solid promotion of the BRI humanities exchanges and cooperation, the highlights have brought people's hearts and minds closer together in the exchanges, gaining the support and active participation of more and more countries, consolidating the foundation of public opinion and building a firm social foundation for the construction of the BRI, as well as bringing convenience and opportunities for friendly exchanges, trade and commerce among people in countries along the route.

Humanities exchanges are an important and indispensable fulcrum for promoting the co-construction of the BRI meticulous "brushwork". With firm cultural self-confidence and adherence to the civilizational concept of equality, mutual appreciation, dialogue and tolerance, the BRI humanities exchange further coalesces the consensus on cooperation with the joint efforts of all parties and usher in a wonderful and promising vision of development, injecting more positive energy and warm power to promote the construction of the BRI in a deeper and more practical way. It promotes the building of a human community of a shared future and create a better future for the human society together.

The BRI is not only the economic prosperity path. It is also the path to green development. China has always upheld the concept of green development in the construction of the BRI, and has taken practical actions to build green BRI with other countries. Since the introduction of the BRI in 2013, China has always incorporated ecological environmental protection into the overall institutional arrangement for the construction of the BRI, which is evident to the international community. Chinese President Xi Jinping stressed that he would "make green a base color, promote green infrastructure construction, green investment and green finance" and "actively work with other countries to build a 'Green Silk Road'." In recent years, the Chinese government has issued a series of policy documents to strengthen relevant top-level design: the 2013 *Guide to Foreign Investment Cooperation Environmental Protection* requires enterprises to carry out pollution prevention and control in accordance with host country environmental protection laws, regulations and standards; the 2015 *Vision and Action for Promoting the Construction of the Silk Road Economic Belt* and the *21st Century Maritime Silk Road* proposes to highlight the concept of ecological civilization in investment and trade; the 2017 *Guidance on Promoting Green BRI Construction* and *The BRI Ecological and Environmental Protection Cooperation Plan* clarify the ideas and task measures for green BRI construction. China's push to build a green BRI is by no means a passing fancy, but rather a proactive, long-term plan.

At this stage, China is vigorously promoting international cooperation on ecological and environmental protection under the BRI framework, working with all parties to create green international public goods, bringing confidence to global environmental governance. It advocated the establishment of the BRI cooperation mechanism to combat desertification, set up the Climate Change South-South Cooperation Fund, launched the BRI International Alliance for Green Development together with relevant countries and international organizations, issued The *BRI Green Investment Principles*, and implemented the Green Silk Road Messenger Program to train more than 2,000 trainees from various countries (Li Mo, 2021, January 10). China's cooperation with the BRI countries in the field of "environmental protection" has achieved fruitful results, injecting new momentum for the cohesion of global environmental governance and the implementation of the 2030 Sustainable Development Goals, as well as adding new highlights to the co-construction of the BRI.

The economic and cultural development centre of the BRI regards the concept of "Peace and Cooperation, Openness and Tolerance, Mutual Learning and Mutual Benefit" as the spirit of the Silk Road. The idea of "implementing the national strategy of 'One Belt, One Road' proposed by General Secretary Xi and the act for the benefit of the country and the people" are the central mission. The notion of "efficiency, pragmatism, pioneering and piloting" is the central aim. The "interconnection, interoperability, cooperation and win-win" is the central goal.

References

Anonymous. (2016). The 2015 World Investment report analyzes seven major trends of global investment (2016, May 4). http://www.ccpithn.org/index.php?m=content&c=index&a=show&catid=22&id=446.

Jin Qiu. (2019). Xi Jinping's keynote speech at the opening ceremony of the second Belt and Road Initiative Forum for International Cooperation, 2019. https://www.ccps.gov.cn/xtt/201904/t2019 0426_131284.shtml

Li Mo. (2021). White Paper on China's International Development Cooperation in the New era (2021, January 10). http://www.scio.gov.cn/zfbps/32832/Document/1696685/1696685.htm.

Liu Chang. (2020). The proposal of the Central Committee of the Communist Party of China on the formulation of the 14th Five-Year Plan for national economic and social development and the visionary goals for 2035 (2020, November 3). Xinhua News Agency. http://www.gov.cn/zhengce/2020-11/03/content_5556 991.htm.

The Guidance on Promoting Green BRI Construction and The BRI Ecological and Environmental Protection Cooperation Plan (2017, May 16). http://www.scio.gov.cn/xwfbh/xwbfbh/wqfbh/37601/38609/xgzc38615/Document/1633 106/1633 106.htm; https://www.mee.gov.cn/gkml/hbb/bwj/201705/W020170516330272025970.pdf.

The Guidelines on Foreign Investment Cooperation Environmental Protection (2022, January 5). https://www.mee.gov.cn/xxgk2018/xxgk/xxgk05/202201/t20220110_966571.html.

Unicom leads Development Partners to focus on Cooperation: speech at the Host Partnership Dialogue on "strengthening the Partnership for Connectivity". (2014, November 8). www.xinhuanet.com/polit ics/2014-11/08/c_12 7192119.htm.

Unswervingly continue to make full use of the key trick of reform (2020, May 3). https://jjsb.cet.com.cn/show_515453.html.

Vision and actions of the Silk Road Economic Belt and the 21st Century Maritime Silk Road (2015, March 25). Ministry of Foreign Affairs, https://www.yidaiyilu.gov.cn/yw/qwfb/604.htm

Xi Jinping. (2020). Xi Jinping's speech at the general debate of the 75th session of the United Nations General Assembly (2020, September 22). https://www.ccps.gov.cn/xxsxk/zyls/202009/t20200922_143558.shtml.

Xu Huiyun, Lin Xiaozhao, Zhu Yanran. (2020). The free trade zone has been expanded to 21 provinces, and a higher level of opening-up has promoted a new pattern of development (2020, September 22). https://baijia hao.baidu.com/s?id=1678 468083159865579&wfr=spider& for=pc.

Zheng Chengqiong. (2016). Liu Jieyi: Why is the the Belt and Road initiative included in the General Assembly resolution. http://www.scio.gov.cn/ztk/wh/slxy/31200/Document/1534910/1534910.htm.

VIII

Investment Planning and Financing for Co-constructing the BRI

Introduction

"The Outline of the 14th Five-Year Plan" proposes to "promote the co-construction of the 'BRI' high-quality development". Therefore, during the 14th Five-Year Plan period, it is necessary to accurately grasp the profound connotation of "high-quality development" when determining the investment plan for the BRI countries. The total scale of investment should be appropriately controlled. The structural distribution of investment should be appropriately adjusted. The proportion of investment in relevant projects in New Economy forms such as the digital economy and the green economy should be gradually increased. In addition, the industrial structure and layout of investment in the BRI countries during the 14th Five-Year Plan period should also be adjusted and optimized, with differentiated arrangements for Central Asia, East Asia, Southeast Asia and West Asia in terms of industrial layout, giving full play to the location advantages of the four main cooperation regions themselves, and integrating industrial adjustment based on the international and domestic double cycle in terms of industrial structure, and striving to build a two-way trade posture of "investment + trade".

Specifically in terms of financing, during the 14th Five-Year Plan period, the promotion of high-quality co-construction of the BRI requires a rich variety of investment and financing actors. It gradually forms a hierarchical and differentiated structure of investment and financing actors, and the establishment of a coordinated framework of diversified investment and financing actors. We should actively try various "benefit sharing, sharing of risks" investment and financing models, making good use of traditional financing methods and various types of capital market. It achieves a combination of equity financing, debt financing and investment, construction and operation. We should deepen financial cooperation and risk supervision cooperation, and jointly improve the financial support and guarantee system to serve the construction of the BRI. Through the construction of the investment and financing rules system, the long-term compatibility of the investment and financing cooperation mechanism should be enhanced, so as to provide strong support for the promotion of high-quality practical cooperation among the countries along the route to build the BRI.

1. Investment planning for the BRI countries in the 14th Five-Year Plan period

(i) Scale and structure of investment

The Outline of the 14th Five-Year Plan and Vision 2035 for the National Economic and Social Development of the People's Republic of China was officially released in March 2021 (*The Outline of the 14th Five-Year Plan and Vision 2035 for the National Economic and Social Development of the People's Republic of China*, 2021, March 13). As in the 13th Five-Year Plan, the construction of the BRI is listed in a separate chapter as an important part of the 12th chapter of the outline, "Opening up to the outside world in high standards, develop new situation of win-win cooperation". "The changes in the 14th Five-Year Plan on how to build the BRI are mainly reflected in the overall requirement of 'high-quality development' ". In the outline of the 14th Five-Year Plan, the expression BRI was changed from "promoting the construction of the 'BRI' " in the outline of the 13th Five-Year Plan to "promoting the co-construction of the 'BRI' for high-quality development" (*The Outline of the 14th Five-Year Plan and Vision 2035 for the National Economic and Social Development of the People's Republic of China*, 2021, March 13).We believe that the meaning of "high-quality development" needs to be grasped from three perspectives. (1) From a microscopic point of view, the quality of each project has to be up to standard.

Both domestic and international standards and local standards must be met. The quality of each project must stand the test of history and meet designed life. (2) The construction of the BRI requires a concept of sustainable development. High-quality projects must be made to benefit local social, economic and ecological development needs in the long term. Establishing this concept of sustainable development and vigorously advocating "building a green Silk Road" is of great significance to the enhancement of China's national image and the effective prevention of environmental risks in "going abroad", thus promoting the smooth implementation of the BRI construction. (3) The 14th Five-Year Plan outline "high-quality development" of the overall requirements clearly states the basic principle of the BRI. In other words, we must adhere to the principle of "shared benefits and shared risks"; although the initiative is proposed and led by China, the promotion of comprehensive construction depends on the joint efforts of all parties involved. In this way, the dividends inherent in the BRI can be transformed into real wealth, to be shared by all parties involved in its construction.

Therefore, in the next stage of international cooperation in the co-construction of the BRI high-quality development, based on the above overall requirements and basic principle, combined with the results of the basic theoretical model in Chapter III of this study, we suggest that the scale of total non-financial investment by the Chinese side in the 14th Five-Year Plan period should be controlled at about US$100 billion. Since the implementation of the BRI, the total scale of investment in the previous eight years was US$ 138.2 billion, while the 14th Five-Year Plan period and the period ahead is more important to work on the actual details. It should perform a serious job in each project, rather than continuing to expand the big scale. During the 14th Five-Year Plan period, efforts should be made to coordinate domestic and foreign investment, properly deal with the challenges posed by changes in the external political and economic environment, and effectively respond to the negative impact of foreign pandemic developments on local project construction. Therefore, appropriate control should be exercised on the total scale of investment. The total non-financial investment of US$100 billion during the 14th Five-Year Plan period is a broad guideline scale. It does not include the scale of investment and financing of international capital market and co-build countries. Specifically, during the next five years covered by the 14th Five-Year Plan, the scale of Chinese non-financial investment is limited to about US$20 billion per year. In addition, it is necessary to make a budgetary plan for the financing of non-financial investment and construction during the 14th Five-Year Plan from the total scale of investment in the national economy, so as to ensure that the above-mentioned investment planning arrangements are financed.

At the same time, the investment structure of China's promotion of the BRI construction during the 14th Five-Year Plan period should also be adjusted. Of the US$100 billion investment, the share of investment in digital economy and green development projects gradually increases to about 40%. Digital technology, the latest technological achievement of the fourth industrial revolution, is used in the construction of the BRI through digital transformation. For example, coal-fired power plants are no longer simply copying what is already in place in the country construction project, but are going green through digital measures to prevent pollution and improve coal savings. The 14th Five-Year Plan period can focus on the digital transformation of these traditional projects to promote the further development of digital economy industries. China is already ahead of most countries in terms of application scenarios, and digital economy technologies, including mobile payment. These are promoted in developing countries along with the construction of the BRI project. The application scenarios of new Chinese digital economy technologies are realized in developing countries, so that developing BRI countries can enjoy these new digital technologies even earlier than some developed countries ("Belt and Road Initiative" to develop digital economy to light up the road of innovation, 2017, May 22). In terms of green development, we should continue to strengthen policy communication with BRI countries, play a good role in the BRI international alliance for green development and the ecological and environmental protection data service platform. It promotes environmental policies, planning, standards and technology docking, and share China's ecological civilization construction concepts and practical experience. At the same time, it continues to improve the ecological environmental administration level of the BRI projects, improves policy tools, plays the role of financial institutions to guide green investment, guides enterprises to better fulfill their environmental responsibilities, and prevents the BRI construction ecological and environmental risks. We continue to promote high-quality development through the construction of green BRI and build a beautiful home in harmony with our partners.

(ii) Industrial pattern

1. Development differences in the four main cooperation regions

The BRI differs in its focus and opening strategy towards the four regions of Central Asia, East Asia, South-East Asia and West Asia. These four regions have different bases in terms of political and cultural foundations, strategic resource priorities, and demographic and economic needs, and differ in the manner and timing of their advancement. The East and South-East Asian region is the main area for near-territorial trade and the search for expansion of maritime rights, while the Central Asian region is the main area for the search for expansion of the right of way.

The East Asian region is central to near-domain trade with Japan and Korea and to securing international maritime transport alleyway. From the perspective of near-region relations, Japan and South Korea are China's traditional trading partners, but their product types are similar, competition is greater than cooperation, and the proportion of China's trade structure relatively decline. In terms of long-range links, the eastern coast is bound to remain the main landing point for oceanic trade route for a long time to come, but in the face of the "first island chain" of maritime encirclement, seeking landing points in the northeast and Taiwan to bypass Japan and South Korea is still a strategic consideration for the region in the future. One of the more likely possibilities because of its common roots with Taiwan is that the relationship with Taiwan is not so much about the volume of economic trade as it is about forming cultural and political alliances thus bringing new access to the sea.

Southeast Asia and the South Asian region are important transition areas for near-area trade to ASEAN and linking land-sea alleyway. In terms of near-region trade, ASEAN is the emerging trade partner with the fastest rising proportion in China's foreign trade, with a larger population and fast growth in economic development. Some of the countries are just in the early stages of industrialization, which has a greater complementary relationship with China's economic and industrial system. As China's share of investment in south east Asia region is increasing day by day, it is pushing for further economic-industrial integration and has recently been gradually increasing its influence through the construction of trade integration and investment in industrial parks. The China-Singapore New Economy corridor focuses on further integration of ASEAN, with key implementation projects including road, rail, fiber optic communications, port and park construction, with route selection and construction work starting on the Nanning-Singapore, Kunming-Yangon and Kunming-Bangkok routes respectively. The China-Pakistan Economic Corridor (CPEC) focuses on the political-military-economic-strategic significance of the Gwadar port to form an easy access to the Indian Ocean and an outpost into the Middle East. Because of its close proximity to Iran and the Persian Gulf, the port of Gwadar is the closest overland port of entry for Middle East crude oil that China can fully control.

The Bangladesh-China-India-Myanmar Economic Corridor is China's southwest entry into the connected Indian Ocean alleyway, with a heavy emphasis on opening up the southwest access to the sea. It is strategically important to open up the second alleyway of the Indian Ocean, seeking to provide the nearest southwestern access to the sea at Chittagong, Sonadia, Kyaukpyu and Colombo ports. Key implementation projects include the construction of public and railway projects, and the construction of industrial parks. The opening of this alleyway is of strategic importance for changing China's energy and ore imports alleyway, and also fundamentally change the pattern of China's energy and resource deep processing industries. The difficulty lies in Myanmar's wavering and India's wariness.

The Central Asian region is a major area for securing international land transport alleyway and long-distance trade with Europe and West Asia. Central Asia is relatively less important for near-domain trade and more strategically important for long-distance overland links with Europe alleyway and for control of energy resources in the West Asian countries along the route. The Central Asian region relies mainly on the construction of two land bridges, the New Eurasian Continental Bridge and the China-Mongolia-Russia Economic Corridor. The New Eurasian Continental Bridge focuses on connecting the two ends of the Eurasian continent to form a direct and convenient trade alleyway to Central and Eastern Europe, and to promote the export of high value-added complete sets of mechanical and electrical equipment. The key implementation projects include the promotion of trade facilitation in customs, commercial and sanitary inspection, and investment. The difficulty of which lies in the long distances and the need for cooperation from Kazakhstan and Russia. The strategic significance of the China-Mongolia-Russia Economic Corridor lies in the development of Mongolia's coal and mineral resources as the focus. It aims to improve the freight network and the difficulty of which lies in the different interests of the three countries and the large gap in the understanding of the selection of routes.

The West Asian region is in the early stages of industrialization. With a large population and a small economy, but rich in oil and other mineral resources, and the region's trade focus is not on industrial manufactures but on energy resources. At this stage, China's share of investment and trade in the region is relatively small and its influence is not strong compared to that of European and American countries. Therefore, resource-based and infrastructure-based investments in the region could be enhanced to promote the construction of the China-Iraq-Turkey Economic Corridor. The strategic significance of this alleyway is to stabilize the hydrocarbon resources supply, eliminate the breeding ground for the "three" forces through development, and compete with the major powers for hinterland. Key projects include the development of oil and gas and potash resources, and the construction of overseas Chinese industrial parks. The difficulty lies in the geopolitical complexity along the way, with intense sectarian conflicts and complex state relations

2. Industrial restructuring

In the 14th Five-Year Plan period, the high-quality common BRI faces a profound adjustment of the international and domestic double cycle. The type of national trade shifts from the traditional model of exporting cheap consumer goods to the global market for energy mineral resources to the future model of exporting high-tech products, equipment set and service trade, which inevitably leads to a structural adjustment of China's industrial layout. China's original European and American-oriented global value chain construction has potential safety hazard problems. The BRI through comprehensive opening can not only allow more cities to participate in the international industrial division of labor, but also through the industrial collaboration of different sectors to form a diversified and flexible global value chain, so that more countries can benefit from the global industrial division of labor.

During the 14th Five-Year Plan period, the high-quality development of the BRI continues to promote the development of trade in services with the export of technology, information, services and standards. It continues to drive the transformation of the economic development of the maritime province, especially the traditional industrial advantageous areas of the Pearl River Delta and Yangtze River Delta. It promotes the development of productive services and high-end manufacturing, and accumulate innovation capacity and international functions in core cities such as Beijing, Shanghai, Guangzhou and Shenzhen, which gradually moves from a mere manufacturing base to a global economic hub.

In the context of the BRI, we need to coordinate industrial adjustment based on the international and domestic double cycle. The trade cooperation among countries further shifts from a single pattern of trade in goods to "investment + trade" two-way trade posture. It has increased its outward industrial investment by establishing overseas economic and trade cooperation zones, setting up R&D centers in areas along the BRI route with scientific and technological strength, and building export processing zones, energy and resource production bases, industrial transfer bases and industrial platforms in BRI countries. It also makes the country's western border gateway cities more prominent and prominent in their frontier role in coordinating cross-border cooperation.

2. Financing for high-quality co-production of the BRI in the 14th Five-Year Plan

(i) Subjects of investment and financing

1. Hierarchical and differentiated structure of investment and financing entities

High-quality co-construction of the BRI requires a wide range of investment and financing actors, actively attracting the participation of various international organizations, development agencies and market institutions, and enhancing the transparency and inclusiveness of project finance. According to the source of funds, the main investment and financing agents in the BRI financing structure are divided into the following levels: the first level is the domestic policy banks, which mainly include China Development Bank and the State Export-Import Bank. The second layer is the emerging multilateral development financial institutions, represented by the BRI tailor-made Asian infrastructure investment bank and the Silk Road Fund, both of which are established by international standards, with the Silk Road Fund mainly focusing on equity investment; the third layer is the domestic Commercial Bank, mainly dominated by the four major state-owned Commercial Bank, of which Bank of China and ICBC, with their overseas affiliated agency and mature diversified financing service system, are market share larger in the BRI financing; the fourth layer is the traditional world multilateral financial institutions, with the World Bank and Asian Development Bank closely associated with the BRI; the fifth layer is the auxiliary institutions represented by import and export credit insurance (Bank of China Selected Outstanding Achievements in Industry Development Research Awards, 2019).

According to the financing model, the financing systems of the financing entities differ significantly from each other. There are many domestic Commercial Bank overseas affiliated agency and a wide range of financing products offered, and the financing modes are mainly bank credit (on-balance sheet credit and off-balance sheet credit); syndicated loan; issuance of domestic and overseas bonds; cross-border financial integrated services, etc. The strengths of domestic policy banks are low interest rate and long term loans such as two prime loans and special loans, strategic equity investments through multilateral/bilateral cooperation funds, etc. The emerging multilateral financial institutions represented by the AIIB and the Silk Road Fund are characterized by their specificity to the BRI and provide diversified and innovative financing modes such as credit, bonds, equity investment and insurance for the BRI in accordance with international standards. The traditional world multilateral financial institutions mainly focus on loans, and the interest rate and maturity of loans are favorable to the BRI project. It can be invested and financed jointly with domestic policy banks, Commercial Bank and the Silk Road Fund. The financing auxiliaries are mainly China Credit Insurance, which provides services such as overseas investment guarantee and export credit insurance.

In the context of high-quality co-construction of the BRI development, the investment and financing mechanism should fully reflect the close collaboration between countries along the route and international institutions. Governments along the BRI route need not only to fully communicate and coordinate the different legal norms, tax systems, business environments and regulatory frameworks along the route, but also to take a leading role in financing arrangements with coordinated action by the public sector. In particular, governments along the route should actively direct the flow of funds through development strategic planning docking and cooperation in the field of development finance. At the same time, national development financial institutions and multilateral and regional development banks, as well as market players, should also strengthen their cooperation and coordination, bring into play the advantages of multilateral development banks in providing sustainable financing, institutional know-how and intellectual integration services to countries along the routes, and respect and encourage the decisive role played by market mechanisms in the financial resource allocation. In the process of building the BRI project with high quality, there is more parallel financing, tripartite cooperation, multi-party cooperation and other forms of cooperation. For example, the Ministry of Finance of China, together with the World Bank, the Asian Infrastructure investment bank, the New Development Bank, the Asian Development Bank, the European investment bank and the European Bank for Reconstruction and Development, signed the *Understanding on Strengthening Regional Cooperation Under the BRI Initiative Memorandum Entry*, setting up a coordination framework for diversified investment and financing entities.

(Department of international financial cooperation, 2017, May 14).

2. Specific sources of funding for investment and financing

Specifically, in order to form a hierarchical and differentiated structure of investment and financing entities, there are six main potential sources of funding for the construction of the BRI high-quality development project.

(1) Traditional world multilateral financial institutions

The traditional world multilateral financial architecture for the BRI financing is mainly the World Bank and Asian Development Bank. The main lending institution in the World Bank is the IBRD, and to a lesser extent the IDA. Three of these, IBRD for energy and mining, transport, water, sanitation and flood control, account for about half of the Bank's financing and are closely linked to the BRI promotion industry.

Asian Development Bank to support the development of its members in areas such as infrastructure, energy, environmental protection, education and health through the provision of loans, co-financing guarantees, technical support and grants. Asian Development Bank is headquartered in Manila, the capital of the Philippines, with a clear location advantage. South east Asia region is currently one of the best progressed areas for the BRI, south east Asia region for the BRI financing can be hoped for Asian Development Bank. Asian Development Bank Support the development of its members in areas such as infrastructure, energy, environmental protection, education and health, mainly through the provision of loans, co-financing guarantees, technical support and grants. Moreover, Asian Development Bank as a regional development agency, it can cooperate with domestic Commercial Bank in joint secured financing to improve credit enhancement and reduce financing costs.

(2) Emerging multilateral development finance institutions

Emerging multilateral development finance institutions include the Asian Infrastructure investment bank, the BRICS New Development Bank and the Shanghai Cooperation Organization Development Bank. Among other things, the AIIB is committed to promoting the development of infrastructure and other production facility in the Asian region. The advantages of the AIIB are: (1) its specialization in the BRI financing; (2) it can form a complementary relationship with the Asian Development Bank and the World Bank in the Asia-Pacific region to jointly promote infrastructure investment; (3) it can combine with the "Silk Road Fund" and other emerging multilateral financial institutions to maximize the potential of financing through credit, bond issuance, concept shares, public offerings, insurance, aid, and various types of asset securitization and other financial innovations to generate greater effectiveness.

BRICS China Development Bank invests mainly in infrastructure development in BRICS countries and other developing countries, with the founding members being the five BRICS countries, and is committed to developing green finance, with funding sources relying mainly on the issuance of green bonds in addition to BRICS capital funds. The investment areas of interest for the BRICS NDB are currently the BRICS countries, of which Russia, India and China are important nodes of the BRI strategy.

SCO is still actively promoting the preliminary preparations for the SCO Development Bank. The SCO member states and observer states are densely distributed on the important alleyway of the BRI. The SCO provides an important mechanism to support and guarantee the implementation of the BRI strategy, and become an important platform for development planning and mechanism docking in the construction of the Silk Road Economic Belt. The SCO Development Bank primarily finances infrastructure development in SCO member countries. Russia and Kazakhstan are rich in energy resources among the SCO member states, but their development is limited due to transportation conditions and other constraints. The establishment of the SCO Development Bank helps bridge the financial gas of infrastructure development in these countries and promote regional economic cooperation among them.

(3) Domestic policy financial institutions

Domestic policy banks include China Development Bank (CDB) and China Export Bank, especially CDB is the largest foreign investment and financing bank in China. Both provide not only traditional credit, but have also established a Sino-foreign cooperation fund. It is through commercial loans (individual bank credit/syndicated loan), preferential buyer's credit, foreign aid loans, export credit insurance, establishment of country/industry funds, etc., that it serves the low-cost financing support for domestic and foreign enterprises, large-scale projects, etc.

The main sources of financing for policy banks are financial bonds, deposit-taking, and borrowing from government and other financial institutions. Policy banks provide financing for BRI projects in the following modes: (1) providing commercial credit (preferential credit, etc.); (2) setting up investment (cooperation) funds for equity investment, and investing in infrastructure, energy and minerals, high-tech, manufacturing, agriculture and financial cooperation in countries or regions along the BRI route through mergers and acquisitions, holding investments, important minority share holding investments, debt and other forms.

(4) Commercial financial institutions

The main advantage of Commercial Bank participation in the BRI is the extensive overseas affiliated agency and mature experience in going global. The four largest state-owned Commercial Bank dominate the BRI investment and financing, with the Bank of China and Industrial And Commercial Bank of China playing a major role in the BRI financing. Commercial Bank financing is mainly regulated by bank credit, international syndicated loan and bond issuance abroad.

Commercial Bank also actively launch diversified cross-border financial services, providing investment bank, offshore insurance, financial consulting, risk management and other innovative businesses for domestic cross-border investment enterprises.

(5) Dedicated investment funds
The Silk Road Fund is a medium- and long-term development investment fund tailored to the BRI and a medium- and long-term development investment fund around the BRI strategy. It mainly provides financing support for economic and trade cooperation and multilateral connectivity within the BRI framework. The Silk Road Fund carries out investment business in accordance with the principles of marketization, internationalization and specialization. It can provide investment and financing services in various ways, such as equity, debt, fund and loan, etc. It can also initiate and establish joint investment funds with international development institutions, domestic and foreign financial institutions, etc. It conducts asset trustee management and external entrusted investment, etc.

(6) Export credit insurance agencies
China Export Credit surety company is the only state-owned policy surety company that undertakes export credit insurance business in China. With the state fiscal budget arranging its funding sources, it aims to promote domestic exports and facilitate the development of foreign trade and investment. The way credit insurance institutions participate in the BRI is mainly to provide export credit insurance for enterprises' overseas investment, which is equivalent to the "peace of mind" for domestic enterprises to invest in overseas projects. As the focus of the BRI development is mainly on infrastructure, energy and other projects with long construction period, large volume and large capital demand, and some countries investment climate are still unstable, so enterprises need export credit insurance to provide protection. However, since such insurance is risky and difficult to measure, export credit insurance for Chinese enterprises is mainly provided by the government-backed China Export Insurance Credit Corporation.

China Credit Insurance provides export credit insurance services such as medium and long-term export credit insurance, overseas investment insurance, short-term export credit insurance, domestic credit insurance, credit guarantee and reinsurance related to export credit insurance, accounts receivable management, and information consultation and insurance fund utilization. This is dominated by short-term export credit insurance, followed by overseas investment guarantees and medium- and long-term export credit insurance business.

(ii) Investment and financing modalities

In the 14th Five-Year Plan period, in the context of high-quality co-development of the BRI, it is necessary to develop diversified financing methods, not only using traditional financing methods such as development finance and bank credit instruments, but also actively trying to finance through more transparent and efficient capital market, especially bond market. Specifically, for the optimization of capital financing, it is necessary to combine equity financing and debt financing, investment, construction and operation, and to actively try various "benefit-sharing, sharing of risks" investment and financing models.

1. Combination of equity and debt financing, investment, construction and operation

The BRI construction project have multiple financing needs, including loans, equity, bond financing, leveraged lease and aid. The Chinese Government can promote a multi-level financing system for the construction of the BRI by providing loans, equity financing, bond financing, development assistance and other financing arrangements. In terms of lending, this includes concessional loans from multilateral development banks and China Development Bank in addition to the usual Commercial Bank loans (Sun Rong, 2019, June 12).

Syndicated loan and bank credit are the main forms of bank loan business lines. Development banks at the national and multilateral levels have lower interest rates and longer maturities than the traditional Commercial Bank. For example, China Development Bank has set up the BRI special loans for infrastructure, the BRI special loans for production capacity cooperation, the BRI special loans for financial cooperation and other more targeted special loans. The Import-Export Bank of China also has a flexible and diverse range of special loans such as the BRI special loan value degree and the BRI special loan value degree for infrastructure (China Development Bank official website).

Equity investment is another common form of financing. The market is more likely to be private equity investments with shorter investment horizons, but these liquid investment often do not meet the BRI construction project medium to long-term capital needs. In this regard, the Chinese Government has already made some attempts. During the 14th Five-Year Plan, it can continue to explore the establishment of some special medium- and long-term funds, such as relevant industrial funds, regional or country-specific funds and special BRI construction funds. In continuing to promote the quality development of the BRI project, further attempts are needed in these areas to meet the funding needs of the BRI project.

Bond financing is a weak link in the development of high-quality co-financing BRI. Asian countries have now made the advancement of Asia bond market a goal of regional financial cooperation. The Asian Infrastructure Bank's practice of issuing long bond provides lessons for the development of bond market in the region. BRI countries have reached a consensus to vigorously promote the development of local currency bond market and expand the BRI construction project sources of medium- and long-term funding. At the same time, how to attract private capital has been a challenge for infrastructure financing. Strengthening public-private partnerships and leveraging the entry of private investment through public sector funding are important means of supplementing the financing gap for infrastructure development. "In the 14th Five-Year Plan period, the role of government and social capital cooperation in the high-quality construction of the BRI project finance should be actively enhanced to realize the financing pattern of joint participation of public and private funds. The government-social capital cooperation model (PPP) can optimize the cost-sharing and sharing of risks mechanism for cross-border project construction and reasonably avoid various risks. Multilateral development agencies, national development banks and early investments from special national funds should actively play the role of catalytic agent and attract more social capital to participate.

2. Exploring a "benefit-sharing, sharing of risks" investment and financing model

On the one hand, it is necessary to innovate the design of financing structures, explore investment and financing methods that share the benefits and risks, and flexibly arrange the corresponding financing models. Most BRI countries are developing economies, in a different historical development stage and facing constraints in their macroeconomic policies and financial resource allocation. When allocating financial products and instruments, financial institutions need to fully consider local physical condition and innovate the design of financing structures to provide flexible and diverse financing models that are economic benefit oriented and take into account social benefits (Liu Wei, 2017, May 30).

For some countries that are lagging behind in economic development and urgently need to launch relevant projects to break development bottlenecks, it is suitable to adopt a direct and convenient financing model to enhance financing efficiency and realize early returns. There are countries that have made great strides in energy, power and infrastructure after years of economic development. They are in a position to achieve higher levels of financial governance and factor allocation efficiency. In these countries then, the legal and regulatory environment of project finance should be explored to improve, encourage the free flow and moderate competition of various types of social capital in a regulated and orderly market environment, and guide the financial resource allocation to be more refined and professional. International syndicated loan and asset securitization should be fully utilized to diversify and manage risk in a larger market context. Actively develop various types of financial derivatives to help enterprises hedge various exchange rate and interest rate risks in transactions and enhance capital security. When participating in large-scale projects, we actively use international financial organizations' guarantees, export credit insurance, international commercial insurance and other means to mitigate risks. In certain markets with high country risk, risk management channels could be expanded by introducing multilateral investment guarantee agencies.

On the other hand, it is necessary to innovate investment and financing models. It promotes innovation in investment and financing products and systems, and promotes the sustainable development of the BRI financing. First, we should accelerate the layout of financial institutions along the BRI route. Encourage Chinese financial institutions to use the BRI as an axis to plan for global network topology. It further improves the support policy for offshore CNY loans. It studies the formulation of tax incentives. By setting up a "white list", it gives preferential policies to reduce increment tax, differentiated management of the scale of offshore CNY loans, low-cost funding support to banks on major projects by means of collateralized supplementary loans (PSL) and other means to improve pricing competitive power. Second, we should continue to deepen the medium- and long-term investment and financing advantages of China's development and policy financial institutions. Reasonably determine the target of external debt structure regulation and control, and appropriately increase the scale and proportion of medium- and long-term external debt. Simplify the procedures for Chinese enterprises to obtain currency swap funds for overseas projects, and gradually increase the proportion of overseas CNY loans to be used as a pre-set condition to be explicitly proposed in the early stage of project development. Finally, we should explore the establishment of an investment and financing service system covering international syndication, green finance, PPP models, project revenue bonds, asset securitization (ABS) and other market-based methods to meet multi-level financing needs. At the same time, it establishes a good sharing of risks mechanism with the governments of cooperating countries, financial institutions and Chinese and foreign enterprises to achieve long-term, stable, sustainable and risk-controlled financial support for the BRI.

(iii) Investment and financing cooperation

Through the previous eight years of exploration and practice, the financial cooperation mechanism between China and the BRI countries at the national level has basically taken shape, and the overseas layout of Chinese financial institutions has also tended to be perfected. Some new multilateral cooperation institutions have been established to provide financial guarantee for the infrastructure construction and international production capacity cooperation of the BRI countries together with domestic and foreign financial institutions and international organizations. Nonetheless, financial cooperation in the BRI has lagged behind other areas of cooperation. Cross-border financial cooperation that has taken place, whether at the official level or through private action, has been focused on particular industries and regions. For example, China and ten Southeast Asian countries have constructed a regional financial stability framework, and major Chinese Commercial Bank have established affiliated agency in major countries in Southeast Asia and Central and Eastern Europe, while Chinese financial institutions are not yet fully operational in countries along the BRI in West Asia and North Africa. During the 14th Five-Year Plan period, the co-construction of the BRI high-quality development requires further innovation in financing channels, giving full play to the role of Asian infrastructure investment bank, Silk Road Fund and other platforms, deepening financial cooperation, carrying out risk supervision cooperation, and jointly improving the financial support and guarantee system for the construction of the BRI. Specifically, investment and financing cooperation can be deepened in the following five areas.

1. Extending the breadth and depth of cooperation at the national level

In terms of currency swaps, although China has signed currency swap agreements with 22 countries and regions along the route, these countries are mainly concentrated in the Asian region where there are close economic and trade relations with China, while other BRI countries have basically not carried out currency swap cooperation with China. China's currency swap quota is also mostly allocated to developed regions such as Hong Kong, South Korea, etc., which cannot yet meet the urgent demand of many developing countries for CNY. In view of this, China should respond more actively to the relevant demands of other regions along the route and developing countries. On the basis of strict risk assessment, it carries out more extensive and in-depth currency cooperation with more BRI countries and other emerging market countries, in order to increase the weight of the CNY in global foreign exchange reserves and thus expand the influence of the CNY in the international Mombasa-Nairobi Standard Gaugeary system. With regard to the construction of a financial stability framework, China is actively promoting the reform of international institutions such as the IMF Petroleum Fund organization and striving to gain a greater voice for developing countries in international affairs. In this regard, the establishment of a framework for Mombasa-Nairobi Standard Gaugeary cooperation among BRI countries to enhance regional resilience is a pragmatic option, based on the characteristics of developing countries themselves. The Chiang Mai Agreement, despite its good intentions and institutional arrangements, is limited to Southeast Asia and East Asia, while no corresponding safeguard mechanism has been established in the areas along the BRI route and in Latin America, where financial risks are accumulating. In addition, the Chiang Mai Agreement has a dollar swap limit of only US$240 billion, of which only about US$72 billion is freely available for borrowing by member countries, while the rest is subject to financial support from the IMF Petroleum Fund organization. This is a drop in the bucket compared to the trillion dollars available to be borrowed by the IMF Petroleum Fund organization. The existing regional financial stability framework is still far from meeting the needs of the developing countries along the route, and the cooperation between China and the countries along the route needs to be made real and deep.

2. Advancing capital market construction

China should vigorously develop its domestic capital market, develop a wider range of financial products and provide more open and transparent financial services to respond to the investment needs of international markets. At present, China's financial support for the construction of the BRI is mainly through indirect financing channels, while the role of capital market as a direct financing platform has not been fully exploited. Looking to the future, China could consider becoming a BRI financial centre. It takes New York, London and other major international financial centers as reference, and practice "internal skills" in market size, product diversity, market efficiency and regulatory level to promote the development of domestic capital market.

First, Chinese stocks and bond market could be an important source of financing for the BRI construction project. This requires our stock exchanges and interbank bond market to strategically dovetail with the capital market of BRI countries. It promotes open information sharing, and actively pursue the framework design and institutional arrangements for capital market interconnection. By launching quality cross-border financial products in a targeted manner, it actively promotes the inclusion of China's stock and bond indices in international indices, and builds a cross-border risk prevention network. It facilitates more convenient and inexpensive financing for enterprises in China and countries along the route, and enables investors from all countries to share the dividends of infrastructure development and regional economic growth. Second, it actively develops an interconnected bulk commodity trading platform. This is of great benefit to China and the BRI countries as well. The BRI countries have diverse resource endowments, and by strengthening cooperation with countries along the route, China's commodity spot market can be significantly expanded and the participation of cross-border investors greatly enhanced. Therefore, commodity trading platforms should be actively built to create interconnected market conditions and actively develop futures contract that match the bulk commodity characteristics of countries along the route. In addition, it promotes the establishment of regional bond market. It adheres to the concept of green development and integrate green financial elements into the investment of the BRI. By the end of 2018, the Import-Export Bank of China issued CNY 2 billion of green financial bonds to global investors to promote the green development of the BRI construction (The Import-Export Bank of China official website). We should continue to promote the opening up of domestic bond market to the outside world, establish a regional bond market covering BRI countries, issue CNY-denominated bond investment and financing instruments, and provide diversified, long-duration financial support for the construction of large-scale BRI projects. Finally, China also needs to increase cooperation with BRI countries to jointly build an interconnected capital market. On the one hand, China can encourage financial institutions from along-route countries to come to China to participate in secondary market transactions and welcome high-quality along-route crown corporation to come to China for direct financing. On the other hand, China can also participate in the capital market construction of along-route countries through equity participation and technical consultation to better protect investors' rights and interests and facilitate financing for enterprises from various countries in these along-route countries.

3. Building a diversified investment and financing framework

In 2017, China established a BRI PPP working mechanism and signed a cooperation understanding with the United Nations Europe ECE memorandum entry to jointly promote the better application of the PPP model in BRI construction cooperation projects. The PPP model can not only compensate for financial gas problems in project construction, but the introduction of private sector supervision and management mechanisms can improve project construction quality and increase project construction efficiency. Looking at the development trend of building a diversified framework of investment and financing system, in addition to financial platforms such as the AIIB and the Silk Road Fund that can play a greater role in financing infrastructure, it is also possible to issue long bond or set up various innovative financing instruments, especially more private sector funds should be mobilized. The PPP model should be actively promoted to enhance the attractive force of infrastructure projects for private sector investors.

In order to solve the problem of insufficient funding and financing for the BRI connectivity, we cannot rely solely on public finance and international capital to play a role, but must also give full play to the enthusiasm of private and social capital. Public and private sectors should be combined. The potential of governmental funds should be fully exploited, and more than two dozen funds such as China-ASEAN, Eurasia and Central and Eastern Europe, as well as the newly established Silk Road Fund, should be used. Encouraging private capital to participate in the construction of the BRI, it combines public-private partnership (PPP) and build-operate-transfer (BOT) models effectively, enhancing private capital participation by increasing long-term potential returns, encouraging private capital, surety company, social security funds, pension funds and other institutions to participate in the construction of the Silk Road Fund. It guides the joint participation of commercial funds and social funds. Carrying out innovation in financing models, it encourages bond issuance and equity financing. It encourages the government, financial institutions and enterprises to issue medium long bond in various financial markets. It combines the characteristics and risks of the BRI interconnection project, providing diversified and multi-level direct financing tools and financial innovation products that meet the specific requirements of the project, encouraging the combination of investment and lending, and developing a new model of promoting lending through investment.

4. Promoting the internationalization of the CNY

It compared to major international currencies such as the US dollar and the euro. The level of internationalization of the CNY is far from adequate. At the end of 2018, the stock of international bond in circulation denominated in CNY was only US$107.5 billion, well below the stock of US$11.3 trillion in bonds and US$9.3 trillion in eurobonds (Wu Shuyu and Hu Biling, 2019). Trade and investment transactions between China and BRI countries are still mostly settled in the US dollar and other third party currencies. Promoting the widespread use of the CNY in cross-border trade and financial transactions in markets along the BRI route does not only help enterprises in China and along the route to reduce transaction costs, but also make investment and financing more convenient for producers and force domestic financial reform and financial opening.

The internationalization of the CNY and the co-construction of the BRI are complementary to each other. Expanding the use of the CNY in cross-border settlements reduces trade costs and makes investment activities more convenient. The people of the region benefits from smoother trade. In addition, increasing the share of CNY in international foreign exchange reserves can also reduce the dependence of BRI countries on the US dollar, thus reducing regional financial risks. In order to enhance the level of CNY internationalization, at the national level, central banks can increase Mombasa-Nairobi Standard Gaugeary cooperation with BRI countries, expand the scale of currency swaps, and work together to build regional financial safety barrier networks. At the institutional level, China needs to strongly support interconnectivity with countries along the route Commercial Bank and promote the construction of offshore CNY markets. By encouraging China Commercial Bank to "go global" and establish affiliated agency overseas, the accessibility of the CNY is significantly enhanced, creating convenience for importers and exporters, investors and residents of China and countries along the route. In addition, China needs to develop diversified currency derivatives targeted at the needs of enterprises in different countries and regions to hedge corporate currency risks and promote the use of CNY as a settlement and investment currency by enterprises and investors in more countries.

(iv) Rules for investment and financing

In the construction of the BRI investment and financing rules system, full consideration should be given to the different needs of the cooperating countries (regions) in terms of their levels of economic and financial development and degrees of marketization. It aims to enhance the long-term compatibility of the investment and financing cooperation mechanism, to coordinate the collaboration in the fields of financial institutions, financial markets and financial supervision, to promote the three-dimensional layout of financial institutions and financial services of various countries, and to guide the formation of a convenient information disclosure and sharing mechanism, so as to provide strong support for the bilateral and multilateral economic and trade exchanges and economic and social development of the cooperating countries (regions) (BRI Economic Development Report, 2019).

First, it actively promotes the construction and improvement of financial market infrastructure and guides the coordinated and orderly development of diversified financial institutions. The current level of financial development of BRI countries varies, and the financial infrastructure of many countries is very weak, which weakens the international flow of funds and service efficiency. Therefore, strengthening the construction of financial infrastructure such as international currency payments, clearing and currency swaps can effectively reduce transaction costs, improve capital service efficiency, guarantee capital security and expand capital financing. This is both an important area of financial cooperation for the participating countries and an inherent requirement of the principle of openness. In the construction of the BRI investment and financing rules system, we promote the formation of a networked development pattern among banks, surety company, securities companies, credit rating companies, accounting firms and consulting firms in the countries along the route with complementary advantages, mutual coordination and active expansion. At the same time, it effectively coordinates the coordination and cooperation of policy and commercial finance, communicates and cooperates on cooperation plans, interest rate arrangements and guarantee methods involving projects along the route. It provides strong financial support for project construction. In addition, the role of investment funds in the construction of the BRI can be actively brought into play. Industrial funds co-sponsored by relevant countries can be set up according to specific investment projects to guide the inflow of funds and provide support for the implementation of projects.

In addition, it promotes the disclosure and sharing of the BRI construction project information, ensures consistency of rules, strengthens financial regulatory cooperation and works to avoid financial regulatory arbitrage. With reference to the common practice of the World Bank, IMF Petroleum Fund organizations, Asian Development Bank and other international institutions in providing project credit, formulate minimum information disclosure standards, frameworks or directories for the BRI construction project investment and financing, etc. It makes timely disclosure of relevant business information, research and data pool information and financial information, and builds a platform for the publication and exchange of information, effectively address the problems of misunderstanding, misinterpretation and asymmetry of information, and fully realizes the sharing of information, so as to promote the formation of a normative system for information disclosure and sharing that is open and transparent, with both outlines and objectives, clear in its hierarchy, easy to operate and fairly implemented. Efforts should be made to promote the docking of the financial regulatory systems of BRI countries. It actively promotes cooperation among the financial regulatory agencies of various countries to jointly address possible risks and hazards. At the same time, financial regulatory arbitrage makes financial market transactions and activities more hidden, volatile and fickle, reinforcing information asymmetries and increasing systematic risk. In this regard, it is actively advocating for all participants to jointly promote the construction of consistency in financial rules, establishing and implementing an accountability system, an inspection system and a social monitoring mechanism, enhancing coordination among various rules, avoiding the improper use of discretionary powers by regulators and eliminating rent-seeking behaviour, so as to achieve the construction of a fair system of financial rules.

References

"Belt and Road Initiative" to develop digital economy to light up the road of innovation (2017, May 22), http://www.scio.gov.cn/31773/35507/35510/35524/Document/1552 823/1552 823.htm.

China Development Bank. (2019). UNDP converging investment and financing rules for BRI Sustainable Development, BRI Economic Development Report.

China Development Bank Institute. (2019). "Policy reflections on the BRI investment and financing models and cooperation mechanisms", Bank of China Selected Outstanding Achievements in Industry Development Research Awards.

Department of International Financial Cooperation. (2017). The Ministry of Finance and six multilateral development banks jointly signed the memorandum of understanding on "the Belt and Road" cooperation. (2017, May 14). http://www.mof.gov.cn/zhengwuxi nxi/cai zhen gxin wen/201 705/t20170 514_2600 067.htm.

Liu Wei. (2017). The "the Belt and Road" diversified financing mechanism provides solutions for infrastructure financing. (2017, May 30). http://m.canka oxiaoxi.com/column/20170530/2059266.shtml.

Sun Rong. (2019). He Dexu: How to build a diversified investment and financing system for the "the Belt and Road". (2019, June 12). https://www.financialnews.com.cn/zgjrj/201906/t20190612_161668.html.

Wu Shuyu, Hu Biling. (2019). "China Foreign Exchange, BRI financial cooperation: Status, assessment and improvement", China Foreign Exchange, Issue 10.2019. http://www.chinaforex.com.cn/index.php/cms/item-view-id-48478.shtml.

IX

Risk Control and Project Management of Co-constructing the BRI

Introduction

During the 14th Five-Year Plan period, the BRI faces many potential risk factors in the economic, social, political and financial spheres. These potential risks are partly due to uncertainties arising from the business environment, resource environment, cultural differences and political situation of the BRI countries. It partly due to the complex and volatile nature of the international trade environment brought about by the current global economic slowdown and the rise of trade protectionism. In the process of promoting the further high-quality development of the BRI, and in the face of such a complex and volatile international trade situation, countries along the route need to always uphold the concept of global governance based on mutual consultation, construction and sharing. Constantly, it promotes the construction and deepening of inter-country trade security dialogue and cooperation mechanisms, and strives to build a coordination mechanism for the policy system and governance among countries along the route. The participating countries need to adhere to the correct concept of righteousness and benefit, put righteousness first and benefit together, flexibly use various financial and reciprocal tools, negotiate and build various policy cooperation frameworks, and continuously promote a high level of openness and innovation in economic and trade activities. Under the BRI framework, we work together to promote a comprehensive and coordinated development of the economy, culture and environment in each of the countries along the route, with national leadership, social response and public participation.

At the same time, the achievement of the goal of high-quality co-construction of the BRI is ultimately dependent on the successful implementation of various cooperation projects under the framework of the BRI. To optimize the management of the BRI projects, it is necessary to avoid the problem that traditional project planning may be disconnected from the current situation of local economic development and future planning. It needs to prepare general planning and action plans for cooperation projects in a scientific and reasonable manner by the relevant participating countries. It aims to establish a special BRI organization and coordination body with the joint participation of countries along the route from the perspective of promoting project implementation, supervision and management. It apt to adhere to the principles of cooperation and construction of equal consultation. It shares responsibility and common benefit. It establishes and improves various project risk management and benefit coordination mechanisms. It attempts to establish a dynamic assessment mechanism for the implementation of planning and to ensure that cooperation projects are implemented in accordance with the planning system. It pays attention to the ecological environment and climate effect of relevant projects, and to establish a graded and classified management system for projects environmental impact. It tries to establish an information reserve platform for major projects, to pool resources to give appropriate guidance and support to the classification and grading of projects. It aims to establish a project in construction information sharing platform and a quality surveillance platform to provide a reliable grasp for tracking and assessing the progress of the implementation of relevant projects.

1. Strengthening risk control in the BRI

(i) Potential risk factors for co-constructing the BRI

1. Economic risk

At the macroeconomic level, the current global economic growth is sluggish, trade protectionism is on the rise and financial risks remain high. As our current BRI project is dominated by basic investment, with large capital investment and longer processing cycle, it is vulnerable to economic fluctuations. Against the backdrop of sluggish global economic growth, some of the countries along the route have experienced greater economic volatility and a clear downward trend in their economies, with many countries also experiencing significant currency depreciation. The macroeconomic declines in these countries pose a challenge to the sustainability of our overseas investment and also add to the financing pressures facing the country. Over the past eight years, most of the landmark projects of the BRI have been infrastructure projects, which are heavily financed and require continuous follow-up funding. Therefore, there is an urgent need to broaden financing channels, further diversify financing methods and vigorously promote cooperation and synergy in financing among governments.

At the level of the business environment of the countries along the route, the overall international business environment has deteriorated, and the business risks of enterprises participating in the BRI construction project have increased. The global spread of the pandemic has triggered a global adjustment in fiscal Mombasa-Nairobi Standard Gaugeary policy and trade policies. The business environment of some countries along the route has deteriorated, and the risks of enterprises operating overseas have increased (Office of the Leading Group for the Promotion of BRI Building the BRI, 2019). (Joint "Belt and Road Initiative": progress, contribution and prospect, 2019, April 22) This is reflected in the following four aspects: (1) With the rise of counter-globalization, populism and protectionism, some countries have strengthened their censorship restrictions on foreign investment and foreign enterprises. In the past few years, the gap between the rich and the poor has been widening in some countries, counter-globalization, populism and protectionism are prevalent, populist forces have come to power in a few countries and passed laws that are unfavorable to foreign enterprise, and the risk of Chinese enterprises operating in the relevant countries has increased significantly. (2) The complex and volatile international political situation, coupled with frequent domestic political turmoil or regime change in the countries concerned, has led to less stable policies on the use of foreign investment in some countries. Some of the BRI countries have faced internal and external problems in the past few years, with domestic political turmoil and war risk, leading to greater uncertainty in their business environment and greater risks to the investment and operation of Chinese enterprises in the relevant countries. (3) The laws and rules of some countries are not in line with international norms. There is excessive regulation of foreign capital and excessive protection of domestic capital. For example, in terms of market access standards, hidden thresholds and excessive barriers to investment reviews have greatly increased the cost of entry for foreign firms. The quality criteria of imported products changes frequently, and the opinions of foreign enterprises are less taken into account in the formulation of standards, which increases the difficulty and cost of production adjustment by foreign enterprises. (4) Some countries have a certain bias against China due to competition and ideological differences. It retains a wariness and defensive mentality towards China's rising national power, so there are cases of preferentially enforcement against our enterprises.

2. Social risk

The regions along the BRI route have huge social and cultural differences, complicated historical problems, very different languages and customs, and sharp ethnic and religious conflicts. At the same time, the smearing and alienation of our country by a few countries has not ceased. In such a complex cultural background, the construction of the BRI should avoid falling into the trap of clash of civilizations and clash of values. China needs to avoid misunderstandings among countries along the route due to differences in cultural values in its foreign propaganda, leading to alienation and misinterpretation of China's culture and image.

Environmental impact is one of the important social risks in promoting quality co-construction of the BRI. In terms of the impact of economic development on the environment, most of the regions along the BRI route are economically backward developing countries, and the current urbanization process is generally accelerating. In terms of development methods, many countries rely more on the exploitation of water, oil and gas, and mineral resources exploitation. In the process of exploitation, the protection of the environment is insufficient, which seriously affects the sustainable development of the region. Power and energy cooperation and investment is one of the current priority areas of the BRI. However, in recent years there has been a growing global consensus to tighten controls on greenhouse gas emissions, a growing call to address the growing urgency of climate change, and an increasing number of voices from the BRI host country of "sustainable energy investments". Against this backdrop, making coal power investments overseas is facing increasing controversy and concern.

Some large BRI construction project faced greater social resistance in their implementation. The infrastructure projects implemented by Chinese enterprises overseas are "multiple points, long lines and wide areas". In the process of implementation, they need to solve the problems of land acquisition and resettlement, establish good communication mechanisms with local people, NGOs, mainstream media and tribal and religious forces. It properly deals with the disputes over interests arising from the implementation of the projects, which is a difficult problem to be overcome during the construction of the BRI. In the course of the construction of the BRI, there have been cases of threats and losses to the lives and property of our enterprises and our citizens. Terrorist activities are rampant in some countries. Hijackings and attacks targeting our citizens and the assets of Chinese enterprises have occurred from time to time. The deteriorating internal security situation in some countries is a serious threat to our overseas projects and personnel safety.

The social risks associated with religious and cultural differences are risks that companies must consider when entering the BRI business. Many countries on the BRI have their own cultural and religious characteristics, unlike those of Europe and the United States. Of these, it is particularly noteworthy that many of the countries, both in the Southeast Asian and Central Asian directions, are Islamic. Some of these countries are also more conservative and have a considerable number of cultural taboos unfamiliar to Chinese companies. Due to the problem of sectarianism and sectarian conflict in Islam, there are often significant differences between different Islamic countries. The experience of Chinese companies in some countries is often not adapted to the countries where they are located.

3. Political risk

First of all, the domestic political conflicts of the BRI countries are a destabilizing factor for the high-quality co-construction of the BRI. BRI countries have different political systems. Some of them have extremely uneven economic and social development, serious domestic political differences and are in the process of institutional transition, which inevitably brings various political risks. In recent years, some BRI countries have been in constant political turmoil, with frequent changes in ruling parties and repeated adjustments in foreign policies. It is necessary to strengthen the political risk assessment of such countries, and to make a full assessment and follow-up study of the ethnic, religious and partisan conflicts in their countries.

Second, the issue of strategic checks and balances, and coordination of interests among the major countries along the route. The area along the BRI has traditionally been a place where the interest of great powers compete. Some major countries have put forward their own strategic visions for the areas along the BRI route according to their own geopolitical conditions, such as the "New Silk Road Plan" and the "Indo-Pacific Economic Corridor" proposed by the United States, the "Eurasian Economic Union" proposed by Russia, the "Spice Road Plan" and the "Monsoon Plan" proposed by India, etc. The strategic visions of the above-mentioned powers all place themselves as leading players in shaping the regional order, intending to strengthen ties with neighboring countries and increase their leadership in regional affairs. The contents of these strategic visions are objectively competitive with the construction of the BRI. The promotion of high-quality co-construction of the BRI requires proper handling of the issues of strategic checks, balances and coordination of interests among the major countries along the route.

Lastly, the "three forces" are vigilant in their attempts to erode the security of our borders through the construction of the BRI. Nine provinces and regions in China share borders with BRI countries, which used to be the backbone of opening up to the outside world but are now the forefront of the construction of the BRI route. The stability of the borders is crucial to the smooth progress of the high-quality joint construction of the BRI route. In recent years, China's border security situation has been improving. Cases of violence and terrorism have dropped significantly, but the three forces of separatism, extremism and terrorism have been unusually active in China's neighboring countries in recent years due to the world's political environmental impact. There are still greater uncertainties and risks in the border security situation, so while comprehensively promoting the opening of border areas to the outside world, we must remain highly alert to the infiltration activities of the "three forces".

4. Financial Risks

The developing BRI countries are predominantly economically backward in general, capital market lagging behind in development, with a single industrial structure, and the domestic economic situation is seriously affected by international resources and energy price fluctuation. Financial risks that exist at the macro level include, first, interest rate risk in project finance. In the process of using the PPP/BOT model for project finance, the value of the project is reduced or revenue is lost due to changes in interest rates, either directly or indirectly. The second is exchange rate risk. Most of the BRI countries belong to countries with high exchange risk grade, poor international liquidity of currencies, high exchange risk and lack of exchange rate risk hedging tools. Third, the risk of non-availability and non-transferability of foreign exchange. Difficulties in the smooth conversion of local currency into the required foreign currency and its shipment out of the country to service external debt and other external payments. Fourth, currency inflation risk. Currency inflation problems are common in BRI countries, such as Ukraine, Iran and Turkey currency inflation rates as high as 11.5%, 11.2% and 10.1% respectively in April 2017 (BRI Economic Development Report, 2019).

At the level of the debt structure of the countries along the route, some of the countries along the route have high debt levels, which are prone to trigger emerging market debt default risk. The January 2020 issue of the *World Bank's Global Economic Prospects* report shows that government debt as a share of GDP has risen from 30% to 50% in low-income countries over the past five years. They have to spend an increasing amount of government revenue on interest payments. (Zhou Wuying, 2019, January 11)

If borrowing costs increase at the expected levels over the next few years, the pressure on debt service in low-income countries only increase further. If financing conditions are suddenly tightened, low-income countries may face the dilemma of fight of capital, "borrowing to pay off old debts". Objectively speaking, debt risk is already one of the key challenges in the construction of the BRI. During the second BRI Summit, the Chinese Ministry of Finance released the *BRI Debt Sustainability Analysis Framework*, (BRI Debt Sustainability Analysis Framework, 2019, April 30) indicating that the Chinese government is aware of the potential debt risks associated with the construction of the BRI. As an international cooperation initiative with infrastructure construction as its core element, the BRI is characterized by a long construction cycle. It is not impossible for developing countries to run the risk of debt repayment in the process. It is also worth paying attention to the debt problems that may arise for Chinese enterprises in overseas construction.

(ii) Measures to prevent the risk of co-constructing the BRI

1. Economic risk prevention

(1) Promote BRI financial innovation. (1) We need to deepen innovation in financial products. Innovative financing products based on the characteristics and risks of infrastructure investment projects, such as securitization of underlying proceeds, development loans, large equipment assets, etc. Development of investment insurance products with different risks, maturities, premiums and premiums according to the characteristics of the BRI investment activities. Research and development of corporate overseas assets value preservation and risk hedging financial products and derivative products, and provision of financial advisory, investment, risk management and other financial services. (2) It promotes innovation in fund settlement business, and provides multi-channel and multi-product services for enterprises in investment and financing management on the basis of traditional settlement products such as payment and collection and account management. (3) It promotes the construction of new financial markets and strengthens the construction of electronic network platform in investment and financing, and makes good use of "big data" and "cloud technology"

to provide a platform for sharing information resources for suppliers, investors, commission merchant and other participants on supply chain. (4) Optimize the governance system of financial institutions. It promotes the establishment of a mutually trusting and inclusive governance structure for financial institutions, distinguishes between effective and absolute demand, and conducts proper risk assessment. (5) The BRI financial innovation to the rule of law throughout the whole process attaches great importance to legal risk prevention. It gives full play to the fundamental role of rules and the rule of law in building consensus, balancing interests, safeguarding development and resolving disputes, building a multi-dimensional long-term mechanism in a more comprehensive and systematic manner at the level of the rule of law, and establishing a stable, sustainable and risk-controlled system of institutional safeguards for the BRI financial innovation.

(2) Establish a mechanism for coordinating macroeconomic policies in countries along the route. The construction of the BRI provides a new platform and channel for strengthening macroeconomic policy coordination and communication among participating countries. The platform is used to form policy orientations that are convergent, synergistic and conducive to the development of the world economy. Countries along the route should take this opportunity to promote global macroeconomic policy coordination and communication. It should integrate reform policies in terms of finance, currency, employment, industry and structure. It reduces uncertainty, discontinuity and unevenness in the policies of the countries concerned and minimize negative policy spillover effects in order to support sustainable global economic growth and address potential risks.

2. Social risk prevention

The problems encountered in development can be solved only through development. To crack the risks and challenges in the process of building the BRI high-quality development, it is necessary to strengthen coordination, establish a mechanism of communication and collaboration among the central government, local communities, enterprises and society, and form a "chess" pattern with a clear division of responsibilities, each with its own strengths and leveraging each other's strengths.

(1) At the governmental level, the central government, with regional economic cooperation as the core, continues to highlight the role of infrastructure development as a bridge in the high-quality co-building of the BRI. It plays the pioneering function of humanistic openness, deepens

the consensus on development concepts and values with countries along the route. It promotes the construction of a sense of a human community of a shared future. We continue to strengthen security cooperation with countries along the route, enhancing the sharing of intelligence information and strengthen security control at border crossings. Local governments at all levels should accelerate the construction of government public service platforms and improve the policy and regulatory system. It increases the collection of market information, improves the transparency of information, and provides good information supporting services to help enterprises fully understand the overseas investment risk sources. It discovers risk points in advance, fabricates a timely prevention and control network, and high limit reduces or avoids negative factors.

(2) At the enterprise level, it should strengthen technological research and development, cultivate the core competitive power of the enterprise, and realize the transformation and upgrading of the enterprise in the process of co-constructing the BRI high-quality development. In business thinking, China's overseas enterprises need to attach great importance to the construction of corporate image, in addition to the work of the local government, comprehensive consideration host country of the various political forces, ethnic and racial, religious beliefs, non-governmental organizations influence, and establish extensive contact and normal communication mechanism with them. In the course of business, it respects host country religious beliefs, customs and traditions, consciously protects labor legal interests and the local environment, and reduces the risk of business operation. In terms of talent pool, China's overseas enterprises should focus on training a professional risk management talent team to avoid losses due to unfamiliarity with the political, legal and social culture of host country. In terms of safety and security, our overseas enterprises need to strengthen their own security and protection capabilities. On the one hand, in accordance with the requirements of the National Development and Reform Commission, the State-owned Assets Supervision and Administration Commission and other relevant ministries and commissions, we improve the compliance management system, strengthen compliance training and enhance the compliance awareness and capability of enterprises. On the other hand, it organizes the establishment of overseas private chambers of commerce or overseas branches of industry associations to build a platform for information and resource sharing, save enterprises' costs and efforts, and enhance their

ability to defend rights overseas. In some politically unstable countries along the route, our enterprises can explore a new model of overseas security that suits the conditions of the host country, and equip them with the necessary protective facilities to effectively protect the safety of their personnel and property abroad.

(3) At the public level, as the construction of the BRI develops in the direction of refinement, the bottleneck of insufficient talent pool facing China is further highlighted. In the process of building the BRI together with high quality, China needs technical talents who are engaged in standardization research and standard setting, management talents who are familiar with the language and culture, laws and regulations, and investment policies of the countries along the route, as well as consultant talents who can assess risks and formulate strategies. As the most important link in the training of national human resources, universities can offer courses, disciplines or specialties related to the BRI to train complex and high-level talents. As an important organization that advises the country, think tanks can take advantage of the platform to enhance early warning on potential risks in countries along the route and form high-quality research results to provide services for the country and enterprises. Civil interaction is an important channel for spreading excellent Chinese traditional culture and building a positive and active image of China. As a "heart-to-heart" project, the BRI high-quality development can be promoted in culture, education, scientific research, tourism, sports, overseas Chinese affairs and other aspects of work, forming a synergy to enhance our soft power and national image through civil contacts.

3. Political risk prevention

(1) Decision maker needs to have a sense of developed and strengthened geo-security, and routine geopolitical risk assessments should be conducted when approving major overseas projects. There is an urgent need for policymaking authorities to become more aware of geo-security and to set specific criteria for constituting geopolitical risks as a reference for establishing projects when approving major overseas projects. The specific geopolitical criteria include political, economic and cultural aspects. (1) Consider if the project triggers geopolitical conflicts between major powers, with particular attention to whether the critical infrastructure construction project is in the middle of a fierce game between maritime and land power forces and whether it significantly alters the established geopolitical interest structure. (2) Consider if the project triggers big scale

geo-economic competition, with particular attention to whether it significantly detracts from the economic advantage of major developed economies such as the United States and Europe, leading to containment and retaliation policies against China by these developed economies to maintain their dominant global economic position, and whether it will significantly weaken competitive power and employment in other developing countries, thereby affecting solidarity among developing countries. Again, consider if the project triggers big scale a clash of religions, civilizations, and ethnicities, and whether it leads to a rise in non-traditional security threats. With regard to the identification of geopolitical risks, there should be a lead department and a joint interdepartmental mechanism to formulate effective strategies and tactics for the prevention and control of geopolitical risks on the basis of an accurate understanding of geopolitical risks and their generation mechanisms, to assess geopolitical risks in advance, and to be able to creatively rebalance the structure of geopolitical interests and strive to achieve a new balance in the dynamics of the interests of all parties, including the interests of major powers.

(2) Establishment of a mechanism for security dialogue and cooperation among the BRI countries. The fierce geopolitical and military conflicts in some areas along the BRI route, the serious problem of terrorism and the rampant separatist forces pose a threat to the security and stability of the BRI. Countries along the route should jointly establish a security dialogue and cooperation mechanism, form an institutionalized common dialog box framework, communicate and consult on key security issues that affect the high-quality co-construction of the BRI, and, on the premise of fully respecting the interests of all parties, gather the greatest consensus of all parties and gradually expand the point of convergence of interests. Countries along the route should jointly establish a regular security cooperation mechanism, through regular or irregular joint anti-terrorism exercises, maritime cooperation patrols, major disaster rescue drills, regional peacekeeping exercises, network security emergency drills, etc., to further deepen mutual trust among countries, constantly strengthen the level of security cooperation and establish an effective security force to safeguard the construction of the BRI.

4. Financial risk prevention

(1) Make full use of existing mature financial instruments to prevent and mitigate financial risks. (1) It is to prevent and control interest rate risk. We protect against interest rate fluctuations through loan guarantees from fixed interest rate and interest rate guarantees from the government, use a combination of currencies to settle project expenditures or revenues to prevent shocks to project investments

from interest rate changes, and use hedging techniques such as capping, interest rate banding and underwriting to reduce the impact of interest rate changes. (2) It is to avoid exchange rate risk. In addition to stipulating in the PPP/BOT royalty right agreement the respective responsibilities of the project company and the host country government for exchange rate risk, financial instruments such as swaps and forwards are fully utilized to protect against the risk of exchange rate fluctuations. (3) Offset the risk of non-availability and non-transferability of foreign exchange. The risk is mitigated by requiring the government of the country where the project is located to provide a foreign exchange free conversion guarantee and by contractually agreeing to settle part of the amount due in local currency and part in foreign currency. (4) Control currency inflation risks. Provide for corresponding provisions in the royalty right agreement as a basis for subsequent verification of prices, followed by price adjustments at accepted currency inflation rates, or corresponding increases in tariffs, or extensions of the concession period, and provide for gradual price increases in the product purchase agreement to guard against the investment and financing risks associated with currency inflation.

(2) Co-construct a high-quality financing policy framework for the BRI, and work with the countries concerned to implement the *BRI Financing Governing Principle* (BRI Financing Governing Principle, 2017, May 15). Most of the projects along the BRI route adopt the project finance model, which requires higher requirements to accurately assess the economic benefit of the project, monitor the implementation of the project throughout, and effectively control legal risks. The key to effectively mitigating and avoiding legal risks is to explore ways of intergovernmental cooperation and concessions: (1) Promote the signing of relevant intergovernmental agreements at the macro level. Promote the signing of intergovernmental agreements on individual projects between our Government and the host country Government and royalty right agreements that are conducive to safeguarding the interests of the parties concerned, and actively seek sovereign guarantees from the BRI countries. (2) At the middle level, it promotes the amendment of laws or the enactment of new acts in the countries where the projects are located. Where there are prohibitions in the law, or where the progress of the project involves changes in the law, the amendment of legal provisions should be promoted and the adoption of a new bill specific to the project should be taken as the project construction condition precedent. (3) The choice of investment cooperation models for specific projects at the micro level is based on the legal risk profile. In countries with better legal policies, the "public-private partnership" (PPP) or "infrastructure royalty right" (BOT) models may be considered,

while in countries and regions with high legal risks, the "design, purchase and construction" (EPC) model should be carefully chosen, and construction work should be carried out in strict accordance with the payment schedule.

(3) Properly guard against debt risks and respond positively to international public opinion. First, domestic financial institutions, countries along the route and international institutions should be encouraged to support the use of the *BRI Debt Sustainability Analysis Framework* (2019, April 30) to improve the science of investment and financing decisions and strengthen debt management capacity. Help countries along the route to improve project risk assessment and economic risk resistance, focus on summarizing successful experiences and cases, and respond positively to international public opinion in a reasonable and measured manner. Integrate the sovereign debt security of the BRI countries with the CNY, and develop international capacity cooperation around enhancing the quality development of the host country economy and debt sustainability.

Second, take the lead in designing debt relief crash program. For the BRI countries with the greatest exposure to Chinese debt, such as Djibouti, the Republic of Congo, Laos, Kyrgyzstan and Angola, China should take the lead in designing debt relief crash program in addition to implementing the *Debt Service Moratorium Initiative* under the G20 Initiative. China should give further concessions in terms of debt or debt relief to provide more fiscal space for these countries to fight the Covid-19 pandemic and set clear terms for the use of these funds to support green recovery. For countries with manageable debt risks, China should coordinate with multilateral organizations such as the World Mombasa-Nairobi Standard Gaugeary Fund and the G20 to debt restructuring avoid debt defaults and unsustainable debt, and to build consensus among multiple investors. China should clarify the role of Chinese policy banks such as China Development Bank and The Import-Export Bank of China in debt restructuring and relief.

Finally, standards for official loans abroad should be developed to improve debt governance. During the second BRI International Cooperation Summit to be held in 2019, the Ministry of Finance released the *BRI Debt Sustainability Analysis Framework* (2019, April 30), a debt sustainability analysis tool developed in the context of the actual situation of the BRI countries, which is encouraged to be used on a voluntary basis by financial institutions and international institutions in China and co-builders of the BRI countries. During the 14th Five-Year Plan period, domestic debt transparency should be strengthened and uniform norms should be established for major Chinese official lending institutions such as China Development Bank and The Import-Export Bank of China; while in the international community, multilateral lending mechanisms should be adopted, for example, through the AIIB to increase lending. At the same time, new debt mechanisms, such as debt-for-natural-resource swaps, should be developed and applied, but with a view to avoiding, as far as possible, the creation of disputed debts as a result of resource swaps.

2. Optimizing project management in the BRI

(i) Preparation of project planning and action programmes

The economic development stage and socio-economic and cultural conditions of the BRI related countries are very different from those of the early years of China's development. The conditions in China at that time required that "to get rich, first build roads", because the socioeconomic and cultural conditions in China supported the construction of infrastructure projects that would make it possible to get rich with roads. However, the socioeconomic culture of many BRI countries may not be supportive.

In addition, some of the current investments planned project in the BRI have an objective tendency to involve construction companies wishing to build larger scale projects. For participating construction companies, big scale project decisions are fixed with handling cost the larger the scale, the higher the relative profitability, so it is an objective and realistic condition that the contractor wants a larger project scale. At the same time, although the BRI countries have a strong desire to develop their own economies, most have relatively limited capacity for economic planning and decision-making. The combination of internal and external factors may objectively lead to a state where the project negotiated and designed by both parties exceeds the actual needs of the particular country in the particular economy development stage, resulting in a state where the project does not contribute to the sustainable development of the local economy.

In the 14th Five-Year Plan period, in order to build BRI projects with high quality and avoid the above-mentioned problems of disconnect between traditional project planning and the current situation and future of local economic development, the BRI related projects general planning and special planning should be based on scientific and reasonable feasibility studies and on the basis of the economic development of the BRI related participating countries general planning. It adheres to the people-centred development ideology and follow the path of coordinated economic, social and environmental development. At present, the project-centered and big scale investment-oriented project decision-making mechanism objectively needs to be adjusted, and efforts should be made to build a "people-centered" BRI project general planning and special planning system.

At the press conference of the second BRI International Cooperation Summit, President Xi Jinping (2019) proposed "practicing the concept of high standards, benefiting people's livelihood and sustainability, actively connecting to universally accepted international rules and standards, adhering to the people-centred development ideology, and taking the path of coordinated economic, social and environmental development". A major adjustment needed to implement these concepts and standards is the preparation of the BRI project general planning and special plans, which should be transformed from purely project-centric planning in the past to a project planning process and system based on the economic development general planning of the target country. The first task of the BRI to achieve the goal of a human community of a shared future and high-quality co-construction of various projects should be to move away from mere project planning and replace it with project general planning and special planning in the context of economic development. Even if it is an infrastructure or assistance project, it should be considered and planned in the light of the possible needs and capacities of the local economy of the host country development stage and, in particular, the possibility of integrating the implemented project into the process of sustainable and stable economic development of the country.

Promoting the construction of the BRI in an orderly manner, it is necessary to plan first, and the countries along the route should do their best to quick start the BRI the construction of general planning the preparation of cooperation. Planning should adhere to the preparation of common ground while reserving differences. It fully respects the views and suggestions of countries along the route, and resolves differences of opinion through communication and coordination; adheres to clear authority and responsibility, clear division of tasks among countries, and major investment construction project to clarify construction tasks, construction subjects, construction duration, and investment methods; adheres to orderly promotion, and take the lead in promoting a number of key alleyway, key nodes, and key areas of major projects to form demonstration effects and global momentum. According to the BRI construction general planning, countries along the route should jointly formulate special plans in the fields of transportation, energy, finance, science and technology, culture, etc., enriching the planning level and refine the planning content. And furthermore, according to the BRI general planning and special plans, countries along the route jointly carry out the preparation of action programmes and clarify the timetable and roadmap for the implementation of the relevant plans.

(ii) Optimizing project team coordination mechanisms

1. Clarification of project inclusion criteria

The BRI, as a China-centric narrative system, upholds the inclusiveness and malleability of Chinese culture. This gives a great deal of flexibility to the BRI cooperation and makes it very difficult to strictly define whether a project can be managed under the BRI. At the operational level, the key points that distinguish the high-quality co-construction of the BRI project from the sundry item are Chinese participation, international cooperation and connectivity. Specifically, we believe that the following four points can be used to determine the criteria for the inclusion of quality co-production BRI projects.

First, high-quality co-construction of the BRI projects should involve two or more countries or regions, including China, and have the deep participation of Chinese enterprises or institutions. The project sponsor, investor, contractor, counterparty or business and management party should be from two or more countries or regions and the parties have reached consensus on cooperation on a voluntary and lawful basis. Considering that for a certain period of time, the BRI is still "led" by China, the deep involvement of Chinese enterprises or institutions is appropriate. Accordingly, the BRI project can be a "going out" project or a "coming in" project; it can be a Chinese led project or a project led by a foreign party with Chinese participation. In terms of location, with the exception of HR development project, the project should be physically located along the BRI route or in the territory of the partner, or be a component of a cross-border project in China. Most of the projects should be located in countries along the Silk Road Economic Belt and the 21st Century Maritime Silk Road, with a small number of representative projects in partner countries around the globe.

Secondly, the implementation of the high-quality BRI project should be conducive to promoting connectivity among countries (regions) along the route. The construction content of the relevant projects should be in line with the general direction of policy communication, infrastructure connectivity, smooth trade, financial integration, and civic exchanges. It helps to promote the connection and interconnection among countries along the route in terms of infrastructure, industrial systems, market rules and social public. In this regard, the BRI quality development projects can be either government-led projects under planning or proposed or project in construction projects of a certain scale and representativeness initiated by market players legally; they can be physical infrastructure construction project or non-physical projects such as standards, rules alignment and cultural, scientific and technological exchanges.

Third, the Chinese Government and enterprises should be potential beneficiaries of the implementation of high-quality co-construct BRI projects. The BRI is a mutually beneficial and win-win cooperation initiative, not a unilateral assistance plan by China. As the first initiator, practitioner and promoter of the quality development of the BRI, China should also be one of the direct or indirect beneficial owner implementers of the relevant projects. China does not seek to dominate all BRI projects, but the construction of related projects must not undermine China's sovereignty, security and development interests. The planning and selection of the BRI projects should not only look at the economic benefit of the projects themselves, but also at the potential impact of project implementation on the Chinese side. Projects that are potential hazard to China's national interests, or that are commercially viable but have a serious imbalance between inputs and outputs when China participates, cannot be considered as a quality co-construct BRI projects.

Fourth, high-quality co-construction of the BRI projects should be critical, nodal and demonstration projects. Market dominance means that the relevant projects is very broad in terms of project size to project type. The number of projects that meet the above criteria remains very high. It makes little sense in practice to treat the projects in question indiscriminately as high-quality co-production BRI projects. At the same time, the scale of investment in individual projects varies widely across industries. Even projects of the same type and similar size vary in importance for different countries at different times. The criteria for assessing whether a project is significant should also be differentiated due to the different industry attributes, technical attributes, geographical location, implementation method and financing model of the project. It should start from the common characteristics of like terms projects, combine qualitative and quantitative, and realistically screen out projects with critical, nodal and exemplary characteristics. In this regard, whether a project is a quality development project under the BRI should ideally be determined through an institutionalized selection process involving all key parties involved in the project, in particular the competent authorities.

These four points together outline the broad contours of a high-quality co-production BRI project. Based on this, the BRI cooperation project is further subdivided in terms of the need to gather information on the project, promoting its implementation, and supervise and manage it. Depending on the difficulty and importance of the construction of the project, it can be divided into systemically important projects, demonstration projects and general projects. Systemically important projects are major projects that play an important role in supporting the "Five Links", and most of them are inter-governmental cooperation projects that usually require resource input and policy support beyond market forces; the demonstration projects are generally commercial projects of typical demonstration significance and a certain scale in a certain industry, sector or region.

2. Establishment of an organizational coordinating body

With the continuous promotion of high-quality co-construction of the BRI, it is particularly urgent to build an institutional mechanism for transnational coordination. It is urgent to learn from the international experience of developed countries in setting up dedicated institutions in similar international cooperation. On the one hand, the continuous promotion of the BRI has put forward higher requirements for the improvement of top-level design, government development assistance management and service guidance; on the other hand, the experience of some developed countries also shows that the establishment of institutions dedicated to the implementation of international cooperative development effectively drives the development of foreign regional economic cooperation work and trade.

In order to better promote the construction of the BRI with high quality, countries along the route should jointly set up a special organization and coordination body, which should be composed of representatives sent by the governments of the countries along the route. They should mainly responsible for such organizational and coordination functions as the formulation of the planning and implementation plan of the BRI, the evaluation of the progress of project construction, the selection of major demonstration projects, the publication of relevant information statistics and the consultation on relevant major issues, and the overall formulation of the annual construction work arrangement of BRI with high quality. The agency could establish an institutional secretariat in our country to form a permanent joint working mechanism. At the same time, our government may also consider setting up a full-time international cooperative development agency, integrating the relevant functions to form an agency specializing in the implementation of international cooperative development and achieving a high degree of dovetailing with the BRI organization and its secretariat in China, to play a pivotal role in serving national strategies, coordinating the efforts of all parties, guiding and serving enterprises, coordinating foreign aid funds and promoting international development cooperation.

3. Establishment of a "Risk-sharing and Benefit-sharing" mechanism

In the process of promoting the BRI high-quality co-construction, we should uphold the concept of global governance of co-business, co-build and share. We should always adhere to the correct concept of righteousness and benefit, taking righteousness as the first, and righteousness and benefit at the same time. We should take the development of China as an opportunity to let more countries ride on the development express of China and help participating countries to achieve their development goals. To promote equal participation and full consultation by means of co-consultation, we should take into account the interests and concerns of the partners. The co-construction promotes each person's strength and ability, and form a joint effort to promote them consistently. It also makes the fruits of construction more equitable for the people of the participating countries.

During the 14th Five-Year Plan period, the promotion of high-quality development of the BRI should continue to adhere to the principles of equal consultation, shared responsibility and mutual benefit in cooperation and construction. It takes into account each other's interests and concerns to achieve the goal of win-win cooperation. (1) It is necessary to strengthen communication and coordination with countries along the route, integrating the common interests and divergent interests of China and countries along the route, finding more interests plotted point, mobilizing the enthusiasm of countries along the route for cooperation, enhance consensus on mutually beneficial cooperation, solidifying the basis for cooperation, expand the areas of cooperation and upgrade the level of cooperation. (2) In the spirit of mutual respect, equal treatment, seeking common ground while reserving differences, openness and transparency, and win-win cooperation, we should further strengthen third party market cooperation with countries that have the ability and willingness to promote the complementary advantages of Chinese enterprises and enterprises of other countries, so as to achieve a win-win situation of "1+1+1>3". (3) We should follow the guidelines of host country-based and market-oriented operation. Also we should refer to international best practices and experience, raise the level of multilateral cooperation in the construction of the BRI, promote mutual political trust, interconnection of facilities, mutual financial integration and interpersonal exchanges among countries along the route, and join hands with all parties to build a more open and inclusive platform for high-quality cooperation in the construction of the BRI.

4. Improving project risk management mechanisms

As investment in the construction of the BRI entails risks, it is important to promote a high-quality co-construction of the BRI projects and to manage the risks of the BRI projects within reasonable limits. One of them is to improve debt risk management, enhancing debt management capacity and encouraging multi-channel participation in infrastructure financing through innovative financing means, such as equity investment and PPP cooperation. Second, a risk compensation mechanism should be established to reduce the impact of country risk and credit risk on the implementation process of the BRI. On the one hand, a financial risk compensation fund is established to compensate financial institutions according to a specific rate when country-specific risks occur and the interests of financial institutions are damaged; on the other hand, the insurance industry gives full play to its role in risk compensation and encourage insurance institutions to join in the co-construction of the BRI high-quality development, providing insurance coverage for the personal and property of residents of countries along the route as well as for some infrastructure construction.

(iii) Developing a dynamic system of project evaluation

1. Assessment of planning implementation

Planning and evaluation is an important mechanism to ensure the implementation of the high-quality BRI planning system and to promote performance accountability. It is also an indispensable part of the whole life cycle of the implementation of the BRI plan. The first is to monitor the implementation of the plan dynamically during the process of its implementation, to identify deviations and make corrections and adjustments in a timely manner; the second is to evaluate the results of the plan's implementation after its completion, to summarize the effects of the plan's implementation and to provide a basis for the preparation of a new round of high-quality plan for co-constructing "One Belt, One Road". The dynamic evaluation mechanism of high-quality co-constructing the BRI general planning, special plans and action programmes should initially establish a set of planning evaluation system combining "mid-term evaluation + summary evaluation + annual monitoring and evaluation", "self-evaluation of project subjects + commissioned third party evaluation", "quantitative evaluation + qualitative evaluation", etc. Every two to three years, a scientific assessment of the implementation of the previous phase of the BRI plan for high-quality co-construction and the progress of its construction is conducted, with a view to summarizing experience and identifying problems in a timely manner, and making necessary amendments and adjustments to the plan and the implementation programme.

"In the 14th Five-Year Plan period, the evaluation of the implementation of the plan for high-quality co-production of the BRI should focus on the following two areas. On the one hand, it is important to focus on conducting independent, professional third party assessments. Who conducts the planning assessment has a significant impact on the objectivity, impartiality and independence of the findings. Third party evaluation agency should be independent in its identity, independent from the preparation and implementation departments of the high-quality co-production BRI plan, without direct organizational or personnel relations with the preparation and implementation departments of the plan, and should have a high level of professionalism and comprehensive research strength. The evaluation of the implementation and effectiveness of the BRI quality development plan shall be carried out autonomously by the third party institution, which is independent from the main body of the plan preparation, implementation or beneficiary, and the independence of the third party evaluation process shall be fully guaranteed, and the project participants shall not interfere with the evaluation conclusions of the third party institution.

On the other hand, it is important to ensure openness and transparency in the assessment of high-quality co-production BRI planning. In addition to the implementation of the planning assessment, the scientific and democratization of the pre-programming assessment should be further improved, and public participation in the assessment of the plan for high-quality development of the "Belt and Road" should be enhanced. It is necessary to further improve the transparency of the preparation and implementation of the plan, so that the public and professional third party institutions can play a greater role in the evaluation stage of the planning scheme. Social organizations should be guided to actively participate in the evaluation activities of the BRI plan in a standardized and rule-of-law environment, and more social organizations should be introduced as evaluation subjects in the annual monitoring, mid-term evaluation and summary evaluation. The results of the evaluation of the BRI plan for high-quality co-production should be made public in a timely manner to accept social supervision.

2. Assessment of overall project benefits

Social economic benefit is a comprehensive indicator of all economic activity. In economics studies, economic benefit refers to the comparative relationship between the consumption of factors of production and useful outcomes in economic activity, with indicators such as economic growth rates, efficiency of labour, and resource savings rates; social benefits are the contributions made to society as a result of project implementation. Ecological environment is the general term for the quantity and quality of water, land, biological and climatic resources that affect human survival and development. It is related to the complex ecosystem of sustainable social and economic development. Ecological benefits can be divided into ecological benefits, which refer to the impact of human economic activities on ecological system architecture and functions, and environmental benefits, which refer to the impact of economic activities on physical environment functions.

Building an ecological civilization and achieving green development are important concepts that is upheld during the 14th Five-Year Plan period in the construction of the high-quality BRI. China never follows the original model of international cooperation led by developed countries in the construction of the BRI, i.e., exchanging valuable resources, environment and ecological values for low economic advantage, but fully consider the expectations of people in various countries for a good ecological environment and explore with its partners a cooperation model that gives equal importance to economic benefit and ecological benefits. The BRI strategy advocates full respect for the history and realities of countries along the route, adheres to the principle of "common consultation, common construction and common sharing", and strives to transform green development into the common welfare of all countries, rather than the private interests of a few. The BRI is a green development path, which safeguards the ecological interests of countries along the route and is conducive to enhancing the overall negotiating capacity of developing countries on major international issues such as environmental protection and addressing climate change, thereby promoting the formation of a more just and reasonable global governance system (Hillhouse industry and Innovation Research Institute, Beijing Green Finance and Sustainable Development Research Institute, 2022, March 17).

As the main body of the BRI project planning and design, and construction management, the basic pursuit of the governments of each country along the route is to form a high level of project construction and operation results, and promote the improvement of the level of economic development of the country or region, while the deep-seated pursuit is to solve the sustainable development of the economic, social and ecological environment of the host country. In order to achieve the BRI high-quality development goal, under the leading role of the governments of the countries along the route, high-quality project construction results are formed through the innovation of institutional mechanisms, with the ultimate goal of improving the living standards of the people of the countries along the route and improving the ecological environment of the countries along the route, while improving the development level of the countries or regions along the route. It can be said that by enhancing the overall effectiveness and sustainable development capacity of the BRI project, the expected social economic benefit and ecological and environmental benefits can be achieved, which is the strategic pursuit of the countries along the route to jointly participate in the high-quality development of the BRI. Therefore, it is important to focus on the overall assessment of the social economic benefit and ecological benefits of the BRI project.

(iv) Building a platform for project information dissemination

1. Major Project Information Reserve Platform

Many of the major projects related to the overall situation of the BRI construction, its investment scale, investment cycle, etc. far beyond the affordability of a single or a few enterprises, and even beyond the affordability of some countries along the route. As a result, progress on these projects has been slow to materialize, despite their undoubted importance in promoting connectivity. In the long term, it is necessary to establish a reserve of major projects for the BRI led by the Leading Group for Promoting the Construction of the BRI, to include in it projects of systematic importance and market demonstration for high-quality co-construction of the BRI during the 14th Five-Year Plan period, and to pool government resources to give appropriate guidance and support to the projects included in the reserve, so as to provide effective support for the construction of major projects for high-quality co-construction of the BRI. It provides a reliable grasp for tracking and evaluating the progress of the implementation of the relevant projects. At the same time, through the construction of identification criteria and identification processes, various types of information on existing projects are gradually integrated into the database, and new major projects are directly planned and initiated according to the needs of development strategy alignment, and supported by tailor-made policy tools.

In addition, countries along the route should participate in building a reserve of major projects for the BRI, including major projects in various fields such as transportation, energy, communication, culture, people's livelihood, international assistance and international capacity cooperation, which involve high-quality co-construction of the BRI, making it the main menu for the selection of the BRI projects. The project pool may adopt a proposal system, and countries along the route may individually or jointly submit applications for major projects to be included in the pool. It may be entered into the pool after pre-proposal and scientific evaluation by professional institutions. The projects in the library should become priority projects and key projects for the construction of the BRI.

It can be opened to Asian infrastructure investment bank, the Silk Road Fund and policy financial institutions of various countries, and policy funds should be given priority to the projects in the library. High-quality co-construction of the BRI of the major project reserve can take a rolling implementation mechanism, of which the project may start with, plan and reserve a portion of the work, year by year to adjust, and to regularly assess the progress of the construction of major projects, economic benefit, social benefits, etc., to repeatedly fail to assess the project should be removed from the project pool.

2. Project in construction information sharing platform

On the one hand, it is necessary to explore the establishment of a reporting and statistical system for the BRI cooperation projects. The establishment of a full-caliber reporting and statistical system for the BRI cooperation projects is an important prerequisite for grasping the progress of the BRI construction and giving policy support to related projects. During the 14th Five-Year Plan period, the sources and types of high-quality co-construct the BRI projects becomes increasingly complex, overlapping with overseas investment projects, foreign investment projects, permanent assets investment projects and even PPP projects in the traditional sense. As the transformation of China's governmental functions progresses, most foreign-related projects no longer require government approval, and only need to report information to the local authorities, which makes it more difficult to gather information on BRI projects (Bottom BRI project list, 2015, May 29). To this end, it is necessary to integrate the BRI project statistics into the official national statistical system, to establish a BRI project reporting and statistics system led by the Office of the Leading Group for the Promotion of BRI and the National Bureau of Statistics, and to develop an online project management information system to support long-term reporting and statistics.

On the other hand, the implementation of the BRI project information sharing mechanism should be explored. The positive response of foreign countries to high-quality co-construction of the BRI is considered in terms of the focus on the BRI project cooperation, especially the great interest of extraterritorial partners such as developed countries in how to participate in the BRI projects. It is necessary to study and determine the scope of information on the BRI projects that we can grasp and that can be made public according to the law on the basis of the BRI major project reserve and statistical work, and establish a corresponding information-sharing mechanism. For example, enterprises can choose whether to agree to partially disclose the relevant information when the project statistics are compiled and entered into the database, and specify the corresponding disclosure process and channels for disclosable project information (such as opening a special section and interface on "China BRI website"), which not only demonstrates China's sincerity in cooperation, but also facilitates the participation of social forces in the reserve and implementation of relevant projects.

References

BRI Financing Governing Principle (2017, May 15). http://www.gov.cn/xinwen/2017-05/15/content_5194 23.htm.

China Development Bank, UNDP converging investment and financing rules for BRI Sustainable Development, BRI Economic Development Report (2019). https://www.yidaiyilu.gov.cn/wcm.files/upl oad/CMSydy lgw/201911/2019 11061030039.pdf

Hillhouse industry and Innovation Research Institute, Beijing Green Finance and Sustainable Development Research Institute (2022, March 17). https://www.sohu.com/a/530806109_484815.

Joint "Belt and Road Initiative": Progress, contribution and prospect (2019, April 22). http://www.xinhua net.com/world/2019-04/22/c_1124400071.htm.

Ministry of Finance of the People's Republic of China. (2019). BRI Debt Sustainability Analysis Framework (2019, April 30). Website: http://www.mof.gov.cn/zhengwuxinxi/caizhengxinwen/201904/P020190426412371032816.pdf.

Shanghai Securities News BRI research report team special report "Bottom BRI project list" (2015, May 29).

Xi Jinping. (2019). Second BRI International Cooperation Summit.

Zhou, Wuying. (2019). The data of the World Bank's Global Economic Outlook report were transferred from the Economic Information Daily (2019, January 11). http://www.jjckb.cn/2019-01/11/c_137736714.htm?from=single message.

X

Legal System and Implementation Mechanisms for the Co-construction of the BRI

Introduction

The BRI, as an international regional economic cooperation initiative proposed by China, has become an important platform for global economic governance and component part while tapping new momentum for world economic growth and creating a new platform for international regional economic cooperation. Moreover, these two aspects are the same and complementary. On the one hand, the construction of the BRI can give a strong impetus to the economic and social development of China and the countries along the route. It builds a more solid foundation for improving the global economic governance system, ensuring the stable development of the international political and economic order and enhancing China's participation in global governance; on the other hand, the high-quality development of the BRI and the achievement of its construction goals certainly requires a good economic governance system to provide a stable order and environmental support. The essence of the contemporary international economy is an economy based on the rule of law, which is rule-based and dependent, and rules are an essential element of international economic governance. The core of the international economic governance system is also the legal governance mechanism. The establishment of a good legal governance mechanism, including a system of legal rules and an enforcement mechanism, is necessary to promote the construction of the BRI in a high-quality manner and to guarantee the successful implementation of the BRI. In addition, good legal governance mechanisms and strict compliance with laws and international rules are conducive to eliminating the doubts and resistance of countries along the route to the co-construction of the BRI; at the same time, it is also conducive to eliminating the accusations of some Western countries that the BRI is intended to focus only on China's own development and undermine the existing international economic order, and the hesitant attitude of many countries to join the BRI (He Xiaofeng and Wu Junya, 2019).

The BRI is a public product provided by China to the international community in a new period of development, aiming to promote regional and international economic cooperation, and an inclusive and balanced global governance system. The essence of which lies in the basic legal principles and order followed by the participants. It also lies in the formation of a more stable legal framework and a relatively fixed governance platform for the relevant governance issues, as well as the clear definition of the rights and obligations of the participants and the effective settlement of various disputes. All of which must incorporate elements of the rule of law. The international cooperation required in the supply of this international public good can only be better safeguarded by raising it to the level of international law and binding the behavior of governments and participants through legal mechanisms. Legal interests in particular, under the situation of protectionism in international trade and the prevalence of counter-globalization, as well as in post-pandemic era, we should pay more attention to the use of international legal mechanisms, strengthen the rule of law at home through the internationalization of the BRI cooperation principles and governance models, realize the effective supply of international rules at home and abroad, and gradually promote the change and development of the international economic legal system and governance system (Shi Jingxia, 2021).

The rule of law, as a common civilizational achievement of humankind, is an important language of international dialogue. Only by adhering to the rule of law, building a scientific international economic and legal mechanism, and effectively playing the main role of the legal governance mechanism in the implementation of the initiative, can we create a good and lasting external development environment for the BRI and ensure the long-term, stable and healthy development of the Belt and Road construction. The construction of the international legal governance system of the BRI is a complex and systematic project. In terms of content, it mainly includes the legal system and enforcement mechanisms. In the process of constructing both, several key issues need to be considered and addressed, taking into account the lessons learned from the historical international legal governance system and the BRI's own situation and existing construction practices. The first is to gain an in-depth understanding and comprehensive knowledge of the complex legal environment and major issues facing the construction of the BRI; the second is to deal with the relationship between making full use of existing international rule of law resources and the new rule of law system in the process of establishing the legal rule system and implementation mechanism. It also aims to achieve a positive interaction between the domestic rule of law and the international rule of law, avoiding as far as possible the creation of another stove and reducing the construction costs, obstacles and cycles; the third is to consider how to sort out and set up important legal service systems and platforms related to economic activities such as infrastructure construction project and enterprise trade and investment, so as to guarantee the smooth implementation of specific transactions, reduce the uncertainty of cooperation and enhance the risk response capacity.

1. Perspectives on the legal environment of the BRI

The legal systems established in each country are distinctive. The attitudes and practices towards the rule of law vary widely. The legal environment along the BRI route, which covers many regions and more than 60 countries, is even more complex. The various countries and regions along the route are prone to conflicts between legal systems and rule of law mechanisms due to differences in regional customs. Moreover, the advancement of the BRI is deeply influenced by the complex situation at home and abroad. Therefore, in the process of jointly building the rule system and implementation mechanism of the BRI, it is necessary to pay full attention to and grasp in depth the complexity of the legal environment along the route, and to conduct targeted and systematic consideration of the interlocking legal risks and countermeasures therein.

(i) Greater conflict of laws and legal risks in the BRI domain

The legal differences are enormous. There are many BRI countries, which have different historical, cultural, ethnic and religious backgrounds. In addition, the BRI countries have different levels of economic development and are uneven. There are large emerging economies like China and India, as well as smaller economies including Montenegro, Macedonia and Moldova. Therefore, the legal culture along the BRI route is clearly diversified. The countries along the route belong to civil law, common law and Islamic law systems, and some countries have mixed legal systems, which are a collection of common law, Islamic law, and the Common Law, such as India, Thailand and Nepal, etc. The laws and rules differ greatly from each other, and legal conflicts are common (Jiang Xinmiao and Zhu Yanni, 2018). When actually making investments, many Chinese companies have encountered various problems and caused huge losses because of their inexperience. They are lack of sufficient research and understanding of the local legal environment and risks, and unfamiliarity with the local legal system. This is because we need to conduct a lot of in-depth and highly targeted research. It upholds the "principle of common negotiation, common construction and sharing", promotes the convergence or unification of major legal rules along the route, helps enterprises eliminate or reduce the legal risks in transactions, and thus, ensures the stability and success of the BRI.

Political risks are prominent. As the countries and regions along the BRI route are infiltrated by cultural, religious, political and economic forces, various forces, including religious extremism, political groups and terrorists, converge together. Conflicts break out frequently and political instability poses a great risk to China's overseas investment. For example, the outbreak of armed conflict in Libya in 2011 forced China to abandon its investments there, leading to a great loss. A number of events in Thailand and Myanmar in recent years have also shown that the biggest risk to the region is political instability. In addition, countries along the route may be passively drawn into the geopolitical game with China at the instigation of other geopolitical stakeholders. For example, Myanmar shelved the Sino-Myanmar railway project in July 2014 and Sri Lanka suspended the Colombo Port City project in March 2015. Both directly related to the geopolitical games of the countries concerned. These circumstances make the implementation of the BRI even more complex and serious due to geopolitical problems or obstacles. According to the *2016–2017 Research Report on Chinese Enterprises Going Global* (2017, April 9) jointly published by LexisNexis, a global legal information provider and the China multi-national corporation Law Institute, it states that about 40% of the companies interviewed mentioned that they encountered political risks such as "host country inadequate laws", "host country government corruption, unstable policies and low administrative efficiency" in their overseas investment projects, especially in BRI countries (Law Business Connection, 2017).

Inadequate legalized business environment. One is irrational nationalism. In the construction of the BRI, due to the more frequent exchanges and contacts between countries, the differences between nationalities may be infinitely magnified, which in turn breeds irrational nationalism. This irrational nationalism is shown through narrow "trade protectionism" sentiments. *The 2016–2017 "Going Global" Research Report of Chinese Enterprises*, jointly published by global legal information provider LexisNexis and the China multi-national corporation Law Institute (*The 2016–2017 "Going Global" Research Report of Chinese Enterprises*, 2017, April 9), shows that 44% of the companies surveyed have encountered local government scrutiny in overseas investment. Among them, 36% of the respondents said that they encountered "localization" requirements in the overseas investment projects, mostly in Central and West Asia. Second, the business environment is hostile. Most of the host countries and regions along the BRI route are emerging development economies, mostly facing difficulties in national political, economic and social transformation, with low barriers to investment entry and serious government corruption, which greatly undermine the impartiality of justice. With the exception of a few countries such as Singapore, Poland and Bulgaria, the overall business environment in other countries is poor and poses a greater risk to investment by our companies. According to the World Bank's *Global Business Environment Report 2020* (World Bank's Doing Business 2020 Global Report), countries and regions along the BRI route are underperforming in a number of areas, including starting a business and enforcing contracts. In addition, conflicts arising from the lack of a rule of law business environment continue to have a serious impact on the progress of business investment, as enterprises are unable to keep abreast of local legal policies and accurately and comprehensively predict the local business environment.

(ii) The complex system of bilateral and multilateral rules relating to the BRI

International economic and trade rules along the BRI route are highly fragmented. Just as the phenomenon of pluralism in the legal system among the countries along the route, the basic national conditions and demands of the parties to the BRI cooperation are very different. It is not easy to reach consensus on the governance mechanism. The multi-bilateral international rules along the route are overlapping and complex, with the apparent "spaghetti bowl effect". And plus, there is a lack of a universal and applicable rule system, as well as the legal governance mechanism. From a multilateral perspective, there are also countries in the region that are not members of the WTO. They are outside the multilateral rules, which can easily create barriers to trade and investment, and thus cooperation within the influence area. At the same time, there is a wide variety of regional legal governance mechanisms and other international mechanisms in place along the BRI route. In addition to joining WTO, BRI countries have joined other international organizations and international treaties such as ASEAN, the European Union, the Eurasian Economic Union and various regional and subregional economic cooperation efforts. The trade and investment rules and implementation of these international cooperation mechanisms vary considerably. This can, to some extent, increase the likelihood of disputes and the complexity of dispute resolution in the course of project cooperation.

The BRI is not sufficiently binding. Like many regional international cooperation mechanisms in the world today. The cooperation mechanism proposed by BRI does not have a legal character and is therefore very weakly binding on the parties involved in the cooperation (Ding Tongmin, 2018). As a result of this problem, although China has signed more than 200 cooperation intentions or agreements with 171 countries and international organizations since the introduction of the BRI, it has been impossible to properly address the issues and challenges in the process of promoting the implementation of the BRI using laws or binding international rules, resulting in great obstacles to the promotion of the BRI.

The **imbalance in the role of the State within and outside the region is highlighted**. "Regional economic integration is itself a union of the countries of the region in search of common interests, and if the fate of integration cannot be determined by the countries of the region, then it largely loses its value of existence."(Li Xiangyang, 2018) For reasons of history and political and economic power, important extraterritorial powers, especially the United States, still have great influence, and even a dominant role, in the process of economic integration of BRI countries and many regions. In recent years, the anti-globalization philosophy and the "America First" policy orientation of the United States have posed a major challenge to the multilateral trading system. On the one hand, it has continuously demonstrated its reluctance to assume the role of a public goods provider for global economic integration. On the other hand, it has made unfounded accusations against China's BRI, a new international public good, and even planned to launch a so-called high-standard alternative, launching a non-goodwill strategic and rule-based confrontation. This, in turn, largely increases the competition and conflict of rules along the BRI route, as well as the complexity of the legal environment and governance mechanisms.

(iii) The international dispute settlement mechanism of the BRI needs to be improved

Limited channels and means of intergovernmental dispute resolution. In terms of existing international practice, the main legal settlement channels for the handling of disputes between two or more States concerning measures affecting international trade, in addition to direct consultations, are the dispute settlement mechanisms under the WTO and under bilateral or regional FTAs. Among them, the WTO dispute settlement mechanism is known as the "pillar of the multilateral trade mechanism", with the characteristics of unity, efficiency and compulsion. It provides for a series of relatively complete and strict procedures and monitoring mechanisms to ensure compliance with the substantive rules of the WTO and to ensure the stability and predictability of global trade rules. Although there are still many unsatisfactory aspects, the WTO dispute settlement mechanism has in practice become an important means of resolving trade disputes between members in a fair and efficient manner. In particular, the Appellate Body, as the final arbiter of trade dispute settlement, is known as the "crown jewel of the WTO". Given that China and most of the BRI countries are members of the WTO, the WTO Dispute Settlement Body is likely to become an important channel for us to resolve trade policy issues in the future. However, it is well known that the United States has deliberately obstructed the selection of the members of the Appellate Body for its own interests, at the expense of crippling the international trade dispute settlement system that it has built with its own hands, thus overshadowing the "crown jewel of the WTO". Once the mechanism ceases to be effective and appeals are not finally decided, the "safe and predictable" international trade environment that members have enjoyed for more than 20 years disappears. International trade policy could be plunged into disorderly and vicious competition, with international trade in danger of falling under the "law of the jungle". This is a problem that we have to consider and may face in promoting the construction of the international dispute settlement mechanism of the BRI. The dispute settlement mechanism under FTAs shows that the channels for settlement are even more inadequate. Although as of May 2021, China has signed 20 free trade agreements and preferential trade arrangements with 30 countries and regions, participating in trading partners across Asia, Europe, Latin America, Oceania and Africa. The main ones related to BRI are south east Asia region and ASEAN countries. We have basically not signed any free trade agreements with other BRI countries, and therefore it lacks a corresponding targeted economic and trade dispute settlement mechanism or platform (China Free Trade Zone Service).

There is a tendency for international investment dispute settlement mechanisms to weaken. Most of the BRI projects belong to the infrastructure construction category. These projects have long investment cycles, involving more links and higher risks. They are bound to lead to many investment disputes due to the complex political and legal environment of countries and regions along the route. In the case of investment disputes between private individuals in different countries, investors may use a mutually agreed approach or choose their own appropriate approach to deal with the matter. However, investment disputes between private individuals and host country are more special. The unequal status of the disputing parties cannot be easily resolved by traditional dispute resolution. A fair and efficient dispute resolution mechanism with mutual benefits is needed. For these investment disputes, the most influential settlement mechanism in the world today is the International Centre for Settlement of Investment Disputes (ICSID), which has been regarded as the leader in the field of international investment dispute settlement. Although the design of the system itself leaves much to be desired, it has been praised for its creativity in providing an effective route for the settlement of disputes between States and investors. However, the practical analysis of the countries concerned shows that the acceptance of ICSID in BRI countries is low. There are still more than a dozen BRI countries that are not parties to the ICSID Convention. Although China has signed bilateral investment treaties with more than fifty BRI countries, only 14 of them have explicitly opted for ICSID jurisdiction through the fork in the road clause. For their part, the low level of participation in the ICSID by countries along the route limits the choice of important channels for an international settlement mechanism for investment disputes. In addition, there is also a changing trend in the current ISDS mechanism to return more investment disputes between investors and host country to domestic jurisdiction, with host country holding more regulatory power. Most notably, the United States-Mexico-Canada Agreement (USMCA), completed under the leadership of the United States, (United States-Mexico-Canada Agreement [USMCA] details of the United States-Mexico-Canada Agreement can be found on the website of the Office of the United States Trade Representative) emphasizes the transfer of regulatory authority over investment disputes to the jurisdiction of host country domestic courts, thus weakening the existing international investment arbitration settlement. For our country, however, the weakening of the current approach to international investment dispute arbitration poses a challenge to the BRI. Because the overall development level of countries along the BRI is not high, some countries are politically unstable, and the level of domestic rule of law is also not high. The investment of Chinese enterprises face great risks in terms of political change, nationalization expropriation, restrictions on remittance, etc. Under the investor-State dispute settlement mechanism, a relatively high level of international investment arbitration is relatively secure, but a weakened mechanism would create great uncertainty and pose a huge challenge to China's outbound investment and threaten the stability of China's BRI.

Arbitral decision on the risk of uncertainty in recognition and implementation is high. For international trade and investment activities, arbitration is an internationally accepted and important legal way to resolve economic and trade disputes. It has become increasingly important in dealing with cross-border economic and trade disputes, removing obstacles to trade and investment, and promoting the international rule of law. With the further promotion of the BRI and the increasingly close exchanges between China and countries along the BRI in the field of trade and economy, arbitration plays an increasingly irreplaceable role in resolving commercial disputes and promoting win-win cooperation. However, due to the complex and diversified concepts of international commercial arbitration in the BRI countries, the legal systems of arbitration are in different development stage and the legal regimes of arbitration are also very different. Courts in many countries have been less than accommodating and open to arbitration agreements and arbitral decision, imposing too much interference, resulting in arbitral decision not being finally recognized and enforced. This has resulted in greater uncertainty and risk in international commercial arbitration of related commercial disputes as the BRI progresses (Li Na, 2018).

(iv) Apparent lack of capacity in foreign-related legal services related to the BRI

There is a large demand for foreign legal personnel. With the deepening of the BRI, trade and investment activities are increasingly active. The encountered legal problems and trade disputes are also on the rise. There is urgent need for a proficient in international law, but also the understanding of economic and trade can meet the needs of language diversity of international composite foreign-related legal service personnel. Data reports show that although China's lawyers have long exceeded 400,000, but there is only 1% of foreign lawyers. In some professional fields, qualified foreign lawyers are rare. It is still far from meeting the growing needs of foreign affairs. The imbalance between the supply and demand of professional talents has been sharpened. Although the Ministry of Justice has made new arrangements for the training of foreign lawyers in order to effectively alleviate the shortage of foreign-related legal talents, the training of professionals is a systematic and complex project that requires a lot of human, material and financial resources.

There is a strong demand for networking of foreign-related professional services. The BRI involves dozens of countries in Asia, Europe and Africa. The legal environment in each country and region is complex, with significant differences in legal systems, making it extremely difficult to identify, evaluate and control legal risks (Sun, Hongtao, 2018). It is impossible for our own legal service teams and institutions alone to meet the large and diverse needs of enterprises going out to do business. We need to actively promote domestic foreign-related lawyer service agencies and related organizations to establish cooperation and docking with legal service agencies and other kinds of business associations in BRI countries. We have to build various service systems and platforms, so as to follow up foreign-related legal services wherever Chinese enterprises go, and help enterprises further improve the quality and level of foreign investment cooperation.

2. Building the legal system of the BRI

As mentioned above, the success of the BRI as a programme launched by China to promote global economic development needs to be guaranteed by the rule of law. Only by choosing the path of development based on the rule of law can the BRI strategy fulfil its benign function of promoting global governance and better respond to the general concerns of all parties about its development model and institutional design. At the same time, it strongly guaranteeing the advancement of cooperation projects and the economic development of countries and regions along the route. The first step towards the rule of law in the BRI is the establishment of a legal supply mechanism and a scientific system of legal rules. In order to quickly and efficiently build the legal system of the BRI, we must make full use of existing rule of law resources, improve domestic supporting legislation, sort out and integrate existing international rules, realize the positive interaction between domestic rule of law and international rule of law. On this basis, according to actual needs, it gradually establish a more targeted and regionally applicable rule system with countries along the route.

(i) Improving domestic supporting foreign-related economic and trade legislation

The successful launch and implementation of the BRI, a highly contemporary, comprehensive and holistic project, in itself demonstrates China's determination to comprehensively deepen its reform and opening-up. It further integrate into the new economic globalization in the new era, implying that China has now integrated its own domestic affairs with the world's international affairs as never before. Of course, its effectiveness depends to a large extent on whether China can consciously and effectively integrate international affairs into its domestic development. It can consciously and effectively raise its domestic legal governance to international standards, and thus can proactively adapt to the global governance change and its multidimensional challenges, and lead the global governance change to a more rational and healthy direction (Ouyang Kang, 2018). In order to meet the practical needs of the implementation of the BRI, we should first focus on improving the relevant domestic legislation in two aspects: the legal system for foreign-related trade and commerce, and the legal system for civil and commercial matters relating to the recognition and enforcement of awards (Liu Jingdong, 2017).

The construction of foreign-related economic and trade legal system closely related to the BRI. The orderly and effective implementation of the BRI requires China and countries along the route to pay special attention to the opening up of the trade and investment sectors and the creation of a fair domestic legal environment. China has made great efforts in this regard, adopting the Law of the People's Republic of China on Foreign Investment and its implementing regulations in 2019, while repealing the "Three Capital Laws" (Before the Foreign Investment Law came out, China relied on three main laws for the use of foreign investment: the Sino-foreign Joint Venture Enterprise Law, the Foreign Capital Affiliate Law and the Chinese Foreign Cooperative Venture Law, commonly known as the "Three Capital Laws" or the Three Laws on Foreign Investment), reflecting the cardinal rule and spirit of treating domestic and foreign investors equally and protecting them as one. We have also carried out experiments to deepen reforms or expand the open system by establishing free trade pilot zone in the country. Since the establishment of the Shanghai Free Trade pilot zone in 2013, the number of China's free trade pilot zone has now expanded to 21, including Shanghai, Guangdong, Tianjin, Fujian, Liaoning, Zhejiang, Henan, Hubei, Chongqing, Sichuan, Shaanxi, Hainan, Shandong, Jiangsu, Hebei, Yunnan, Guangxi, Heilongjiang, Beijing, Hunan and Anhui, forming a pilot pattern with distinctive features and focus in a broader field and on a larger scale to promote comprehensive deepening of reform and expansion of opening up. In addition, on June 1, 2020, the Central Committee of the Communist Party of China and the State Council issued the *Program for the Construction of Hainan Free Trade Port* (Program for the Construction of Hainan Free Trade Port, 2020, June 1), and the Twenty-ninth Meeting of the Standing Committee of the Thirteenth National People's Congress voted on June 10, 2021 to adopt the *Law on Hainan Free Trade Port* (Law on Hainan Free Trade Port, 2022, April 14), with an eye on the overall situation of international and domestic development, making a major decision to build free trade pilot zone and a free trade port with Chinese characteristics on the whole island of Hainan, promoting the formation of a highly liberalized, rule of law, internationalized and modernized institutional system, which becomes a benchmark and example for China to achieve socialist modernization. In the future, we should further establish and improve the supporting legal policy system of the foreign investment law and the integrated customs clearance system, so as to provide stronger protection and greater convenience for foreign investment and commodities, including those from BRI countries. It improves the overseas investment insurance system, increase the types of insurance and insurance subjects, and simplify the insurance procedures, so as to provide more adequate protection for Chinese enterprises' investment in BRI countries. It gradually transforms the successful experience of the free trade pilot zone into domestic legislation in relevant fields in due course, constructs a scientific, reasonable and effective institutional support and implementation system for the BRI, creates a good domestic legal environment for the BRI, and forms a model system for countries and regions along the route to improve their foreign-related economic and trade laws, providing Chinese wisdom and Chinese experience.

Improve China's foreign-related civil and commercial legal system, especially legislation related to judicial collaboration and international arbitral decision recognition and enforcement. In recent years, China has been upholding an open and accommodating attitude to innovate the existing foreign-related civil and commercial legal system, safeguard the legal interests of various market players, and protect the interests of Chinese and foreign party concerned equally. The Supreme People's Court issued the *Opinions on the People's Courts' Provision of Judicial Services* (Opinions on the People's Courts' Provision of Judicial Services, 2017, March 20) and *Protection for the Construction of the BRI and the Opinions on the People's Courts' Further Provision of Judicial Services and Protection for the Construction of the BRI* in 2015 and 2019 (2019, December 9), respectively. These two normative documents closely combine the characteristics of BRI construction and the practice of China's foreign-related commercial maritime trials. They draw on advanced international judicial concepts, and make innovative provisions in various aspects such as jurisdiction, judicial reciprocity, application of international treaties and practices, identification of foreign law, judicial review of foreign-related arbitral decision. According to the provisions of the opinion, when the relevant cases involve the application of foreign law, the court will, in accordance with the provisions of *China's Law on the Application of Law to Foreign-related Civil Relations* and other conflicting norms, fully and comprehensively consider the subject, object, content, legal facts and other foreign-related factors of the legal relationship. It fully respects the right of party concerned to choose the applicable law, actively identifies and accurately apply foreign law, eliminates doubts in international trade and commerce between China and foreign party concerned, and creates good conditions for the participants in the BRI construction to freely choose the law applicable to the contract, prompting them to be more willing to choose Chinese courts to settle civil and commercial disputes. This largely reduces the legal risk of the BRI and helps to increase investor confidence (Liu Jingdong, 2017). In the next step, we should further amend and improve the procedural and institutional design of the *Law on the Legal Application of Foreign-related Civil Relations* (2021, October 19) and the *Arbitration Law* (2017, September 12) and other relevant laws. It incorporates more advanced experience in international legislation, expands the space for meaningful autonomy in dispute resolution and governing law choice, weakens domestic judicial intervention, and works to build a friendly legal environment that is more conducive to the recognition and enforcement of international arbitral decision and foreign court decisions. It has a more inclusive attitude towards arbitration agreements and arbitral decision and a more open attitude towards international arbitration, upgrading domestic arbitration institutions competitive power, promoting the development of the domestic arbitration market, enhancing the domestic arbitration market to foreign party concerned attractive force and providing more guarantees for domestic goods, capital and services to go market abroad. At the same time, by actively exploring the mode of legal governance cooperation among the countries along the route, the unity of regional legal rules and systems is improved as much as possible. The goal of mutual benefit and win-win situation of the BRI countries is strongly guaranteed.

(ii) Emphasis on the role of international soft law in promoting the uniformity of rules in the territory

As to the definition of "soft law", the academic community has not yet reached a definite conclusion. Scholars mostly understand and study international soft law in comparison with the "hard law"(The term "hard law" refers to law in the traditional sense, which includes an element of coercion and guarantees its effective implementation by clearly establishing liability for violations and judicial guarantees). However, one of the most widely accepted underlying understandings is that "soft law" is a norm of conduct that is not legally binding in principle, but has practical effects (Francis Snyder, 1994). Strictly speaking, international soft law is not part of the law, but it is well observed in practice. The scope covered by international soft law is so broad that any written document that does not fall within the scope of a convention, treaty or protocol in force, regardless of its name, can be subsumed under it. Since there is no world government, international relations are decentralized. There is no supranational legislative, law enforcement or judicial system. There are huge differences in the areas of concern, interests and development capacities of individual countries. It is very difficult to negotiate a binding international legal document (treaty or convention) between them, often to no avail at great cost of negotiation. It is for this reason that there is great scope for international soft law to play an important role in the international community and to play an important role in regulating and regulating the relations of the international community in practice (He Zhipeng, 2017). Compared with the mainstream rule-based governance model of Western countries, the inherent characteristics of international soft law are well suited to the construction needs of the BRI. It can be used to gradually promote the uniformity of rules in the region with the advantages of international soft law, and then establish a new model of pluralistic regional cooperation and governance.

Establish a system of international soft law and build consensus on cooperation. The BRI is an open and non-exclusive regional cooperation initiative. The countries and regions along the route vary greatly in terms of history, culture, level of economic development and degree of legalization. The United States and Europe as the leading party in the selection of regional cooperation members, often consciously choose a higher price payment capacity or can enhance their geopolitical security and economic advantage countries, through the strength, authority, voice and other ways to achieve the so-called "reciprocity" or "fairness" of rights and obligations. There are clear differences between the two, with the complexity of the BRI governance far exceeding that of the latter, and the potential for longer lead times and higher governance costs to achieve hard law, requiring greater inclusiveness and flexibility in governance models. International soft law, which is relatively simple to conclude and flexible in form, offers a high degree of suitability (Han Yonghong, 2016). By means of a set of principles, norms and decision-making procedures that are not formally legally binding but contain substantive codes of conduct and can have a practical impact on cooperation participants. The sense of community among the countries and regions along the route can be continuously raised, as well as the recognition of the development concept of the BRI. The consensus on the community of a shared future embodied in the BRI, effectively expands the consensus on cooperation. Statistics from the Ministry of Commerce show that as of August 2021, China had signed 206 cooperation documents with 140 countries and 32 international organizations to build the BRI. Among them, soft law documents such as joint statements, memorandum entry and investment guides occupy the majority, and international soft law has become an important component part of the BRI cooperation mechanism.

Build mechanisms for soft law cooperation and promote policy alignment. In order to give full play to the role of international soft law in the construction of the BRI, we also need to build a corresponding international soft law guarantee mechanism and form a fixed platform for dialogue on governance. In addition to the existing forms of the BRI international cooperation summits, summits and strategic dialogues, various forms of cooperation mechanisms can be constructed according to the actual situation and needs of the participating parties, such as the establishment of economic corridors, sub-regional cooperation and other diversified cooperation mechanisms in the context of cooperation with other countries and regions along the route, with reference to the China-Central and South China Peninsula economic corridors and the Lancang River-Mekong River cooperation mechanism. When the conditions are ripe, an economic organization similar to APEC should be established, and regular meetings should be organized to fully communicate the desire for cooperation, to facilitate better quality cooperation projects, and to properly address conflicts through active consultation (Li F, 2017). At the same time, the dialogue content of these platforms should be enriched, with more discussions on topics such as market opening and best practices in legal regulation, so as to promote the exchange and harmonization of trade and investment policies among countries and regions along the route. It reduces the costs and potential risks of cooperation, and thus promotes the development of the BRI strategy in a deeper direction, taking into full consideration the concerns and demands of all parties.

Encourage the participation of multiple actors to enrich the supply of soft law. Unlike the singularity of the subjects involved in the governance of international rules, the subjects of the formulation and implementation of international soft law include not only States and international organizations, but also non-State actors such as business associations, enterprises and even individuals. For the BRI, which has a complex governance environment, the joint participation of multiple actors is particularly important, helping to make up for national shortcomings in terms of information and resources and enriching the supply of international cooperation norms. This does not only help to avoid or reduce the political sensitivity of rules governance and the resulting disputes over rights and obligations, but also help enterprises to reduce the burden of contracting and performance in project cooperation.

(iii) Making full use of the existing international legal system for trade and commerce

As mentioned above, due to the complexity of the political, economic and social environment along the BRI route, there is great scope for international soft law to play a role, but this does not mean that the construction of the BRI can be left out of the international law regime. In order to ensure the long-term, institutionalized and standardized construction of the BRI, we need to adopt a flexible mechanism that combines soft law and hard law, and gradually promote the transformation of soft law into hard law, and establish an effective legal governance mechanism for the BRI. In the long run, China's governance approach to promote the construction of BRI should prefer a legal governance mechanism, using legal norms to draw a roadmap for the construction of the BRI as far as possible, relying on the mandatory binding force of the law to coordinate the implementation of the respective legal responsibilities of countries along the route, thus, enhancing the stability of bilateral and multilateral cooperation (Ding Tongmin, 2018).

Make good use of the WTO rule system. As the Foreign Ministry spokesperson pointed out in response to foreign accusations that China was trying to establish its own rules and guidelines through the BRI. The BRI advocates the principles of common consultation, common construction and sharing. It is open and inclusive and is not intended to create a "small circle", nor does it target any country. It has never sought and will never seek to establish rules dominated by one country (Foreign Ministry: "China will not use BRI to establish rules dominated by one country", 2017, December 20). In fact, the BRI is not a rejection of the established system of international economic and trade rules, but rather facilitates China's use of existing rules to promote connectivity among participating countries, expand mutual trade and investment, and thereby strengthen the multilateral trading system. The use of the WTO's existing and mature rules is conducive to reducing the international community's misunderstanding and concerns about the BRI, creating a favorable international environment for the construction of the BRI, and also helping to avoid the duplicate construction of the legal system and institutional mechanisms, reducing the potential huge cost investment. Among them, the *Agreement on Trade Facilitation* (2015, October 16), as the first multilateral agreement on trade in goods that China has participated in since its accession to the WTO, has a particularly important application value. Because of the complex legal environment along the BRI route, the lack of unified economic and trade rules is a prominent obstacle in the process of promoting and implementing the BRI. The establishment of the relatively unified trade facilitation rules to reduce trade costs has become an urgent task (Zeng Wenge, 2018). *The Agreement on Trade Facilitation*, on the other hand, establishes and improves a series of rules on trade facilitation, including the release and clearance of goods, border agency cooperation, freedom of transit and customs cooperation. Working with countries and regions along the BRI route to promote the application of the agreement on the ground helps reduce trade procedures and trade costs, promotes trade liberalization and market integration on a wider scale, and facilitates the smooth development of trade and investment along the route. While promoting the application of existing WTO economic and trade rules, we can also actively feedback the common interests of countries and regions along the BRI route in the ongoing and future negotiations on plurilateral agreements such as e-commerce at the WTO. We promote more just, reasonable and inclusive international trade and investment rules, and push the current globalization in the direction of openness, inclusiveness, balance, inclusiveness and sustainability.

Make good use of existing economic and trade agreements with countries along the route. In recent years, as the multilateral trading system has been weakened and global trade protectionism has risen, China, while firmly supporting the multilateral trading system, has also accelerated the implementation of its free trade zone strategy and promoted a higher level of opening up to the outside world and the construction of a new system of open economy in response to the new situation of rapid development of global regional economic integration. In 2015, the State Council issued the *Opinions On Accelerating the Implementation of the Free Trade Zone Strategy* (2015, December 6), which further clarified the strategic layout direction of China's free trade zones, i.e., focus on the periphery, circulating the BRI, and facing the world. As of May 2021, the country has signed 20 free trade agreements and preferential trade arrangements. The countries and regions involved are mainly concentrated in the East and Southeast Asia. Trade and investment agreements and various preferential trade arrangements with other countries and regions along the BRI route are also under active construction. The Regional Comprehensive Economic Partnership Agreement (RCEP), signed on November 15, 2020 by ten ASEAN countries and 15 countries from China, Japan, South Korea, Australia and New Zealand, is the largest free trade agreement in the world and is a comprehensive, modern, high-quality and mutually beneficial FTA. The overall number of products with zero tariffs for trade in goods exceeds 90%. The level of services trade and investment liberalization is significantly higher than in the original 10+1 FTAs. These agreements have had a significant catalytic or boosting effect on deepening regional economic integration and stabilizing the global economy. According to statistics from the Ministry of Commerce, in 2020, ASEAN became China's top trading partner for the first time, and China's imports and exports to BRI countries were 9.4 trillion CNY, accounting for 29.1% of total imports and exports. Exports to Saudi Arabia, Turkey, Egypt, Poland, New Zealand and some other BRI countries also saw rapid growth (Ministry of Commerce: Report on China's Foreign Trade Situation, Spring 2021). In the future, we should further leverage the value of these agreements in enhancing regional trade and investment liberalization, promoting the entry into force of the RCEP and the temporarily troubled China-EU Comprehensive Investment Agreement (2021, January 23) as soon as possible. We need to supply the mature and advanced rules therein to future agreements with other participating countries of the BRI, so as to further accelerate the process of trade liberalization along the routes and optimize the surrounding and international legal environment for the construction of the BRI.

Make good use of the advanced institutional arrangements of other international organizations. Given that freedom of transit of goods is an important institutional guarantee for the realization of the BRI connectivity. The transit of goods along the BRI route is faced with legal and practical difficulties such as large differences in systems and inefficient customs clearance, some institutional arrangements on customs clearance facilitation, such as the AEO (Accredited Operator) system advocated by the World Customs Organization, deserve particular attention and promotion, so as to improve the customs control methods of countries and regions along the route and protect the freedom of transit of goods. As of May 2021, China has achieved AEO mutual recognition with 46 countries (regions) in 20 economies, including Singapore, South Korea, EU, Switzerland, Israel, New Zealand, Australia, Japan, Kazakhstan, Mongolia, Belarus, Uruguay, UAE and Brazil, ranking first in the world in the number of mutual recognition countries (regions). In the future, China Customs needs to promote the AEO mutual recognition negotiation with other important node countries of the BRI as soon as possible. It continuously expands the coverage of AEO mutual recognition countries, forms a more convenient customs clearance system arrangement, and helps realize the optimization of China-Europe train transportation organization and AEO enterprises enjoys customs clearance convenience in countries and regions along the route.

(iv) *Active construction of bilateral and multilateral international rules along the BRI route*

Although there are already free trade agreements or preferential trade arrangements between China and 15 BRI countries, bilateral investment agreements have been signed with 56 of the 65 countries along the route. There is still much room for improvement in the scope of application, degree of openness and supply of advanced rules in many agreements. The main problems include the relatively small number of BRI countries that have concluded FTAs, the low level of overall trade and investment liberalization under the agreements, and insufficient attention to the practical needs of the BRI construction in terms of rule provision to promote infrastructure investment. The BRI international rule of law construction also needs to pay close attention to international economic and trade rules recent development and innovatively improve our ability to supply investment and trade liberalization rules for the BRI (Shi Jingxia, 2021).

Upgrading existing economic and trade agreements. Taking the latest economic and trade agreements promoted by China, such as the Regional Comprehensive Economic Partnership Agreement (RCEP), the China-EU Comprehensive Investment Agreement (2021, January 23) or even the Comprehensive and Progressive Trans-Pacific Partnership (CPTPP)(2021, January 11), which China is actively considering joining, as a model, on the one hand, we should upgrade the existing FTAs with countries along the routes, expand regional trade liberalization while increasing the openness of investment and service markets, and cover as much as possible new issues and horizontal rules such as digital trade, competition policy, intellectual property rights and government procurement; on the other hand, we should clean up and revise old bilateral investment agreements, abolish outdated or obsolete regulations and contents, and incorporate new rules related to expanding the level of investor protection and the openness of investment markets, especially increasing the supply of rules to promote infrastructure investment, so as to provide a more stable and transparent environment for enterprises to invest.

Improve the network of economic and trade agreements. At present, there are no FTAs or even ongoing negotiations between most of the BRI countries and China, especially in Central Asia and Central and Eastern Europe. More than a dozen countries have not yet joined the WTO. Since these countries have a low level of acceptance of international rules, it may be more rational to apply international soft law governance first. However, in view of the huge political and legal risks that Chinese companies are bound to encounter as their investments enter these countries in large numbers, China should still take the opportunity to promote economic and trade agreements with them and to prioritize the implementation of access, treatment and investment protection in the infrastructure construction sector in them. Through international soft-law governance, China and these countries are able to unite as a BRI community and a human community of a shared future; through rule-based governance, we promote the extended protection of enterprises' interests in overseas investment and trade (Chen Weiguang and Wang Yan, 2016).

Improving the multilateral mechanism of the BRI. In the long-term development, countries and regions along the BRI have formed a large number of diversified international economic and trade rule systems and governance mechanisms. There are overlapping and conflicting differences among them. In the future, with the further development of the BRI system of legal rules and governance mechanisms, the situation may become even more complex, resulting in a serious "spaghetti bowl" effect. To this end, China needs to maintain good communication with countries along the route. It needs to promote the docking, coordination and integration of these rules and mechanisms based on common interests, and maximize the cultivation of the concept of "community rule of law" and a universal BRI legal rule system and governance mechanism. Of course, this project is so vast and complex that it is impossible to do it all at once or to find solutions to all the problems at once. We need some incremental and regional advancemental approaches. First, a number of countries at important nodes along the route can first form a fixed communication mechanism similar to the G7, and promote their consensus and practices to other participating countries; second, some regional trade agreement arrangements can be formed first, such as China-ASEAN, China-Eurasian Economic Union, China-GCC, China-Central and Eastern Europe or even China-EU FTAs or BITs, and then the universal rules matured under these regional frameworks can be continuously extended to the entire BRI route, continuously enriching the multilateral system of the BRI and institutionalizing the BRI multilateral-style platform. In addition, we can mobilize multilateral mechanisms such as WTO, OECD, World Bank and AIIB to join the framework of BRI and increase the participation in the international decision-making process and the recognition of the BRI, so as to continuously multilateralize it (Wang Huiyao, 2020, November 6). Ultimately, through the BRI, regional cooperation are achieved on a larger scale, at a higher level and at a deeper level. An open, inclusive, balanced and inclusive regional cooperation structure are built together with the countries along the route.

3. Improving the legal enforcement mechanism of the BRI

In addition to improving the corresponding legal rules, the construction of the BRI governance system is also inseparable from a fair and efficient legal enforcement mechanism. China and the countries along the route should make good use of the existing international dispute settlement mechanism, negotiate and establish an innovative dispute settlement mechanism, improve commercial legal services, and "form a set of multi-level, three-dimensional, domestic and international mechanisms to complement each other and interact positively, thus creating a stable and predictable rule of law environment for the construction of the BRI and setting a model for global economic governance in the new century. "(Liu Jingdong, 2017)

(i) Active use of existing international trade and economic dispute settlement mechanisms

Make good use of the WTO dispute settlement mechanism. The WTO dispute settlement mechanism (DSU) is known as the "crown jewel of the WTO" for its unique compulsory jurisdiction, quasi-judicial nature and rule-based approach, which provides stability and predictability to the multilateral trading system and has proven to be effective in practice. Despite the paralysis of the WTO appeals mechanism by the United States, the panel report remains an authoritative determination of the merits of trade disputes under the WTO framework. The compliance of member parties with their relevant commitments and obligations can still provide a strong guarantee for the operation of the multilateral trading system. Since the BRI countries differ greatly in terms of their degree of economic development, political systems, cultures and religions, etc., and we particularly emphasize that the construction of the BRI should respect each country's choice of development path and model. It may not be possible to form a highly unified regional economic and trade dispute settlement mechanism for a longer period of time. In this context, the DSU is an important channel and platform that we can actively use. The application of this mechanism to the BRI regional economic and trade dispute settlement has its unique advantages. The first is that there are many WTO members, and more than 3/4 of the BRI countries are WTO members, so the DSU has a wide scope of application; the second is that the DSU's original principle of "reverse consensus" has greatly improved the efficiency of dispute settlement; the third is that the DSU, as a dispute settlement mechanism under the multilateral mechanism, it is not easily swayed or influenced by a certain member. It is less likely to be biased towards domestic interests than the domestic relief methods of countries along the route, and can be more widely accepted and recognized by participating countries. Therefore, in the event of economic and trade disputes with countries along the route, if there is sufficient support from WTO rules, priority should be given to the use of the DSU to resolve disputes and defend itself legal interests. In addition, most of the BRI countries are developing countries, and certain developed countries are currently trying to restrict the freedom of trade and the legitimate needs of public interest of developing countries under the multilateral mechanism in the name of promoting WTO reform in order to safeguard their own interests. In the face of such a predicament, countries along the route should also gradually reach a consensus, expand cooperation, use the WTO dispute settlement mechanism to advocate common claims and protect the reasonable interests of developing countries.

Make good use of the dispute settlement mechanism under the FTA framework. In agreements such as the *China-ASEAN Free Trade Agreement*, the *China-New Zealand Free Trade Agreement*, the *China-Australia Free Trade Agreement*, the *China-Georgia Free Trade Agreement* and the *Regional Comprehensive Economic Partnership Agreement* (RCEP) reached just last year, a hybrid dispute settlement mechanism is adopted, with the establishment of a panel of experts or the establishment of arbitration court for adjudication when consultations fail, but this is not a mandatory procedure for dispute settlement. Both parties to the dispute are encouraged to resort to conciliation, negotiation, good offices and mediation at any time to resolve their disputes, with more emphasis on friendly consultation and voluntary mediation between countries in the implementation process. Through the flexible use and independent choice of such diversified dispute settlement methods, it not only improves the efficiency of the dispute settlement mechanism and reflects China's traditional ideology of "peace is precious", but also becomes particularly important in the construction of the BRI dispute settlement mechanism. It is worthy of being extended to the future economic and trade agreements with other countries along the route on a larger scale.

In addition, given the low level of rule of law in many BRI countries and the high risk of Chinese outbound investment, we should consider gradually including an investor-state investment dispute settlement mechanism ("ISDS mechanism") in the economic and trade agreements between China and BRI countries.

(ii) *Establishment of a diversified BRI international commercial dispute settlement mechanism*

As the BRI is moving towards the stage of taking root and lasting development, in addition to the need to establish a mechanism for the settlement of economic and trade disputes between governments, enterprises, as major participants in the construction of the BRI, they face the risk of frequent commercial disputes and their failure to be resolved in a timely and effective manner in the course of their trade and investment activities. It is also necessary to establish an international commercial dispute settlement mechanism for the BRI. It is to promote the proper resolution of commercial disputes arising in the course of the construction of the BRI and to serve and safeguard the construction of the BRI. On January 23, 2018, President Xi Jinping presided over a meeting of the Central Leading Group for Comprehensively Deepening Reform, it considers and adopts the Opinions on the Establishment of International Commercial Dispute Settlement Mechanisms and Institutions for the BRI. On March 26, 2018, the General Office of the CPC Central Committee and the General Office of the State Council issued the *Opinions on the Establishment of BRI International Commercial Dispute Settlement Mechanisms and Institutions* (2018, June 28), requiring all regions and departments to earnestly implement them in conjunction with actual practice. The *Opinions* proposed: "It should adhere to the principle of diversification of dispute resolution. It should fully consider the diversity of the subjects involved in the construction of the BRI, the complexity of the types of disputes and the differences in the legislation, judiciary and rule of law culture of various countries, actively cultivate and improve the dispute resolution service guarantee mechanism with organic articulation of litigation, arbitration and mediation, and effectively meet the demand for diversified dispute resolution in China and abroad party concerned. It should create a stable, fair, transparent and predictable business environment under the rule of law through the establishment of the BRI international commercial dispute resolution mechanisms and institutions."

Promote the application of the mediation system. Mediation is an activity to resolve disputes through voluntary and equal negotiation by party concerned under the auspices of a neutral third party mediator, which is a dispute resolution method originated in China and is called the "Eastern experience". In contrast to litigation and arbitration, the party concerned in mediation has control over the outcome of the dispute, and the mediator merely assists the parties in reaching a settlement. This helps to alleviate the conflict-of-laws problems often faced in the handling of international civil and commercial disputes. For foreign-related commercial disputes arising in the construction of the BRI, this advantage is more obvious, because these disputes often involve the laws or international rules of several countries. Conflict of laws is even more unavoidable, for party concerned, the final choice of the governing law to deal with disputes is less predictable, and through mediation can effectively alleviate these conflicts. In addition, mediation is conducive to easing the confrontation between the parties to the dispute, avoiding the intensification of conflicts between party concerned and helping to foster the willingness of both parties to continue cooperation and maintain good business relations. Because of these advantages, mediation systems have received widespread attention from the international community. In recent years many countries have begun to focus on building mediation organizations, systems and rules. Following the adoption of the *Model Law on International Commercial Conciliation* (https://uncitral.un.org/zh/node/3089) by the United Nations Commission on International Trade Law (UNCITRAL) in January 2003, the United Nations General Assembly session considered the adoption in December 2018 of the *Singapore Convention on Conciliation* (known as *the United Nations Convention on International Conciliation Agreements Arising out of Mediation*) (2019), which is intended to address the cross-border enforcement of settlement agreements reached through international commercial conciliation. These conventions can help States and mediation organizations to strengthen legislation and build rules on the use of modern mediation techniques. China can draw on and make use of these international legislative models; on the one hand, it promotes the professional reform of its own mediation system and enhance the international competitive power and credibility of its mediation mechanism; Plus, it actively introduces and promotes China's rich mediation system and practices to countries and regions along the BRI route, creates an international commercial dispute resolution centre and provides important safeguards for enterprises to effectively resolve international commercial disputes and maintains harmonious commercial relations.

Give full play to the unique advantages of arbitration in resolving commercial disputes. In contrast to conciliation, arbitral decision is ipso facto legally binding and, thanks to the existence of the *United Nations Treaty on the Recognition and Enforcement of Foreign Arbitral Decision* (the New York Convention) (1958, June 10), a multilateral convention, international arbitral decision can be recognized and enforced on an almost global scale. Compared to litigation, international arbitration party concerned has greater autonomy in controlling the process and outcome of dispute resolution. Arbitration is final and has the advantages of being quick, efficient, confidential, inexpensive and more neutral. As a result, enterprises from various countries engaged in international trade and investment increasingly tend to choose arbitration mechanisms to resolve transnational commercial disputes. According to statistics, approximately 80% of economic disputes of Global 500 companies are resolved through arbitration each year. Therefore, China should actively promote the application of international arbitration mechanisms along the BRI route, encourage competent domestic arbitration institutions to go out, gradually establish regional permanent institutions for international arbitration at important nodes of the BRI route, and cooperate with arbitration institutions in countries and regions along the route to establish joint arbitration mechanisms for international commercial disputes, so that arbitration can become the actual mechanism for dispute resolution in the construction of the BRI route.

Improving judicial support for the settlement of commercial disputes. Litigation is the most authoritative means of dealing with disputes and the last barrier to preserve legal interests strong judicial support is essential for the handling of commercial disputes in the BRI construction. The Supreme People's Court, as the lead unit in establishing the BRI international commercial dispute settlement mechanism and institution, has conscientiously organized the implementation of the Central Government's *Opinions on the Establishment of the BRI International Commercial Dispute Settlement Mechanism and Institution* (2018, June 27) and set up the First and Second International Commercial Courts of the Supreme People's Court in Shenzhen and Xi'an respectively on June 29, 2018. The International Commercial Court mainly hears major, difficult and complex international commercial cases, broadening the avenues for the settlement of international commercial disputes for party concerned without altering the international commercial dispute settlement mechanism provided for in existing international treaties. It implements first instance and final adjudication, thus achieving a quick and efficient settlement of disputes. The International Commercial Court has also pioneered the system of the International Commercial Experts Committee, which is composed of Chinese and foreign legal experts who are well versed in international law and skilled in national law, with extensive practical experience and a high international reputation. They are broadly representative both professionally and geographically, and can voluntarily mediate cases and provide advice on foreign law issues in cases in accordance with party concerned. In addition, the International Commercial Court is also a "one-stop" dispute resolution platform. As an important task in the construction of the BRI international commercial dispute resolution mechanism and institutions, it has built a diversified dispute resolution mechanism that effectively links litigation with mediation and arbitration. At present, the Supreme People's Court has selected a number of arbitration institutions and mediation institutions to be included in the "one-stop" mechanism for the diversified settlement of international commercial disputes. Within this mechanism, the International Commercial Court provides strong support to international commercial arbitration institutions and international commercial mediation institutions in accordance with the law in terms of procedures and the enforcement of awards or mediation, so that litigation is organically linked with mediation and arbitration to form a synergy and jointly create a "one-stop" multi-dispute resolution platform, providing quality and efficient legal services for partyconcerned (Zhang Xinghua, 2018, June 28).

In the future, we should fully build and make use of this platform, which brings together the most quality resources in the field of litigation, arbitration and mediation. We should encourage and attract parties to international commercial disputes at home and abroad to consensually choose the jurisdiction of international commercial courts. We expand the scope of international commercial expert members, expand the construction of "one-stop" multi-disciplinary dispute resolution platforms, and appropriately introduce extraterritorial international commercial arbitration and mediation institutions to enhance international credibility and better serve the resolution of international commercial disputes in the BRI and even globally.

(iii) Improving the BRI foreign-related commercial legal service system

The BRI Rule of law building is a system ensemble project. Among them, a high-quality and efficient national commercial legal services network and service system is a necessary support and guarantee. It is an important factor in determining the effectiveness of the operation of the future BRI legal system. In May 2016, the *Opinions on the Development of Foreign-Related Legal Services* (2017, January 9) of the Central Leading Group for Comprehensively Deepening Reform put the *Development of Foreign-related Legal Services* (http://www.gov.cn/xinwen/2017-01/09/content_5158078.htm) on the important agenda. In December 2016, the Ministry of Justice, the Ministry of Foreign Affairs, the

Ministry of Commerce and the Legislative Affairs Office of the State Council jointly issued the Opinions on the Development of Foreign-Related Legal Services, clearly proposing to provide legal services for the BRI and other major national development strategies. They jointly promote the development of foreign-related legal services and serving the construction of the BRI to a new level, which has achieved phased results in recent years (Xiong Xuanguo, 2017). In order to further build a good BRI foreign-related commercial legal service system, the following areas of work need to be focused on.

Cultivate and reserve a pool of foreign-related legal service personnel. Owing to the late start of China's foreign-related legal services industry as a whole, there is currently a serious shortage of foreign-related lawyers, foreign-related arbitration and mediation practitioners, etc., resulting in the loss of most commercial disputes related to China to be resolved outside China and often with the help of foreign professional teams. In the universities, international law is cross-disciplinary in nature and difficult to learn, the scale of international law personnel training is small. The reserve of foreign-related legal talents is completely unable to meet the realistic needs of serving the construction of the BRI and foreign relations in the new era. Therefore, on the one hand, as proposed by a deputy to the National People's Congress at the 2021 session, international law is upgraded to a first-class discipline, so that theoretical innovation and normative system construction of international law can fully reflect the frontier, independence, multidisciplinarity and intersectionality. It plays a leading role in the construction of China's international law discipline system and academic theory, strengthen the strategic choice of training foreign-related rule of law talents, innovate the training mechanism of foreign-related rule of law talents and continuously improve the quality of talent training (Huang Jin and Yang Song, 2021, March 7); on the other hand, foreign-related lawyers is incorporated into the national and key local medium and long-term talent development plan and key talent project plan. Under the guidance of the Ministry of Justice, it closely focuses on the needs of the "One Belt and One Road" strategy, continues to strengthen the training of foreign-related lawyers "leading talents", and accelerates the training of a high-quality talent team that is well versed in international rules and adept at handling foreign-related legal business.

Fostering professional foreign-related legal service institutions and platforms. In addition to nurturing a number of law firms with strong professional capabilities that can effectively help enterprises deal with foreign-related commercial disputes, it is also necessary to vigorously develop business associations with comprehensive foreign-related commercial legal support services to provide enterprises with "one-stop" foreign-related commercial legal public services such as early warning of economic and trade frictions, foreign-related legal training and assistance. It represents enterprises to express their demands abroad, organizing enterprises to conduct overseas class action lawsuits, helping enterprises to build branding systems and compliance management mechanisms, guiding and helping enterprises to conduct business in accordance with the law, and providing enterprises with accurate and customized international trade compliance information consultation and risk screening.

Building a network of foreign-related legal services. The demand for foreign-related commercial legal services in the construction of the BRI is very rich and complex. It is necessary to build a huge service network with the help of legal service resources from countries and regions along the route before it can be completed. We should actively promote Chinese legal service institutions and organizations to go out, cooperate and dock with relevant institutions and other business associations outside China. We should open up information and communication channels, promote legal service exchanges and cooperation and the establishment of joint operation and business alliances in countries along the route, and organize and coordinate legal service workers in participating countries to provide quality and efficient legal services for the construction of the BRI. Wherever the enterprise goes, foreign-related legal services shall follow up. It needs to enhance the quality and level of foreign investment and cooperation, protect the BRI construction in the rule of law track to run, and actively maintain regional security and stability, and promote regional economic development and prosperity.

References

2016–2017 Research Report on Chinese Enterprises Going Global (2017, April 9). http://china.cnr.cn/gdgg/20170409/t20170409_523698701.shtml.

Agreement on Trade Facilitation (2015, October 16). http://sms.mofcom.gov.cn/article/wtofile/201510/20151001138374.shtml.

Chen Weiguang and Wang Yan. (2016). 'Building BRI together: An analytical framework based on relational and rule-based governance', World Economy and Politics, (6), 93–114.

China Free Trade Zone Service. http://fta.mofcom.gov.cn/.

China-EU Comprehensive Investment Agreement. (At present, the text of the agreement has not yet been made public by the Chinese side. The European Union has released the main contents, and some domestic scholars have collated them) (2021, January 23). https://www.sohu.com/a/446285138-120220684.

Comprehensive and Progressive Trans-Pacific Partnership (CPTPP) (2021, January11). http://www.mofcom.gov.cn/article/zwgk/bnjg/202101/20210103030014.shtml.

Ding Tongmin. (2018)."Reflections on establishing a sound legal governance mechanism for BRI", Journal of Henan Normal University(Philosophy and Social Science Edition), (6), 45–50.

European Commission. (2020). Key elements of the EU-China Comprehensive Agreement on Investment. European Commission website. https://ec.europa.eu/commission/presscorner/detail/en/IP_20_2542.

Foreign Ministry, "China will not use BRI to establish rules dominated by one country" (2017, December 20). http://news.china.com.cn/world/2017-12/20/content_50115 163.htm.

Francis Snyder. (1994). "Soft law and institutional practice in the European Community", In Stephen Martin (ed.),The construction of Europe: Essays in honour of Emile Noel, Kluwer Academic Publishers, 199–200.

Han Yonghong. (2016). "An outline of the Soft Law Guarantee mechanism for International Cooperation in BRI",Contemporary Law. (4):151–160.

He Xiaofeng and Wu Junya. (2019). "Reflections on the sound legal governance mechanism of BRI". Legal Exposition. (10), 138–139.

He Zhipeng. (2017)."The trend of counter-globalization and International Soft Law", Journal of Wuhan University(Philosophy and Social Science Edition), (4):54–69.

Huang Jin and Yang Song. "Proposal to elevate international law to a first-class discipline", (2021, March 7). https://www.sohu.com/a/454606000_618422.

Jiang Xinmiao and Zhu Yanni. (2018). "Research on the legal supply mechanism of BRI', Journal of Northwestern University (Philosophy and Social Science Edition), (3):83–89.

Law Business Connection. (2017). "A first glimpse of the BRI legal risk map",The Import and Export Manager, (5):66–67.

Law on Hainan Free Trade Port (2022, April 14). https://www.fadada.com/notice/detail-3819.html.

Law on the Legal Application of Foreign-related Civil Relations, (2021, October 19). https://www.sohu.com/a/495907482_121117451.

Law on the Legal Application of Foreign-related Civil Relations (2021) and the Arbitration Law, (2017, September 12),http://www.npc.gov.cn/npc/c30834/201709/c8ca14070ead4c6d904610eea0f535fb.shtml.

Li F. (2017). "The construction of BRI legal cooperation mechanism and its risk prevention", Journal of Hebei University of Economics and Business (Comprehensive Edition), (1):20–24.

Li Na. (2018). "Implications of Egyptian arbitral decision recognition and enforcement practice on the rules of legal application of arbitration procedures and BRI legal risk prevention", Legal Application, (12):60–63.

Li Xiangyang. (2018). "The "absence" of regional economic integration in Asia and the development orientation of BRI", Chinese Social Science, (8):33–43.

Liu Jingdong. (2017). "Study on the construction of the rule of law system of BRI", Political and Legal Forum, (5), 125–135.

Ministry of Commerce of the People's Republic of China. (2021). Upgrade the free trade agreement, and China and Singapore jointly send a "New Year gift package".China Free Trade Zone Service Network. http://fta.mofcom.gov.cn/article//fzdongtai/202102/44421_1.html.

Ministry of Commerce: Report on China's Foreign Trade Situation (Spring 2021). http://zhs.mof com.gov.cn/article/cbw/202006/20200602974110.shtml.

Opinions on Accelerating the Implementation of the Free Trade Zone Strategy. (2015, December 6). http://www.gov.cn/zhengce/content/2015-12/17/content_10424.htm.

Opinions on the Development of Foreign-Related Legal Services (2017, January 9). http://www.gov.cn/xinwen/2017-01/09/content_5158080.htm.

Opinions on the Establishment of BRI International Commercial Dispute Settlement Mechanisms and Institutions (2018, June 28). https://baijiahao.baidu.com/s?id=1604476287894263226&wfr=spider& for=pc.

Opinions on the Establishment of the BRI International Commercial Dispute Settlement Mechanism and Institution (2018, June 27). http://www.gov.cn/zhengce/2018-06/27/content_5301657.htm?trs=1.

Opinions on the People's Courts' Provision of Judicial Services, (2017, March 20), https://www.waizi.org.cn/law/17621.html.

Ouyang Kang. (2018). "'One Belt, One Road' in the changing landscape of global governance", Chinese Social Sciences, (08):5–16.

Program for the Construction of Hainan Free Trade Port (2020, June 1). https://www.ccps.gov.cn/xtt/202006/t20200601_141319.shtml.

Protection for the Construction of the BRI and the Opinions on the People's Courts' Further Provision of Judicial Services and Protection for the Construction of the BRI in 2015 and 2019, (2019, December 9),http://gongbao.court.gov.cn/Details/e42b145854505083e6ba738fb78b71.html.

Shi Jingxia. (2021). "The 'Belt and Road' initiative and international law: An analysis based on the perspective of International Public Goods Supply", Chinese Social Sciences, (01), 156–179+207–208.

Sun, Hongtao. (2018). "Legal risk management of BRI for Big Petroleum and Petrochemical Engineering Services",SME Management and Technology (Zhongguo Xian), (2), 99–100.

The 2016–2017 "Going Global" Research Report of Chinese Enterprises, (2017, April 9). http://china.cnr.cn/gdgg/20170409/t20170409_523698701.shtml.

The United Nations Convention on International Conciliation Agreements Arising out of Mediation (2019). https://uncitral.un.org/sites/uncitral.un.org/files/media-documents/uncitral/zh/mediation_convention_c_0.pdf.

United Nations Treaty on the Recognition and Enforcement of Foreign Arbitral Decision, (1958, June 10). http://www.law-lib.com/law/law_view.asp?id=15259.

United States-Mexico-Canada Agreement [USMCA] details of the United States-Mexico-Canada Agreement can be found on the website of the Office of the United States Trade Representative: https://ustr.gov/usmca.

Wang Huiyao, (2020, November 6). "BRI can be linked to multilateral mechanisms such as WTO". http://www.ccg.org.cn/archives/60191.

World Bank's Doing Business 2020 Global Report. https://chinese.doingbusiness.org/zh/data/doing-business-score.

Xiong Xuanguo (Vice Minister of Justice). (2017). "Promoting the BRI legal services is our common responsibility", Chinese Lawyer, (07): 6–7.

Zeng Wenge. (2018). "On the construction and improvement of the land trade facilitation rule system of BRI", Journal of Law, (11):16–22.

Zhang Xinghua. (2018). The responsible person of the Supreme People's court answered reporters' questions on the opinions on the establishment of international commercial dispute settlement mechanisms and institutions under the "the Belt and Road". (2018, June 28). http://www.gov.cn/zhengce/2018-06/28/content.5301709.htm.

XI

Cultural Communication and Green Development for the Co-construction of the BRI

Introduction

The aim of the BRI is to focus on connectivity, deepen practical cooperation, coordinate to address the various risks and challenges facing humanity, and achieve mutual benefits and common development. By adhering to the principle of co-business, co-construction and sharing, China advocates multilateralism and purports the concept of openness, greenness and cleanness. It takes the ecological "green" as the underlying color, promoting green infrastructure construction, green investment and green finance. It aims at achieving a high standard development, people's livelihoods benefit and sustainable goals. It adheres to the ideology of a people-centered development, so that the results of the co-construction of the BRI can benefit all people. It makes a real contribution to the economic and social development of countries along the route, while ensuring commercial and financial sustainability. Cultural dissemination and green development are the long-term goals of the co-construction of the BRI. They are also the determinant factors for the formation of a sustainable, stable and good development of the countries along the route.

From the perspective of cultural communication, communication and exchange at the cultural level among BRI countries can effectively improve mutual understanding and awareness among countries, regions and people. It enhances mutual understanding and plays an important role in promoting the common interests of countries along the route. In the context of the development of the BRI, it is of great significance to strengthen the overseas dissemination of the Chinese culture in order to achieve the great rejuvenation of the Chinese nation. Through a multi-participation path led by the government and with the collaboration of all sectors and the participation of all people, it encourages the dissemination of culture among the BRI countries, so as to establish a good inter-personal relationship.

From the perspective of the green development, the sustainable development of the countries and regions concerned is enhanced through the agreement reached by the BRI countries in terms of green development models, development concepts and development directions. Making the BRI a green development path is the original intention of China's initiative. It is a concept that expects to achieve the goal of sustainable development in practice, together with the countries along the route. In order to jointly build the BRI high-quality development, green is the color of development. It can reach a consensus on green development of countries along the route through clear requirements and standards. It advocates the path of green development of countries along the route through optimizing and innovating development structures and methods. It escorts the green development of countries along the route through systematic protection, governance and integration of industrial structures.

The Belt and Road cooperation pursues development, aims at mutual benefits, and conveys a message of hope (Xi Jinping, 2021, 20 April). Taking the opportunity of co-constructing the BRI, we tell a good Chinese story, spread a good Chinese voice, show a real, three-dimensional and comprehensive China. It forms an international discourse that matches our comprehensive national power and international status, creating a favourable external public opinion environment for China's reform, development and stability. It makes a positive contribution to promoting the building of a human community of a shared future (Xi Jinping, 2021, May 31). Meanwhile, China has made cooperation in the field of ecological civilization a priority in the co-construction of the BRI. It launches a series of green action initiatives and takes a series of green initiatives to continuously benefit the people of all countries participating in the co-construction of the BRI (Xi Jinping, 2021, 22 April).

1. Multi-party participation in the cultural dissemination of the BRI

Since ancient times, BRI countries take different forms and styles of civilization. They have formed a colorful and splendid "Silk Road Culture" through cultural exchanges. The world today is a pluralistic and multipolar one. Different countries have chosen different social systems and development paths based on different historical traditions, cultural genes and religious beliefs, resulting in a colorful and diverse modern civilization. On the basis of the spirit of interaction along the ancient Silk Road, countries along the route should further strengthen exchanges and cooperation among their cultures and create new civilizations. During the 14th Five-Year Plan period, China strengthens top-level planning and concrete implementation. It joins hands with various BRI countries to elaborate a number of BRI cultural exchange projects, carrying out in-depth exchanges and cooperation in such fields as culture and art, science and education, religion and history, sports and tourism, and health and hygiene. It jointly organizes diverse and colourful cultural forums, exhibitions and performances, holds cultural years, art festivals and film weeks, supports youth exchanges, academic exchanges, volunteer dispatches and exchanges among non-governmental organizations from various countries. It extensively carries out various public welfare and charitable activities in education and medical care, poverty alleviation and development, ecology and environmental protection. The ultimate aim is to promote interpersonal contacts through cultural exchanges.

(i) Government-led promotion

1. Implementing an extroverted cultural development strategy and focusing on the top-level design of cultural communication.

It needs to implement an extroverted cultural development strategy to effectively promote the international dissemination of Chinese culture. In the process of the international dissemination of Chinese culture, it is necessary to implement an extroverted cultural development strategy that focuses on the dissemination of Chinese culture in order to expand its influence and win a broader space for development in the world. The promotion of Chinese culture and the strengthening of cultural exchanges and cooperation with BRI countries must follow the policy of "we are the main focus" and "government-led". It shall be in line with the general policy of national politics and diplomacy, as well as the overall economic development of the country, the construction of domestic socialist spiritual civilization, and the combination of official and private sectors.

It focuses on top-level design of cultural communication and adheres to the strategic layout of cultural opening to the outside world. It plays a leading and coordinating role in government, strengthens cultural exchanges with the governments of countries and regions along the BRI route, focuses on establishing a long-term cooperation mechanism, gives full play to the advantages of domestic provinces, autonomous regions and municipalities, and encourages the active participation and co-construction of social forces. Taking into account the interests and concerns of all parties, it follows international rules and the laws of the market, giving full play to the important role of the market in cultural resource allocation, mobilizing the enthusiasm of all parties, closely integrating culture with diplomacy and economic and trade, forming a coordinated development of cultural exchange, cultural dissemination and cultural trade, and achieving mutual benefit and win-win results. It explores the establishment of a foreign cultural industry fund and strengthens financial cooperation in the field of culture in the BRI. Focusing on key cultural industries and projects, it promotes the effective allocation of cultural resources, the rational flow of production factors and the deep integration of cultural markets, and form a reasonable layout for the development of the Silk Road cultural industry.

2. Optimize the Chinese cultural industry system and promote the "going out" of the cultural industry

On the one hand, we should continue to cultivate a Chinese cultural industry system with socialist core values. The promotion and cultivation of the national spirit is the spiritual force on which our cultural construction is based. It is also an inevitable requirement for Chinese culture to participate in the process of world cultural development in the context of economic globalization. Globalization and modernization are holistic changes driven by multiple factors. Economic, social and cultural aspects must be developed in a coordinated manner. It seems necessary to cultivate a Chinese cultural industry with socialist core values, which not only has the role of providing spiritual pillars and value guidance, but also has the value of psychological satisfaction and behavioral adjustment. The Chinese cultural industry system should be a system composed of mainstream culture, subculture, local culture, elite culture, and popular culture. The Chinese culture that combined with the BRI builds a new pattern of positive, healthy and innovative development of Chinese culture, creating a new image of "cultural China" and establish a multi-level cultural industry system with socialist core values.

On the other hand, we should continue to vigorously promote the cultural industry to go global. It creates cultural media for the international market as a breakthrough, and vigorously develops the international cultural industry. In recent years, China has put a lot of effort into building influential media platforms and personnel, investing a lot of money, upgrading equipment and establishing awards. The effect, it should be said, is clear but limited. While adhering to the roots of excellent traditional culture, we should follow the characteristics of the era of globalization and information technology. It strengthens the building of humanistic exchanges by enhancing foreign-related media, education and cultural industries. We should actively learn from the successful experience of non-English-speaking media in going global and build China's world-oriented media. In this regard, the experience of Al-Jazeera in Qatar and Russia's *Russia Today TV* is worth learning from.

The Chinese media should consciously establish a communication discourse system of equality, mutual appreciation, tolerance and respect. It conveys to BRI countries the peace aspirations of our traditional culture, the concept of new international relations, and the concept of a human community of a shared future. This requires the creation of new concepts and expressions that connect China and abroad. The use of words are appealing and credible so as to tell the Chinese story well. In the development of foreign cultural industries, we should adhere to the model of government promotion, enterprise main body, market-led and commercial operation. The government should formulate and improve policies related to cultural innovation, increase financial support for international cultural industries, and build multiple channels of cooperation to integrate cultural resources. Enterprises should learn from the successful experience of international cultural industries in other countries, making efforts to train international specialized personnel, operating according to market needs for business investment and other operations, and significantly improving their legal awareness, management level, cultural literacy, etc. for transnational operations.

Specifically, the work can be carried out in three ways. First, it needs to raise the awareness of innovation among practitioners in the international cultural industry. In the context of the BRI, if cultural industries want to "go global", cultural industry practitioners must continuously deepen their knowledge and understanding of the BRI international cooperation strategy. The concepts they hold should keep up with the times. Practitioners in China's cultural industry should change their traditional concepts, gradually build up cultural concepts and innovative ways and means that are compatible with the BRI, so that more of China's cultural works can "go out". Second, the policy of the BRI culture industry needs to be strengthened before promotion. Our government should strengthen the promotion of the policy, especially in areas with special affinity to the BRI, such as Quanzhou, Xinjiang, Xi'an, etc. The policy on the development of the cultural industry BRI is brought to the minds of the workers, so that they are inspired to create more excellent works and go abroad. Third, it has to broaden the financing channels for the cultural industry. The financing channels of China's cultural industry mainly rely on government support and bank loans. However, in the context of the BRI international cooperation, relying solely on policy financing can no longer meet the needs of cultural works going abroad. Therefore, our government should, while investing more in the cultural industry, formulate some regulatory documents to provide tax concessions, financial help and other preferential terms for social capital that invests in cultural works. It should also actively encourage them to participate in the work of exporting cultural works.

3. Building a cultural exchange platform and creating the BRI cultural exchange brand

Good cultural exchange platform. The existing dialogue mechanisms, such as the meetings of the Ministers of Culture of the member States of the Shanghai Cooperation Organization, the China-Central and Eastern European Countries Forum on Cooperation among Ministers of Culture, the China-Arab Forum of Ministers of Culture and the China-ASEAN "10+1" Meeting of Ministers of Culture, as well as the bilateral and multilateral cultural tourism cooperation mechanisms, such as China-ASEAN, China-Central and Eastern Europe and China-Russia-Mongolia, play an important role in cultural exchanges. It further promotes the signing of bilateral and multilateral cooperation documents in the field of culture and tourism with BRI countries and relevant international organizations. It jointly promotes the establishment of multi-level humanities cooperation mechanisms and opens up more cooperation channels. It strengthens international theatres, museums, festivals, libraries, art galleries and alliances in the field of tourism. Encourage regional cooperation between domestic places and regions along the BRI route in such areas as cultural exchange, heritage conservation, literary creation and cultural tourism. It actively builds cultural exchange platforms between Hong Kong, Macao and Taiwan, and countries and regions along the BRI route.

It creates a brand of cultural exchange. The international influence of the BRI theme festivals, such as the Maritime Silk Road International Art Festival, the Silk Road (Dunhuang) International Cultural Expo and Dunhuang Walk, and the Silk Road International Tourism Festival, has been enhanced. The connotations of cultural and tourism brands such as the "Silk Road Journey", "Happy Spring Festival" and "Beautiful China" have been deepened. It promotes the translation of Chinese and foreign cultural classics, supports the creation and promotion of stage art works on the theme of the BRI, and co-creates plays and outstanding works on the theme of the BRI with countries along the route. It strengthens the protection of cultural heritage and promote international cooperation with countries and regions along the route in archaeological research, restoration of cultural relics, exhibition of cultural relics, personnel training, museum exchanges, and World Heritage nomination and management.

Establishment of a permanent mechanism for the BRI summit. Taking the successful convening of the first and second Beijing Summits on International Cooperation on BRI in 2017 and 2019 as an opportunity, the Forum continues to be held every two years, bringing together all countries along the route for in-depth exchanges on issues related to the construction of the BRI, making it the highest-level official forum on the construction of the BRI. The first and second Beijing BRI international cooperation summits were successfully held in 2017 and 2019 respectively. After the exploration and experimentation of these two forums, a rich and diverse format of the opening ceremony of the summit, leaders' roundtable summits, high-level meetings, sub-forums and entrepreneurs' conferences has gradually developed, promoting the building of a wider circle of friends, more partners and deeper cooperation in the BRI. The theme of cultural cooperation and integration should be given more prominence in the next few BRI summits. In the short term, consideration can be given to holding them in important domestic cultural node cities along the BRI route. In the long term, consideration can be given to continuing to hold them in key cultural cities of other major BRI countries, adhering to the concept of open, inclusive and transparent cooperation, deepening the docking of development policies and plans, bringing the cultural advantages and potential of each country along the route into full play, and promoting a high-quality co-construction of the BRI cultural cooperation.

4. Integrating and promoting foreign aid cooperation

During the 14th Five-Year Plan period, the high-quality development of the BRI has provided new directions, principles and priorities for China's foreign aid, which is more inclined to countries along the BRI and accelerate the transformation of China's foreign aid. In the context of the BRI, China does not only expand the scale of its foreign aid, but also give more prominence to key areas such as infrastructure, climate change, poverty reduction and sustainable development, security capacity and peace-building. During the 14th Five-Year Plan period, China's foreign aid practice should be guided by President Xi Jinping's foreign aid theoretical system, appropriately expand the share of regional aid and multilateral aid. It takes on more responsibilities and obligations within its capacity, innovate foreign aid models, improve the effectiveness and quality of aid projects, and focus on foreign aid projects in areas such as infrastructure improvement, health care, agriculture and education, so as to bring tangible benefits to local people.

Specifically, the first step is to identify medium- and long-term country assistance action plans for countries along the route. With the BRI as a guide, China's foreign aid policy framework and strategic planning is treated as a direction, and China's diplomatic priorities as a guide/ China should expeditiously define country-specific aid action plans for some countries along the BRI, clarifying the challenges and opportunities of the countries along the route in terms of policy measures, economic conditions, climate change and development priorities, in order to play its leading and exemplary role. In the 14th Five-Year Plan period, the high-quality construction of the BRI should be based on the neighboring countries, with the diplomacy of the countries along the route as a foothold, and the interpersonal contact as the key. In terms of China's diplomatic priorities, relations with Central Asian countries, China-Mongolia relations and China-Pakistan relations should be the priority directions. For example, Sri Lanka is in an important position in China's neighboring diplomacy. Fiji is the first Pacific island country to establish diplomatic relations with China. In this regard, we believe that China should first identify five country assistance action plans for Mongolia, Kazakhstan, Sri Lanka, Fiji and Pakistan. It promotes the policy measures and key areas of the BRI to strengthen interpersonal ties.

Second, foreign aid projects should be carried out evaluation of performance. With the further promotion of the BRI, the scale of China's foreign aid is certainly further expanded. In accordance with the *Performance Targets for Central Sector Budgets Management Method* (2015, May 21) issued by the Ministry of Finance on May 21, 2015, budget execution should be organized in accordance with the performance targets set at the time of the foreign aid budget in the future. A corresponding performance indicator system should be established, and performance self-assessment and evaluation should be carried out according to the performance targets and performance indicator system. As for the evaluation of performance organization structuring issue of foreign aid work, China has not yet clarified it. The Ministry of Commerce's regulations generally provide for the Ministry to carry out monitoring and evaluation, but do not refine it. As seen in developed countries, countries pay much attention to the evaluation of the effectiveness of aid extravehicular activity and evaluate it through independent expert juries or by setting up specialized foreign aid evaluation agencies. The evaluation of foreign aid performance is actually a comparison of the "input-output" of foreign aid funds, which can effectively prevent the waste of national resources and further enhance China's soft power and discourse by strengthening the management and evaluation of foreign aid projects.

Finally, it is necessary to deepen tripartite development cooperation. Triangular development cooperation has become an important new trend in the field of foreign aid, helping to improve aid effectiveness and to make full use of the resources, knowledge and experience of different aid providers. In the construction of the "Silk Road Economic Belt", China should take the China-Russia-Mongolia Economic Corridor, the China-Pakistan Economic Corridor, the Bangladesh-China-India-Myanmar Economic Corridor and the Central Asia-West Asia Economic Corridor as the basis to carry out tripartite development cooperation between China and countries along the route, especially with Russia, India, Kazakhstan and other countries. With regard to the construction of the Maritime Silk Road, China needs to strengthen tripartite development cooperation with ASEAN and Indonesia and other member countries based on the South China Peninsula Economic Corridor. It deepens tripartite cooperation with New Zealand under multilateral frameworks such as the United Nations and APEC, and jointly helps island countries in the South Pacific region to address climate change. Given the openness and inclusiveness of the BRI, China also strengthens communication and coordination with the United States, the United Kingdom, the European Union, the Nordic countries and other BRICS countries in the areas of medical and humanitarian assistance under the BRI. They share knowledge and experience, and work together to achieve common goals such as poverty eradication and global development in the 2030 Agenda for Sustainable Development, food security, global health security, humanitarian assistance and multilateral institutions.

(ii) Concerted efforts by all sectors

1. Enhancing the comprehensive competitive capacity of cultural enterprises

(1) Fostering cultural enterprises to equip with international competitive power. Cultivating cultural market players with international competitive power is a key link in promoting Chinese culture "going global" during the 14th Five-Year

Plan period. First, we should promote cultural enterprises to be psychologically prepared to "go global". At the same time, in order to improve the level of localization, it is necessary for cultural enterprises to learn as much as possible about local social customs, interpersonal relations, environmental awareness and other characteristics and features. It uses international language to integrate into and develop international markets. The second is to enhance the industrial level, product quality, technological content, supply chain management and brand building capabilities of cultural enterprises. Cultural enterprises must adapt to the current upward momentum of their position in the international value chain and follow the trend to improve their capabilities and levels in all areas and eventually develop a comprehensive advantage. Third, promote culture industrial restructuring. It is actively fostering new cultural industries, nurturing a number of export-oriented backbone cultural enterprises and large cultural intermediaries, improving the scale, intensification and specialization of cultural industries, vigorously building brands for the export of cultural products and services, and enhancing the cultural competitive power of enterprises "going out".

It strengthens dialogue and communication between Chinese and foreign media, deepens the construction of the BRI journalism cooperation alliance, the BRI journalist organization cooperation platform and the Silk Road TV international cooperation community. It actively carries out cooperation in the fields of journalist organizations and news publishing. In order to make use of such platforms as the China-ASEAN Information Ministers' Meeting, the China-Arab Media Dialogue, the China-Africa Media Summit and the "10+3" Media Cooperation Seminar to carry out a wide range of multi-bilateral exchange activities. It encourages domestic media to strengthen cooperation with the media and social media platforms of countries along the route, tell the "Silk Road story" in the new era, effectively enhance people's sense of affinity, integration and identification with the high-quality construction of the BRI in various countries, and create a favorable public opinion environment for the construction of the BRI.

(1) Focus on the deep integration of culture and technology. Today's society is highly dependent on technological development. Many cultural changes and innovations are often brought about by technological developments and breakthroughs, such as the Creators, "Internet+", "Industry 4.0", etc., all reflect the future trend of economic development and represent the direction of the BRI high-quality construction. Enhancing the integration of culture, science and technology, and improving the power of communication and expression of culture can make cultural products more powerful in influencing the world and benefiting humanity. First, we should promote the transformation and application

of advanced and key technique achievements in the relevant fields of science and technology to the cultural field. We should innovate cultural products and service models, and enhance the scientific and technological content of cultural products. Second, the use of Internet thinking drives the integration of culture and technology. The Internet has changed social life and gradual modification traditional business models. In promoting the integrated development of culture and technology, we should insist that the platform and the content are very important factors. It continuously promotes innovation in business models based on the Internet and mobile Internet, and promotes emerging models such as professional verticals, the O2O model, equity crowdfunding and online participation as the mainstream of cultural industry development.

Along with the development of the Internet economy, the New Economy form represented by "Internet+" has come into being.The integration of the Internet and traditional industries by relying on Internet information technology can help build the BRI with high quality. In terms of means of communication, new media technologies makes full use of the Internet and smartphones. In cultural communication, the Internet and other new media technologies vividly reflect the docking of China's foreign aid, cultural cooperation and local development needs as well as changes in the lives of local people. It encourages Internet companies such as Baidu, Alibaba and Tencent to actively participate in the BRI high-quality co-construction, cooperate with local Internet, media and entertainment companies, and adopt ways to participate in the development of cultural and creative industries by working with local artists cooperative development animation, games and filming movies. The State should also set up a special fund for the development of the Internet to help the BRI and encourage social enterprises, nonprofit organization, religious groups and individuals to participate in the cause of the BRI public welfare and cultural dissemination.

2. Expanding cultural exchange and cooperation channels

(1) Support social organizations in their international engagement. As an unofficial force, private non-governmental organizations should actively participate in the high-quality construction of the BRI and promote the spread of Chinese culture in countries and regions along the route. First, through the direct dissemination of Chinese culture by private non-governmental institutions (such as Confucius Institutes), China's peaceful and tolerant ideas are conveyed to countries along the route; second, through the use of private organizations to understand the customs and cultural charms of countries along the route, cross-cultural exchange activities are carried out, experiences and lessons worthy

of internationalized enterprises and organizations are drawn, the management level of various construction institutions and their ability to consciously resist the invasion of undesirable ideas are improved, and the communication and integration of the cultures of various countries are strengthened.

At the same time, we should improve the construction of the cooperation network of civil organizations along the Silk Road, attracting more social organizations from countries along the route to join and build an effective platform for exchanges between social organizations in China and countries along the route. We further promote the "Silk Road Family" initiative, encourage social organizations in China to actively carry out relevant international exchange activities. We need to enhance the recognition and support of people from various countries for the high-quality construction of the BRI. It guides Chinese enterprises cooperating in countries along the route to entrust social organizations to undertake public welfare projects, improves the level of funding service efficiency and project specialization, and effectively enhances the sense of participation, access and happiness of people in the project sites.

(2) To give full play to the role of religious and cultural exchange. Religious culture is an extremely important element of culture that touches many aspects of social life. The impact of religion is not only reflected in the material, cultural life and spirituality of its followers, but also in the relations and stability of peoples, countries and regions, as well as in political, economic and social development. The trend towards politicization and extremism of religion in today's world has gradually increased its influence on regional politics. Religion has become a resource that international forces are competing for. There are numerous and complex ethnic and religious problems along the BRI route. Terrorism and extremism are rampant in this area, especially in areas dominated or sought to be dominated by hegemonic States, such as Central Asia, the Middle East and North Africa, which has become a common challenge for China and other BRI countries in the context of high-quality co-construction. Therefore, in the process of building BRI together with high quality, it is necessary to realize that most countries along the route have their own religious and cultural beliefs. We need to play the role of religious dialogue and integration on the basis of equal respect and in a peaceful and friendly manner, so as to avoid contradictions and conflicts caused by differences in religious beliefs.

3. Promoting cooperation in the public service

(1) Strengthen cooperation in education and think tanks. It strengthens educational cooperation and enhances the capacity of talent support services. It

strengthens educational exchanges and cooperation with UNESCO, and other international organizations and professional institutions, and promotes the signing of educational Lead City cooperation agreements with more countries along the route. It supports the active participation of relevant department in the mechanisms for humanities exchanges between Chinese and foreign governments and multilateral education ministers' meetings in Central and Eastern Europe, China-ASEAN, the Shanghai Cooperation Organization, APEC and the European Union. It promotes bilateral, multilateral and inter-regional education exchanges among BRI countries, and the development of multi-level and multi-city practical cooperation. It promotes mutual recognition of academic degrees, interchange of standards and mutual appreciation of experience with more countries along the route. It supports qualified higher education institutions and vocational colleges to establish university alliances and vocational education alliances with universities in countries along the route.

It encourages Chinese and foreign think tanks to play an active role in the high-quality co-construction of the BRI. The BRI International Think Tank Cooperation Committee and the BRI Think Tank Cooperation Alliance have the advantage of extensive contact with all sectors of society, including politics, industry, academia and research. They are striving to build a BRI high-end think tank dialogue platform with greater international influence. It supports domestic think tanks to deepen cooperation with think tanks of countries along the route, widely carry out dialogue and seminars, joint research, release results and other forms of cooperation, providing more public products for the co-construction of the BRI high-quality development. It provides intellectual support for practical cooperation between governments and enterprises, and contributing wisdom and strength to promote the building of a community of a shared future.

It promotes the construction of international talent think tanks and strengthen international exchanges and cooperation among think tanks. Up to now, the scope of national think tanks is still mainly domestic. Chinese culture is marginalized in the international arena, which is not conducive to the mastery of national discourse in the process of cultural dissemination. Chinese think tanks should develop their own internationalization strategies. They should deepen close cooperation with think tanks in the United States, Russia and other high-level overseas countries. They learn from their effective experiences, find academic and countermeasure research partners in countries along the route, and jointly conduct strategic research and malefactors that serve local economic and social development goals. Through a series of think-tank platforms, we have established daily cooperation and dialogue mechanisms with overseas think-tanks, gaining public discourse and influence through reciprocal exchanges. They should create an objective and realistic public opinion environment on the cooperative, inclusive and public nature of the BRI in the world.

(2) Promote the spread of the Chinese language in the regions along the route. In view of the real demand during the 14th Five-Year Plan period, the Confucius Institutes and classes, which are mainly aimed at spreading Chinese culture and increasing the popularity of the Chinese language, are still very inadequate and unevenly distributed along the BRI route. From an objective point of view, the weak promotion and low popularity of the Chinese in countries and regions along the BRI route is not just a concern, but a fact. If Chinese remains a language that is not spoken in front of the courts in the countries and regions along the belt and road, the development of the belt and road in the region and its global development projects are greatly hindered. Therefore, it is imperative to promote the spread of Chinese language in countries and regions along the BRI route. First of all, at the national and governmental levels, the construction of Confucius Institutes and classrooms should be speeded up in the regions along the routes with the support of the country's strong economic strength and under the premise of reasonable, scientific and systematic planning, with the ideas of "having priorities", "setting examples" and "leading by points". Specifically, during the 14th Five-Year Plan period, a number of model Confucius Institutes should be built in the regions along the route to drive and lead the development of other Confucius Institutes in the region. In addition, with the help of the BRI, the main media platform for foreign communication, we can set up special columns on Chinese language and choose special names such as the "Window of Chinese Language" to spread the knowledge of Chinese culture in the columns.

4. Strengthening talent bidirectional flow movement

(1) Strengthen the exchange of talents with countries along the route. Promote the BRI two-way exchange of talents and expand the scale of mutual international students. We continue to implement the "Silk Road" government scholarships and the Head of the Wild Goose Project, providing training opportunities for relevant countries to come to China. We help train professionals and build up expert teams in countries along the route. Taking the lead in sending students on state study tours, we are promoting the implementation of a more active, open and effective talent policy. It actively sends more international students, experts and scholars to study and exchange in countries along the routes. It promotes the construction of Confucius Institutes and Luban workshops in countries along the route, providing vocational skills training for those employed in countries

along the route. It also promotes the understanding of traditional Chinese culture among people in countries along the route. It guides domestic universities to set up research institutions and cooperation platforms for countries along the route.

Emphasis on nurturing a reserve of think-tank talent. Under the BRI high quality development strategy, spreading Chinese culture to countries and regions along the route means "going global". A key prerequisite for cultural outreach is the cultivation of talent, which is the primary resource and central strength of think tanks. The circulation of Chinese culture in the world cannot be separated from the flow of outstanding talents. At present, China is actively establishing Confucius Institutes and creating Confucius Classrooms in various countries and regions to actively teach foreign students Chinese language and other Chinese cultural knowledge. At the same time, foreign language teaching activities are also in full swing in China and have made remarkable achievements, both in terms of professional settings and teaching materials, as well as in terms of faculty strength and enrollment scale, all of which have shown strong vitality, but this is far from enough for the international dissemination of Chinese culture. The high-quality construction of the BRI is a new historical opportunity for China's cultural communication, which requires high-quality talents with global vision and the ability to participate in international affairs and cope with international competition, especially in the areas that need to be improved in the countries along the route. In the 14th Five-Year Plan period, we need to cultivate a group of scholars and experts who are well versed in both Eastern and Western cultures, so that Chinese culture can be disseminated overseas. At the same time, it is developed in an innovative manner, so as to form a cooperative research and development force with the local people and accelerate the pace of quality development of the BRI together.

(2) Establishment of the BRI National Study Abroad Fund to promote international talent mobility. The establishment of a study abroad fund to strengthen the education of foreign students in China and to encourage domestic students to "go out" and "tell the Chinese story" is of strategic importance, especially in view of the shortage of high-end talents in small languages, the lack of strategic foreign language talent reserves and the lack of understanding of the public sentiment of partner countries. The destination countries of Chinese students are mainly English-speaking countries, but BRI countries and other countries with economic activities with China are gradually gaining attention. It is important to clearly see that the current shortage of non-general language personnel in our country is far from meeting the development needs of a high-quality common BRI. The country can make the problem go away through student exchanges, teacher

transfers, and even achieve full coverage of the forty or so official languages along the route. The relevant national educational administrative institutions may set up BRI study abroad funds to support student and teacher exchanges with countries along the route, covering at least professional talent programs, joint training programs and language programs.

(iii) Broad participation of all staffs

1. Playing the role of a bridge for students studying in China

(1) Overseas students in China are the inheritors of the Silk Road spirit and the spreaders of the Chinese story

International students from China have the advantage of "connecting China and abroad". It can bring Chinese language and culture to the world, and tell "Chinese stories" to suit the tastes of their compatriots. Many of them grows up to be pillars of the society. The Chinese stories they tell have a large audience and a great influence, and the effect of dissemination increases geometrically. For the general public, the direct channel to learn about China is to be surrounded by people who have been there or are studying there. The construction of the BRI straddles Europe, Asia and Africa, involving more than 60 countries and 4.4 billion people along the route. There are many developing countries along the route, and it is the common dream of the people in these countries to "get rid of poverty, improve people's livelihood and make the country rich and strong". China's peaceful rise has demonstrated to the world the Chinese solution and Chinese wisdom. In response, the eyes of the world have been envious and uncomprehending. The international community also has some doubts and misunderstandings about the BRI. The global pandemic of 2020 and China's outstanding performance in fighting the pandemic not only made China the safest place to be, but also reached out to the international community to help people in other countries fight the pandemic, which is itself a manifestation of the Silk Road spirit. International students who experienced the pandemic were touched by the heroes of China's pandemic fight and the story of the great love of the Chinese people. They supported China's pandemic fight by donating masks, volunteering in the community and making videos to cheer for China. They became part of the "China's pandemic fight story" and the inheritors of the Silk Road spirit.

(3) Cultivating BRI builders through education for study in China

Young people from BRI countries who came to China to study and grow up in China in the context of the high-quality construction of the BRI are the builders and successors of the future BRI. The youth of today leads the world of tomorrow. Their correct understanding of the BRI is an important foundation for the stability and success of the BRI. It is possible to cultivate people's hearts and public opinion by "educating people with culture". It set up relevant BRI knowledge courses in Chinese language teaching, tell stories about the history and culture of the Silk Road, the stories of Silk Road people, the culture and customs of various countries, as well as international economic and trade cooperation and traditional friendship, so that international students can understand the contribution of Silk Road culture in the history of human civilization and its far-reaching impact on contemporary economic development. It can help students gain a deeper understanding of the Chinese and world cultures, cultivating their understanding of diversified cultures and their sense of identity with BRI, enhancing their awareness of the human community of a shared future, gather more consensus and confidence, and make them the builders of the BRI who know and love China. At the same time, it improves the quality of education for students coming to China and cultivate high-quality talents of international students so that they can play a greater role and influence after returning to their home countries. It deepens international students' identity and emotion towards Chinese culture with good quality of education service, and sow the seed of friendship in their hearts so that they are willing to advocate for China's education and culture after graduation.

2. Leveraging the "localization" energy of the Chinese diaspora

(1) Chinese and overseas Chinese are the "liaison" for Sino-foreign friendship and cooperation

The Chinese diaspora has skillfully navigated between Chinese and foreign cultures, bridging the cultural exchange between BRI countries. Overseas Chinese are the inheritors of fine Chinese culture and should take up the responsibility and mission of spreading Chinese culture. They are unique in telling Chinese stories, spreading Chinese culture and promoting Chinese brands. The establishment of Chinese education by overseas Chinese, the spread of Confucius Institutes all over the world, and the popularity of the Chinese language around the world have made it possible for the profound and colorful Chinese culture to spread overseas with cross-cultural communication. It greatly enhances the influence of Chinese culture in countries around the world. Chinese brands are an important carrier of Chinese culture. China certainly creates a series of brands in the process of building the BRI with high quality. Overseas Chinese media, associations and major enterprises founded by overseas Chinese are the main positions of Chinese brand publicity and promotion; on the one hand, they can publicize the design concept, development history and cultural connotation of Chinese brands overseas. In addition, they can also promote the brand experience of other countries, playing a positive role of mutual exchange and common development. In short, the BRI countries are culturally diverse and complex. The unique cultural integration and identity of the Chinese diaspora makes them comfortable in participating in the construction of a high-quality culture in the BRI.

(2) Chinese diaspora as a "guide" for overseas talent building

At present, the international competition for talents is intensifying. The flow of talents is a major trend of globalization, while talents are an important support and basic guarantee for the in-depth promotion of the high-quality development of the BRI construction. Only by cultivating a large number of international composite talents with international vision, excellent quality and strong skills can we meet the requirements of the construction of the BRI and provide a more solid guarantee for the construction of the BRI to move steadily and far. As a result, countries have introduced various measures to attract foreign talent. The overseas Chinese, by virtue of their diligence, hard work, courage, solidarity and mutual assistance, have rapidly emerged in the world of science and technology, culture and education, and other fields. According to the survey, there are more than 4 million overseas Chinese professionals. The vast majority of them work in foreign enterprises, mainly involved in the education industry, financial industry and high-tech fields, etc. About 19% are in educational institutions, and a considerable part is devoted to new bio-engineering, new medicine, new material, cultural creativity, energy saving and environmental protection and other emerging industries (Chen Shuisheng, 2016). The increasingly diversified career paths and gradually increasing scientific and technological strengths have given strong impetus to the building of China's overseas talent pool and can provide strong intellectual support for the promotion of the BRI. On the one hand, overseas Chinese can make full use of their professional advantages to deeply interpret the connotation of the BRI related policies. They build exchange and cooperation platforms and hold a series of activities. They build bridges and pave the way for promoting international exchange and cooperation, carrying out project finance, and personally participate in guiding the implementation of some projects; on the other hand, the brilliant achievements of the overseas Chinese in various fields are inseparable from their unity and mutual assistance. They work hand in hand to advance together. They share the common bloodline of being born from the same root. They uphold the ideal belief of working together for development and set up a series of social groups. All of these assembles the elites of overseas Chinese through the force of mutual assistance and provide intellectual support to better participate in the construction of the BRI.

(3)The Chinese diaspora is the "paver" for the smooth flow of trade in the BRI
The smooth flow of trade is an extremely important element of the high-quality co-construction of the BRI, which is in line with the objective law of economic development of China and BRI countries. At a time when the world economy is facing downward pressure and trade protectionism is on the rise, the high-quality co-construction of the BRI provides a strong guarantee for the promotion of smooth trade flow of countries along the route. The Chinese diaspora, with their extensive and convenient trade channels, as well as their deep knowledge of bilateral trade rules and investment, play an important role in promoting high-quality trade facilitation in the BRI region. By expanding trade exchanges, promoting two-way investment cooperation and advancing platform construction, overseas Chinese have continued to promote trade facilitation and common economic prosperity in countries along the route. More importantly, the Chinese diaspora provides a vane for Chinese enterprises' investment development, which is conducive to China's enterprises going global while combining local needs and better integration with local enterprises, thus promoting the internationalization process of Chinese enterprises. In addition, the Chinese diaspora can promote the establishment of a favorable international business climate and play an irreplaceable role in removing trade barriers, optimizing the trade structure and promoting trade and investment facilitation. Therefore, overseas Chinese should give full play to their role as "pavers" and contribute their fair share in promoting high-quality trade facilitation in the BRI.

3. Cultural self-awareness of expatriates

(1) Strengthening national image awareness training for personnel going abroad
Today, with thousands of Chinese people going abroad, the image of ordinary nationals has become an important component part of the country's image. In addition to government officials, there are four main categories of Chinese expatriates: (1) those sent by Chinese enterprises to engage in economic and trade activities and infrastructure construction; (2) teachers and students sent at public expense and a large number of self-funded students studying abroad; (3) general tourist shoppers organized by Chinese travel agencies; and (4) those who go to foreign countries to participate in various conferences and events. With their words and actions, these personnel naturally become the purveyors of Chinese culture and the shapers of the country's image. They are a large but weakly organized

actor in China's public diplomacy activities. All levels and types of expatriates organizational structure should pay great attention to the issue of national image awareness of expatriates, providing acute and targeted training to expatriates, and effectively improve the national image awareness of our expatriates.

In addition to entrepreneurs and corporate employees, a large number of party and government cadres, experts and scholars, as well as a large number of tourists travel to BRI countries every year, so that all people going abroad are aware that they are China's business card and the image ambassador of Chinese people. Targeted training for expatriates is an emergency measure to make up for the "lack" of Chinese people in foreign countries. The training of personnel going abroad should not be limited to foreign affairs discipline, but should also enable them to understand local history and culture, customs and manners, religious habits, and to overcome all uncivilized habits in themselves. To this end, the training of personnel going abroad should be shared by the Ministries of Foreign Affairs, Commerce, Culture and Tourism. It is important to carry forward the good tradition of pre-departure training for public service expatriates and to extend it to all expatriates as far as possible. Whether it is an enterprise sending personnel to work abroad or a travel agency organizing people to travel abroad, all types of expatriate personnel at all levels organizational structure should include pre-training for expatriates in their schedules. The training content should be highly relevant in order to achieve better behavioural and image-building results. In addition to the general contents of our foreign policy (especially the policy towards the target country) and the history, culture and customs of the target country, technical contents such as intercultural communication skills and foreign affairs etiquette should be added. The working regulations, habits and practices of the target country and partner companies are very important elements for staff going abroad. For the average tourist, the basic values and cultural preferences of the people in the target country are also indispensable elements.

Cultural exchanges should be geared towards cultural identity, in which our values are spread, rather than talking about ourselves and entertaining ourselves. In view of the fact that the people of the BRI countries still lack a comprehensive, objective and correct understanding of China. We have to tell the "China story" and spread the "good voice of China" in the BRI countries. The is also the expectation of President Xi Jinping. The arts and cultural groups we send to BRI countries, as well as the departments and units that organize exhibitions and activities, should be consciously aware of the need to promote high-quality and jointly build the BRI culture. It pays attention to the research and study before performances and exhibitions. It looks for the cultural elements we share with the people of the countries along the route, and produces programs or exhibition contents that can create the greatest cultural identity for them.

(2) Raising the self-awareness of cultural communication among students studying abroad

Study abroad visitors are the best civil ambassadors for cultural dissemination. They go out to see the world and let the world know China through them. International students themselves are the name cards and spokespersons of the Chinese culture. They are an important bridge and link for spreading Chinese culture. Residents of the countries where they study form an impression of China in their daily communication and interaction with them. As China's exchanges with the world become closer and closer, more and more domestic study visitors are getting valuable opportunities to work or visit abroad. Then, after "going abroad", how these "overseas Chinese" should make use of their local advantages to master the cross-cultural communication skills of "introducing China" and better take up the mission of cultural communication is an issue that deserves great attention.

(c) Use learning opportunities to demonstrate "cultural awareness". Chinese students working and visiting overseas should develop a high degree of "cultural consciousness" and make full use of their research and study opportunities to take up the mission of cultural dissemination. If we can make good use of our local research advantages and resources, we can not only introduce Chinese culture through scientific cooperation, but also gain insight into the perception of Chinese culture in our host countries. Moreover, the host country's perception of Chinese culture is a valuable research information for Chinese study visitors to take back to their home country, as it helps the country to better grasp the international context of Chinese cultural diffusion, especially the pressing issues that need to be addressed.

Participation in social activities to avoid "cultural misinterpretation". Generally speaking, Chinese study visitors have the opportunity to participate in social activities during their time abroad. With the radiating power of these activities, it is not only possible to enhance the effectiveness of Chinese study visitors as "opinion leaders" of Chinese culture, but also to avoid cultural misinterpretation and the resulting cultural gap in a timely manner. For example, as China's international status continues to rise, various exhibitions on Chinese culture are gradually increasing. However, given cultural differences, historical reasons and ideologies, cultural misinterpretations often inevitably occur in the exhibition, sometimes even directly affecting the interpretation of China's basic national policies. At this time, overseas Chinese study visitors are the most suitable cultural communication messengers and cultural translators.

We demonstrate our "cultural philosophy" by devoting ourselves to public service. The cultural concept of "when one side is in trouble, all sides support each other" is what the Chinese culture advocates. It expresses the generosity of the Chinese people who follow the "great love without borders". This is completely consistent with the spirit of fraternity in Western culture. Anyone familiar with Western culture is well versed in the Western focus on the "common good" and "charity". In Western countries, both governments, corporations and individuals care a lot about their performance in public service. In this sense, charity activities is an opportunity for the exchange of ideas between Chinese and foreign cultures. In various public welfare activities, the advantages of overseas Chinese study visitors are their accumulated research, professionalism and familiarity with various public welfare activities, which can precisely complement the rich experience and networking resources of local Chinese. Through their professional knowledge, students abroad can help the Chinese in their home countries to carry out various public welfare activities and truly play the role of "opinion leaders" in spreading Chinese culture.

2. Deepening the implementation of the BRI green development

Green development is a common development concept in the world today. It is also the consensus of the BRI co-construction countries. In the joint communiqué of the second BRI International Cooperation Summit Roundtable in April 2019, the participating leaders reaffirmed that "implementing the commitments of the United Nations 2030 Agenda for Sustainable Development, achieving strong, sustainable, balanced and inclusive growth in the world economy, and improving the quality of life of people are our common goals"; and unanimously reaffirmed their commitment to "co-construct the BRI together with high quality, insisting on co-business, co-construction and sharing, adhering to openness, greenness and integrity, and pursuing high standards, benefiting people's livelihoods and sustainability". It adheres to the people-centered development ideology and follow the path of coordinated economic, social and environmental development. However, green development faces many challenges. On the one hand, in the implementation of the concept of green development, the BRI common countries have large environmental and climatic differences. The ecological environment is sensitive, and the strengthening of ecological environmental protection has not yet received sufficient attention. On the other hand, green development is an all-round development that includes production, consumption, distribution, innovation and finance, and requires, in addition to the implementation of concepts, the synergistic development of governance capacity and public awareness, innovation in institutional mechanisms and technological advances.

In the Boao Forum held in April 2021, President Xi Jinping (2021, April 20) pointed out a concrete implementation path for the green development of the BRI, stating that "to achieve green development, the BRI initiative must first ensure that the BRI projects meet the basic requirements and fundamental standards of ecological protection. Second, the energy structure should be optimized to use and develop clean energy, i.e., wind power, solar power and hydro power, etc., as far as possible. Third, the industrial structure should combine ecological protection with climate governance". Specifically, the BRI high quality development together can be carried out in the following three specific aspects of green development.

(i) Clarifying green requirements and refining development standards

In terms of content, green development includes pollution control, ecosystem management, ecosystem protection, green energy, green production, green living and green finance. In the process of implementation, clear requirements and standards are given for green development, which is conducive to the implementation and promotion of the concept. Around the concept of green development, China and countries along the BRI have now issued a series of relevant requirements and standards. In terms of pollution prevention and control, China finalized and approved the *Airborne Pollution Prevention and Control Action Plan* (2013, September 10) in 2013, the State Council issued the *Three-Year Action Plan for Winning the Blue Sky Defense War* (2018, July 4) in 2018. Israel has made significant amendments to the Water Supply and waste water disposal Company Law promulgated in 2001 to address water reduction and water pollution issues. On the basis of the existing requirements and standards, it is necessary to continue to deepen the green development of the BRI in the following five areas, clarify the requirements for green development and refine the standards for green development.

First, in terms of ecosystem management and protection, the participating units and enterprises are required to highlight the concept of ecological civilization in the construction of the BRI. It promotes green development, strengthens ecological environmental protection and jointly builds the Green Silk Road. It strives to use 3–5 years to build a pragmatic and efficient ecological and environmental cooperation and exchange system. It supports service platform and industrial technology cooperation base. It focuses on the development and implementation of a series of ecological and environmental risk prevention policies and measures for the green BRI construction of a solid foundation; within 5–10 years, it aims to build a more complete ecological and environmental services, support, security system, the implementation of a number of important ecological and environmental projects, and achieve good results.

Promoting the construction of green BRI is an inherent requirement for sharing the concept of ecological civilization and achieving sustainable development. It is conducive to enhancing mutual understanding and support among governments, enterprises and the public along the route, sharing China's ecological civilization and green development concepts and practices, improving ecological environmental protection capacity, preventing ecological and environmental risks, and promoting the common achievement of the 2030 Sustainable Development Goals by countries and regions along the route. In order to further promote the green BRI construction of 13 major tasks, including comprehensive services "five pass", it promotes green development and ensure ecological and environmental security. It integrates existing domestic and international cooperation mechanisms and play the role of a window for international cooperation on ecology and environmental protection. It does the groundwork to optimize the layout of production capacity and prevent ecological and environmental risks. It promotes the construction of green infrastructure and strengthen ecological environmental quality protection. It promotes green trade development for sustainable production and consumption. It strengthens outward investment environmental administration and promoting the development of a green financial system. Strengthening the construction of green cooperation platforms, it provides comprehensive support and services, formulating and improving policy measures, strengthening the coordination of government and enterprises, and guaranteeing the effectiveness of implementation. It leverages local strengths, strengthening capacity building and facilitating the implementation of projects, etc.

Ecological civilization is a new form of human civilization in which human beings live in harmony with nature. It is a stage of civilization of sustainable development relative to the agricultural society and industrialization stage in which human beings demand from nature and the relationship between human beings and land is contradictory and antagonistic. In the face of the ecosystem problems and risks that may arise in the construction of the BRI. First, we should actively advocate the concept of ecological civilization, implementing the United Nations sustainable development goals, and implementing green development strategies, that is, applying green technologies to achieve harmonious and environmentally friendly development of man and nature. Second, a four-tier circular economy model has been established, namely, promoting "enterprise circulation, industrial circulation, regional circulation and social circulation". It actively exploring the road of green industrialization to promote green urbanization. Third, we will work together to build an ecological city. Implementing the "Six Cities" model of building ecological cities, actively building safe cities, recycling cities, convenient cities, green cities, innovative cities and harmonious cities. Fourth, it builds a firm green BRI ecological security barrier, and reasonably control the intensity of development. Fifth, the construction of international environmental cooperation benefit-sharing mechanism helps to promote the BRI effective protection of the ecological environment, achieving a win-win situation.

Second, in terms of green energy, we should develop environmental standards and norms for infrastructure construction, increasing ecological and environmental services and supporting for major infrastructure construction project along the BRI route. It promotes energy-saving and environmental standards, and practices in green transportation, green buildings, clean energy and other industries. It promotes water, atmosphere, soil, biodiversity and other areas environmental protection. It promotes environmental infrastructure construction, and enhances green and low-carbon construction and operation. On the one hand, in the cooperation with the countries along the route, we accelerate the development of special energy resources, build up a stable supply and demand relationship in the green energy market, and form landmark projects for cooperation in the field of green energy among the countries along the route. On the other hand, countries along the route should work together to strengthen the efficient use of traditional energy sources, and promotes in-depth cooperation among multiple parties through the layout of a number of key projects and the use of each party's advantageous links.

Infrastructure is a bottleneck that constrains the economic development of developing countries. Accelerates infrastructure connectivity is a key area and core element of building the BRI together. The building of comprehensive and complex infrastructure connectivity should be a priority direction for the co-construction of the BRI, and efforts should be made to promote the construction of a high-quality, reliable, risk-resistant and sustainable infrastructure. While strengthening infrastructure construction, we should continue to take full account of environmental factors, continue to introduce international environmental standards for green transport, green buildings, green energy and other industries, and upgrade the level of low carbonization, so as to ensure economic benefit, but also minimize the direct and indirect risks that may be caused to the environment and society. Also, we should contribute to the sustainable development of each participating country. In particular, in the construction of key roads, port infrastructure, aviation infrastructure, network of pipelines lines and other priority project, efforts should be made to avoid risks, reduce risks and grasp the balance between economic growth, social progress and environmental protection.

The BRI involves a large number of industrial projects investment. In order to create a green BRI, since green energy is one of the important support content, the government should set up a cooperation and exchange platform. China's enterprises should undertake the construction of industrial environmental protection infrastructure. The construction of the green BRI is guided by the concept of green industry and green development. It adheres to the principles of resource conservation and environmental friendliness and integrates green energy into the whole process of the construction of the BRI industry. By strengthening enterprises green and low-carbon technological innovation, rational development and the use of green energy, it promotes green BRI construction and strengthening ecological environmental protection. It is conducive to enhancing mutual understanding and support among governments, enterprises and the public along the route. Green energy construction promotes the green BRI development. It includes the ecological and environmental protection, pollution prevention, environmental protection technology and industry. It also carries out energy saving and emission reduction, ecological and environmental protection and other infrastructure and capacity construction project. The Government encourages environmental energy enterprises to explore the markets of countries along the route and guides green industrial clusters with green energy development technologies to "go global".

Third, in terms of green production, the ecological environmental element of the BRI participating countries varies widely. Many countries have a relatively weak foundation in ecological environmental protection. Therefore, in the process of promoting the green supply chain management system in China, priority can be given to cooperation in regions and areas with high potential for emission reduction, strengthening green supply chain international cooperation in all aspects from procurement, production, packaging, circulation, consumption and recycling, etc. The promotion of the establishment of the green supply chain management system is a focused and step-by-step manner; it combines the development of green products, green factories, green industrial park and green enterprises. It plays the leading role of core enterprises, amplifying the demonstration effect of ecological environmental protection, promoting cooperation among enterprises in the supply chain system, raising the environmental protection awareness of each enterprise in the whole chain, and promoting the construction of ecological civilization to the breadth and depth.

Central to strengthening green supply chain management and promoting green production are systems and standards. It accelerate the implementation of green manufacturing projects, further playing the standardization and leading role of standards, and promoting the standardization of green manufacturing. *The Construction Guide Emphasizes that the Green Manufacturing Standards System* is a comprehensive and systematic system (2016, October 10) . It promotes the formulation and revision of standards in key areas such as green products, green factories, green enterprises, green parks and green supply chain from design planning to evaluation services. Standardization is an important means to achieve the supply chain, and an important technical guarantee to lead and regulate the development of the industry. Carrying out green supply chain management work is an important initiative to implement the green manufacturing project. Grounded in the producer's responsibility, the management of the whole life cycle of products can be realized in terms of green supply in procurement, production, marketing, recycling and logistics, etc. The urging enterprises along the industrial chain can reduce resource and energy consumption and pollution emissions, in order to better promote green supply chain management work.

In the process of promoting the development of green supply chain, BRI countries should strengthen the construction of green supply chain in terms of policies and regulations, information platform construction, exchange and cooperation, enterprise supervision and financial support. First of all, in the future, we should cooperate with BRI countries to formulate laws and regulations related to the environment, and introduce corresponding policies and regulations on the use of raw material, manufacturing, packaging and transportation, etc., and crack down on violations. Secondly, the construction of green supply chain is promoted in a focused and systematic manner. According to the specific situation of industry development and pollution emission in different countries, the focus of green supply chain construction is determined. Through the whole life cycle concept, green products, green factories, green industrial park and green enterprises are organically linked to form a green closed loop of producers and consumers. Once again, BRI countries should strengthen the environmental social responsibility constraints on enterprises. It requires them to disclose information on their emissions data and the energy-saving and emission reduction measures they have taken in the course of their operations. Finally, in order to reduce the economic burden of enterprises at the early stage of implementing green supply chain, BRI countries must strengthen financial support and green financial services.

Fourthly, in terms of green living, the BRI has been actively practicing the concept of green development since its inception, focusing on deepening environmental cooperation, increasing ecological environmental protection efforts, advocating low-carbon, recycling and sustainable production and lifestyle, and striving to build a green Silk Road. In the ecological environmental protection cooperation, the civil power, mainly private enterprises, private institutions and civil organizations, is indispensable. Strengthening multi-disciplinary communication and exchange among civil forces of various countries and giving full play to the bridging role of civil forces are of strategic importance in promoting green development. At present, China has been working with many of the BRI participating countries to carry out various forms of civil interaction, such as public welfare and charity activities, forums and exhibitions, which have promoted the people of the countries along the route to pay attention to the construction of ecological civilization and promote the formation of a consensus on green development. In the future, the participation of civil society in the construction of ecological civilization in the BRI should be further expanded, so as to form a good situation for all sectors to join hands to promote the construction of ecological civilization.

At the same time, it is necessary to actively advocate the ecological civilization concept of respecting nature, responding to nature and protecting nature. It strengthens policy communication among the governments of the participating countries of the Belt and Road, enhances understanding and mutual trust by improving policy transparency. It forms a consensus on resource conservation and environmental protection; strengthens exchanges among the scientific research institutions of the participating countries, fully discusses green development models, relevant technical standards, quality criteria and so on. It jointly formulates green development plans and measures to promote regional cooperation, and negotiates to solve resources conservation and ecological environmental protection problems in the process of the construction of the Belt and Road. It widely carries out cultural exchanges, tourism cooperation and volunteer services among the participating countries, share the "Chinese experience" and "Chinese wisdom" in the construction of ecological civilization, and lay a solid foundation for further cooperation.

The establishment of a strong green concept and awareness is important in promoting the green development of the BRI, especially the formation of green living thinking and habits. Countries, regions and people along the BRI route should establish green values and green development concepts. They gain a deeper understanding of the ecological wisdom of different human civilizations, promote awareness of protecting the ecological environment of countries along the route, and enhance the green consensus among governments, enterprises and the public along the route. It adheres to bottom-line thinking and clearly establish the concept that human activities cannot exceed the carrying capacity of natural resources. It adheres to the mindset of perpetual development and leave room for future generations to develop.

Fifth, in terms of green finance, increasing financial support for ecological and environmental protection enterprises, it reduces the economic burden of ecological and environmental protection enterprises and provides diversified and sustainable investment and financing support for the relevant enterprises that are important measures to strengthen the construction of ecological civilization and carry out green financial services. It is necessary to carry out research on the investment and financing needs of green environmental protection projects in the participating countries of the Belt and Road. It formulates corresponding green investment and financing guidelines, and explores the administrative standard, identification procedures and methods of green investment and financing. For enterprises that the increase investment in environmentally friendly production technologies and actively transform their crude production methods, consideration can be given to relaxing financial credit conditions and granting preferential treatment in terms of financing limits and loan interest rates to ensure precise matching of green investment and financing, and promote the focus of capital investment on the construction of ecological civilization.

Energy transitions in countries along the route also require green finance support. On the one hand, the fossil energy sources such as coal and oil account for a relatively large share of the energy consumption structure of countries along the route, which not only causes high carbon emissions, but airborne pollution is also becoming increasingly serious. On the other hand, the BRI countries are rich in natural resources and have greater development potential in clean energy. The green finance market can both provide fossil energy-dependent enterprises with the capital needed for a green and low-carbon transition and effectively support clean energy and technologies to obtain financing, helping countries along the route to effectively reduce fossil energy while tapping into their own strengths in renewable energy and promoting energy transition.

In order to address the rising climate and environmental challenges, and to solve the financial bottlenecks of green transformation and sustainable development, there is a huge demand for green finance in countries and regions along the route. In order to continue to enhance the sustainability of the BRI and highlight China's responsibility as an important participant in the BRI, China should firmly promote the development of green finance in the BRI and build a green BRI in the future. Specific paths include, first, withdrawing from high-carbon emission-related projects as soon as possible. Second, strengthen third party cooperation in green finance. Third, capacity-building related to addressing climate change should be strengthened. On the one hand, there is an important role for capacity-building at the official level. On the other hand, we should give full play to the role of a think tank in the academia.

(ii) Optimize energy structure and carry out green innovation

Efforts are being made to build a clean, low-carbon, safe and efficient energy system, particularly by proposing restrictions on coal power. For example, the new green bond catalogue released in 2020 has explicitly stopped supporting any coal-related projects. Actively it builds a green BRI to guide projects and investments from home and abroad into green and low-carbon areas. Energy production should not only optimize the stock, promote the clean and efficient development and utilization of coal as the foothold and primary task of energy transformation and development, but also expand the increment, accelerate the proportion of renewable energy sources such as hydropower, wind, solar and biomass. It develops nuclear energy safely and efficiently, and optimizes energy production distribution. We should grasp the total amount and intensity of dual control, make comprehensive use of economic, legal and necessary administrative means, focus on key areas such as industry, construction and transportation to effectively promote energy conservation and emission reduction. We need to improve energy efficiency by eliminating backward production capacity, accelerate the upgrading and transformation of traditional industries and cultivate new dynamic energy, and promote the formation of a lifestyle and social culture focused on energy conservation. It is now important to integrate the winter residential heating and airborne pollution prevention.

Accelerating technological innovation and institutional reform are key dependencies in promoting sustainable energy development. We should focus on making breakthroughs in renewable energy development and utilization, especially in new energy grid-connected technologies and energy storage and micro-grid technologies, comprehensively building "Internet +" smart energy, improving the regulation capacity of power grid systems, increasing the capacity of new energy consumption, developing advanced and efficient energy-saving technologies, and seizing the high ground in energy science and technology competition. Actively it promotes mass entrepreneurship and innovation, further stimulates the creative passion and innovation potential of the staff of energy industry enterprises and research institutes. It cultivates more energy technology advantages and transform them into economic advantages. It is necessary to deepen the market-oriented reform of energy, and actively support the entry of the private economy into the energy sector by simplifying and decentralizing government, combining management and administration, optimizing services and reforming the oil and gas mining rights system and rationalizing the electricity transmission and distribution chain, while deepening the reform of energy state-owned enterprises. It improves mechanisms and policies to encourage the development of distributed energy, rationalizes the energy price system, restores the attributes of energy commodities, gives full play to the decisive role of the market in allocating resources and better play the role of the government, and builds a fair and competitive energy market system.

Ensuring national energy security requires the integration of both the domestic and international situation. Both domestically based and deepening international cooperation form a diversified and stable supply pattern. It is necessary to consolidate mutually beneficial cooperation with traditional host country of the resources countries, optimizes the energy trade structure and seizes the major opportunity of the BRI construction. It promotes energy infrastructure interconnection and interoperability, increases international production capacity cooperation and drives the export of energy equipment with competitive advantages. It actively participates in global energy governance and promotes the development of the international energy order and governance system in a more reasonable direction.

In optimizing energy consumption structure, industrial restructuring drive energy structure adjustment. Energy structure adjustment helps to upgrade the industrial structure. It guides the industrial system to intensive and high-end upgrade. In terms of energy conservation and consumption reduction, the price mechanism should be used to guide the restructuring of energy production and consumption. Gradually, it eliminates backward production capacity and promotes green energy transformation and upgrading. The energy science and technology revolution is a strategic guarantee to seize the high point of scientific and technological development and ensure the long-term energy security of

China. It is an important grasp for us to get rid of path dependence and grasp the initiative of energy security at present. We need to deeply implement the innovation-driven development strategy, promote the energy technology revolution, and continuously improve the capacity for independent innovation in energy technology and the level of localization of major equipment. The first step is to increase the research and application of advanced energy technologies. Second, it focuses on the "neck" technology to accelerate the pace of attack. Finally, we vigorously explore the development of new models and new business models.

(iii) Systematic protection and governance, and integration of industrial structures

The co-construction of the BRI has responded to the inherent requirements of the change of the global governance system, manifested the sense of a human community of a shared future and responsibilities, and provided new ideas and new solutions for improving the change of the global governance system. This has profoundly clarified the important relationship between the BRI and global governance. It has pointed out the direction for the BRI to promote the improvement of global governance under the guidance of the concept of a human community of a shared future. The global Covid-19 pandemic is a major test of the global governance system. It requires the concerted efforts of all countries to strengthen cooperation. The BRI is a public product offered by China to the world as a solution to global governance issues. It is based on China's own practical experience in development, upholding the principles of co-business, co-construction and sharing, advocating that everyone should discuss and manage their affairs, and insisting on win-win cooperation and mutual benefit.

The BRI, in helping undeveloped or underdeveloped countries to move towards industrialization, should establish better ecological standards and not repeat the old path of "polluting first and treating later". Considering the construction of ecological civilization at the beginning stage save governance costs and improve governance efficiency from the development path. It helps the undeveloped or low-developed countries to enjoy the dividends of green development. With the green BRI platform, China can bring ecological civilization awareness to countries and regions along the route; with the green BRI platform, China can promote and guide Chinese enterprises going abroad to take more responsibility for green development and drive local communities to achieve comprehensive sustainable development; with the green BRI platform, it can attract different subjects such as governments, enterprises, associations and people along the route to learn from China's green governance actions and explore local green development paths based on local conditions; with the green BRI platform, it can promote the building of new international relations and a more representative, inclusive, open and fair global governance system.

The green development of the BRI should systematize the protection and governance. In the process of promoting it, we should start from the actual local industry, finding the right fit for industrial upgrading, realizing the precise force, and continuously improve and integrate the quality of the industrial structure. In building the BRI industry service platform, it needs to integrate into the BRI to achieve industrial transformation and upgrading. The government and enterprises need to play a joint role. In addition to the high level of government attention, businesses need to be actively involved. Therefore, the government needs to establish an industrial service platform to provide scientific help for enterprises to integrate into the BRI. We should encourage enterprises to establish an international mindset. It should have an international vision, taking the initiative to integrate into the BRI and comprehensively promote the process of "going out". Of course it also establishes the concept of innovation. It establishes the concept of active innovation, changes the traditional strategy of low-priced marketing, actively incorporates innovative activities into corporate and industry strategies. It establishes a comprehensive competitive concept, participates in the investment and construction of the BRI countries and regions, takes the initiative to initiate international technical standards certification, actively participates in international competition, and improves global operation capabilities. At the same time, the government should help enterprises to actively integrate into the BRI process. The government should make full use of its advantages to provide enterprises with authoritative information inquiry and retrieval services. It provides high-quality consultation service to help them realize the role change and integrates into the international market as soon as possible. At the same time, the government should actively organize associations to help enterprises grasp the entry point for "going global", while integrating industrial resources, realizing exchanges and cooperation between industrial chains, capital and entities, forming strategic alliance, holding a group for warmth, and enhancing international competitive power.

Protection governance systematization is also based on supporting technological innovation and fostering growth industry, dovetailing with the BRI, the key direction of industrial upgrading is corporate technological innovation and fostering growth industry. From the perspective of the benefits of the expansion of market size, it is necessary to increase financial support to promote technological innovation between China and countries and regions along the BRI route to create a development community. Considering the uncertainty of innovation, the government should give necessary financial support to science and technology innovation. It strongly supports enterprises to participate in technological innovation cooperation in countries and regions along the BRI route. It adopts such means as co-establishing R&D centers, creates technological innovation platforms, implement network technologies for innovation and development, supports enterprises to allocate resources scientifically in countries and regions along the route, promotes the construction of innovation communities, and realizes enterprise technological innovation and industrial structure upgrading.

At the same time, we should increase investment in research and improve the overall capital of innovative talents. Industrial transformation and upgrading requires not only a large amount of financial support, but also the support of innovative talents. On the one hand, enterprises need financial support to carry out technological innovation activities. The Government should provide special research funding support to research institutions, especially for technologies that play a major role in the technological upgrading of the BRI industry, and invest more in them. On the other hand, we should vigorously nurture green and innovative talents. An important factor in green innovation activities is talent. The government should further improve the system of scientific research incentives and distribution of earnings, especially by allowing factors of production in management and technology to participate in the distribution. This fully reflects the value of the creative labour of the researchers.

References

Xi Jinping (2021, April 20). At the opening ceremony of the Boao Forum for Asia Annual meeting 2021[full text].Xinhuanet, http://www.xinhuanet.com/politics/leaders/2021-04/20/c_1127350811.htm.

Xi Jinping. (2021). 'Strengthening and improving international communication to showcase a true, three-dimensional and comprehensive China', May 31, 2021.

Performance Targets for Central Sector Budgets Management Method (2015, May 21). http://www.gov.cn/gongbao/content/2015/content_2937333.htm.

Chen Shuisheng & the Research Team of the State Council Overseas Chinese Affairs Office. (2016). "Analysis of the role of overseas Chinese in the BRI strategy", *The Theory of United Front Research*, Beijing: Academy Press, 194–203. [陈水胜和国务院侨务办公室课题组. 统一战线理论研究成果《华侨华人在 "一带一路"战略中的作用分析》[R], 2016:194–203.]

Airborne Pollution Prevention and Control Action Plan (2013, September 10). http://www.gov.cn/zhengce/content/2013-09/13/content_4561.htm.
Three-Year Action Plan for Winning the Blue Sky Defense War (2018, July 4). http://www.nea.gov.cn/2018-07/04/c_137301079.htm.
The Construction Guide Emphasizes that the Green Manufacturing Standards System (2016, October 10). https://www.sohu.com/a/115755348_465

References

Addessi, W., E. Saltari. (2012). "The perverse effect of debt tax benefits on firm investment decisions", *Economic Notes*, 41(3), 101–114.

An, Shuwei. (2015). 'The impact of "One Belt and One Road" on China's regional economic development and the reshaping of the pattern', *Economic Issues*, (04), 1–4. [安树伟. "一带一路"对我国区域经济发展的影响及格局重塑[J]. 经济问题, 2015, (04): 1–4.]

An, Xiao-Ming. (2016). 'The research lineage and progress of "One Belt, One Road" in China', *Regional Economic Review*, (02), 77–88. [安晓明. 我国"一带一路"研究脉络与进展[J]. 区域经济评论, 2016, (02): 77–88.]

Aschauer, D. A. (1989). 'Is public expenditure productive?', *Journal of Mombasa–Nairobi Standard Gaugeary Economics*, 23(2), 177–200.

Asiedu, E. and A. P. Villamil. (2002). 'Imperfect enforcement, foreign investment and foreign aid', *Macroeconomic Dynamics*, 6(4), 476–495.

Bai, Jie. (2018). Xi Jinping met with Vietnamese Prime Minister Nguyen Chun Phu: Speed up the docking of "Belt and Road Initiative" and "two corridors and one Circle" (2018, November 4). Xinhuanet, https://www.yidaiyilu.gov.cn/xwzx/xgcdt/70604.htm.

Bai, Yunzhen. (2015). 'The "Belt and Road" initiative and the transformation of China's Foreign Aid', *World Economy and Politics*, (11), 53–71+157–158. [白云真. "一带一路"倡议与中国对外援助转型[J]. 世界经济与政治, 2015, (11): 53–71+157–158.]

Bardhan, P. K. (1973). 'On terms of foreign borrowing', *American Economic Review*, 63(3), 458–461.

Brakman, S., P. Frankopan, H. Garretsen, C. Van Marrewijk. (2019). 'The New Silk Roads: An introduction to China's Belt and Road Initiative', *Cambridge Journal of Regions, Economy and Society*, 12(1), 3–16.

Buckley, P. J. (2020). China's Belt and Road Initiative: Changing the rules of globalization[J]. *Journal of International Business Studies*, 51(2), 279–281.

Carlstrom, C. T., T. S. Fuerst. (1997). 'Agency costs, net worth, and business fluctuations: A computable general equilibrium analysis', *American Economic Review*, 87(5), 893–910.

Chatterjee, S., G. Sakoulis, S. J. Turnovsky. (2003). 'Unilateral capital transfers, public investment, and economic growth', *European Economic Review*, 47(6), 1077–1103.

Chen Hong, Yang Chengyu. (2015). 'Study on the international economic effects of the national strategy of "One Belt, One Road"-analysis based on CGE model', *International Trade Issues*, (10), 4–13.[陈虹, 杨成玉. "一带一路"国家战略的国际经济效应研究——基于CGE模型的分析[J]. 国际贸易问题, 2015, (10): 4–13.]

Chen Jiyong, Jiang Yanping, Wang Baoshuang. (2017). 'A study of trade competitiveness between China and countries along the "Belt and Road": A dual perspective based on product and market domains', *World Economic Studies*, (08), 3–14+135[陈继勇, 蒋艳萍, 王保双. 中国与"一带一路"沿线国家的贸易竞争性研究:基于产品域和市场域的双重视角[J]. 世界经济研究, 2017, (08): 3–14+135.]

Chen Shuisheng & the Research Team of the State Council Overseas Chinese Affairs Office. (2016). "Analysis of the role of overseas Chinese in the BRI strategy", *The Theory of United Front Research*, Beijing: Academy Press, 194–203. [陈水胜和国务院侨务办公室课题组. 统一战线理论研究成果《华侨华人在 "一带一路"战略中的作用分析》[R], 2016:194–203.]

Chen Weiguang, Guo Jing. (2016). 'Potential estimation and location selection of Chinese investment in countries along the "Belt and Road" ', *Macroeconomic Research*, (09), 148–161.[陈伟光, 郭晴. 中国对"一带一路"沿线国家投资的潜力估计与区位选择[J]. 宏观经济研究, 2016, (09): 148–161.]

Chen Weiguang, Huang Liangxiong, Cheng Yonglin, Han Yonghui. (2017). 'The creation of "One Belt, One Road" economics and its many directions', *Reform*, (02), 5–16.[陈伟光, 黄亮雄, 程永林, 韩永辉. "一带一路"经济学创立及其诸多向度[J]. 改革, 2017, (02): 5–16.]

Chen Yongjun. (2016). 'A theoretical model of the "Belt and Road" strategy', *Journal of the Party School of the Guizhou Provincial Committee of the Communist Party of China*, (01), 29–35. [陈甬军. "一带一路"战略的理论模型研究[J]. 中共贵州省委党校学报, 2016, (01): 29–35.]

Chen Yongjun, Deng Zhongqi, Zhang Jihuan. (2019). 'An economic analysis of the "Belt and Road" initiative to achieve win-win cooperation based on the perspective of benefit creation and sharing mechanism', *Journal of Xiamen University (Philosophy and Social Sciences Edition)*, (05): 83–97.[陈甬军, 邓忠奇, 张记欢. "一带一路"倡议实现合作共赢的经济学分析——基于利益创造与共享机制的视角[J]. 厦门大学学报(哲学社会科学版), 2019, (05): 83–97.]

Chen Yongjun, Li Yajie, She Tiancheng. (2019). 'A study on debt sustainability of cooperating countries in the construction of "Belt and Road"-based on the assessment of debt risk in Kenya', *Price Theory and Practice*, (07), 30–34.[陈甬军, 李雅洁, 佘天诚. "一带一路"建设中合作国家债务可持续性研究——基于对肯尼亚债务风险的评估风险[J]. 价格理论与实践, 2019, (07): 30–34.]

Cheng L. K. (2016). 'Three questions on China's "Belt and Road Initiative"', *China Economic Review*, 40, 309–313.

Cheng Weihua, Zhang Binhai. (2020). 'Opportunities, challenges and reflections on the development of graduate education in China in the New Era', *Graduate Education Research*, 000(002), 27–33.[程伟华, 张海滨. 新时代来华留学研究生教育发展机遇,挑战与思考[J]. 研究生教育研究, 2020, 000(002):P.27–33.]

Chen Weiguang, Wang Yan. (2016). 'Building BRI together: An analytical framework based on relational and rule-based governance', *World Economy and Politics*, (6), 93–114.

Cui Na, Liu Chun. (2017). 'The impact of institutional environment on China's export efficiency in countries along the "Belt and Road"', *World Economic Studies*, (08), 38–50+135–136.[崔娜, 柳春. "一带一路"沿线国家制度环境对中国出口效率的影响[J]. 世界经济研究, 2017, (08): 38–50+135–136.]

Cui L, Song M. (2019). 'Economic evaluation of the Belt and Road Initiative from an unimpeded trade perspective', *International Journal of Logistics Research and Applications*, 22(1), 25–46.

Cui Riming, Huang Yingwan. (2016). 'Research on the evaluation index system of trade and investment facilitation in countries along the "Belt and Road"', *International Trade Issues*, (09), 153–164.[崔日明, 黄英婉. "一带一路"沿线国家贸易投资便利化评价指标体系研究[J]. 国际贸易问题, 2016, (09): 153–164.]

Dai Xiang, Song Jie. (2021). 'The optimization effect of global value chain of the "Belt and Road" Initiative: A perspective of the improvement of global value chain division of labor in the participating countries along the route', *China Industrial Economics*, (06), 99–117.[戴翔, 宋婕. "一带一路"倡议的全球价值链优化效应——基于沿线参与国全球价值链分工地位提升的视角[J]. 中国工业经济, 2021, (06): 99–117.]

Deng Zhongqi, Chen Yongjun. (2018). 'Capital structure analysis of public-private partnership model of financiers in the context of "One Belt, One Road"', *Industrial Economics Research*, (03), 90–102.[邓忠奇, 陈甬军. "一带一路"背景下融资方公私合营模式的资本结构分析[J]. 产业经济研究, 2018, (03): 90–102.]

Ding Tongmin. (2018). 'Reflections on establishing a sound legal governance mechanism for BRI', *Journal of Henan Normal University (Philosophy and Social Science Edition)*, (6), 45–50.

Dong Hong, Lin Huihui. (2015). 'Changes in China's foreign trade pattern and prevention of trade frictions under the "Belt and Road" strategy', *China Distribution Economy*, 29(05), 119–124.[董红, 林慧慧. "一带一路"战略下我国对外贸易格局变化及贸易摩擦防范[J]. 中国流通经济, 2015, 29(05): 119–124.]

Dong Zhanwei, Li Yuanying, Gao Zhenbo (2018). North star power news (2018, December 20). http://news.bjx.com.cn/html/20181220/950545.shtml.

Du Li, Ma Yaoyao. (2019). 'Green development and its performance assessment in countries along the "Belt and Road"', *Journal of Social Sciences of Jilin University*, 59(05), 135–149+222.[杜莉, 马遥遥. "一带一路"沿线国家的绿色发展及其绩效评估[J]. 吉林大学社会科学学报, 2019, 59(05): 135–149+222.]

Du, Longzheng, Lin, Wei-Fen. (2018). 'A study on the efficiency of Chinese direct investment in capacity cooperation along the "Belt and Road" based on data from 24 emerging and developing countries', *Quantitative Economic and Technical Economics Research*, 35(12), 3–21.[杜

龙政, 林伟芬. 中国对"一带一路"沿线直接投资的产能合作效率研究——基于24个新兴国家、发展中国家的数据[J]. 数量经济技术经济研究, 2018, 35(12): 3–21.]

Fan, Jojun, He, Huan. (2016). 'Strategies for the "entry" of infrastructure interconnection in "Belt and Road" countries', *World Economic and Political Forum*, (06), 129–142.[范祚军, 何欢. "一带一路"国家基础设施互联互通"切入"策略[J]. 世界经济与政治论坛, 2016, (06): 129–142.]

Fukunaga, I. (2002). 'Financial accelerator effects in Japan's Business Cycles', *Bank of Japan Working Paper*, 02–6.

Francis Snyder. (1994). "Soft Law and Institutional Practice in the European Community". In Stephen Martin (ed.), *The Construction of Europe: Essays in Honour of Emile Noel*, Kluwer Academic Publishers, 199–200.

Gao B, Li J, Shi B, Wang X. (2020). 'Internal conflict and bank liquidity creation: Evidence from the belt and road initiative', *Research in International Business and Finance*, 53, 101227.

Gong, Jie. (2017). 'Research on cultural communication under the "Belt and Road" strategy', *Nanjing University of Information Engineering*.[龚洁. "一带一路"战略下的文化传播研究[Z]. 南京信息工程大学. 2017]

Gu A, Zhou X. (2020). 'Emission reduction effects of the green energy investment projects of China in belt and road initiative countries', *Ecosystem Health and Sustainability*, 6(1), 1747947.

Koo, Shengzu, Wu, Qinqin, Chuang, Qinqin. (2017). 'Promoting the construction of "One Belt and One Road" and the countermeasures of enterprises "going out"', *Economic Review*, (02), 1–9.[辜胜阻, 吴沁沁, 庄芹芹. 推动"一带一路"建设与企业"走出去"的对策思考[J]. 经济纵横, 2017, (02): 1–9.]

Hanlin. 'The current development of China-European Liner', refer to Hanlin Logistics, <www.shanghaivast.com/Information/250.html>

Han M, Lao J, Yao Q, Zhang B, Meng J. (2020). 'Carbon inequality and economic development across the Belt and Road regions', *Journal of Environmental Management*, 262, 110–250.

Han Yonghui, Luo Xiaofei, Zou Jianhua. (2015). 'A study on the competitiveness and complementarity of trade cooperation between China and West Asia in the context of "One Belt, One Road" strategy', *World Economic Studies*, (03): 89–98+129.[韩永辉, 罗晓斐, 邹建华. 中国与西亚地区贸易合作的竞争性和互补性研究——以"一带一路"战略为背景[J]. 世界经济研究, 2015, (03): 89–98+129.]

Han Yonghong. (2016). 'An outline of the soft law guarantee mechanism for international cooperation in BRI', *Contemporary Law*, (4), 151–160.

He Lifeng. (2017). Chairman of the National Development and Reform Commission, at the BRI International Cooperation Summit on May 14, 2017, "Strengthening Policy Communication and Making the Four Docking Points".

He Xiaofeng, Wu Junya. (2019). "Reflections on the sound legal governance mechanism of BRI", *Legal Exposition*, (10), 138–139.

He Zhipeng. (2017). 'The trend of counter-globalization and international soft law', *Journal of Wuhan University (Philosophy and Social Science Edition)*, (4), 54–69.

Hu Biliang, *"Eight years of BRI: Strengthening connectivity to achieve win-win cooperation"*, *Guangming Daily*, October 22, 2021, p. 12.

Hu, Junchao, Wang, Dandan. (2016). 'Study on country-specific risks in countries along the "Belt and Road" ', *Economic Issues*, (05), 1–6+43.[胡俊超, 王丹丹. "一带一路"沿线国家国别风险研究[J]. 经济问题, 2016, (05): 1–6+43.]

Hu, Mei, Zheng, Wei. (2019). 'Analysis of the competitiveness and complementarity of trade between China and the "Belt and Road" countries', *Economic Issues*, (02), 101–108.[胡玫, 郑伟. 中国与"一带一路"国家贸易竞争性与互补性分析[J]. 经济问题, 2019, (02): 101–108.]

Hu Zaiyong, Fu Shaojun, Zhang Luchao. (2019). 'A study on the international trade effects of infrastructure in countries along the Belt and Road', *Quantitative Economic and Technical Economics Research*, 36(02), 24–44.[胡再勇, 付韶军, 张璐超. "一带一路"沿线国家基础设施的国际贸易效应研究[J]. 数量经济技术经济研究, 2019, 36(02): 24–44.]

Huang He. (2015). ' "One Belt, One Road" in the perspective of public goods', *World Economy and Politics*, (06), 138–155+160.[黄河. 公共产品视角下的"一带一路"[J]. 世界经济与政治, 2015, (06): 138–155+160.]

Huang X, Han Y, Gong X, Liu X. (2019). 'Does the belt and road initiative stimulate China's inbound tourist market? An empirical study using the gravity model with a DID method', *Tourism Economics*, 26(2), 299–323.

Huang, Xianhai, Yu, Xiao. (2017). 'Reshaping the global value chain with the construction of "One Belt, One Road"', *The Economist*, (03), 32–39.[黄先海, 余骁. 以"一带一路"建设重塑全球价值链[J]. 经济学家, 2017, (03): 32–39.]

Huang Y. (2016). 'Understanding China's Belt & Road Initiative: Motivation, framework and assessment', *China Economic Review*, 40, 314–321.

Jianying X, Jixing C, Yanxu L. (2020). 'Partitioned responses of ecosystem services and their tradeoffs to human activities in the Belt and Road region', *Journal of Cleaner Production*, 276, 123–205.

Jiang Xinmiao, Zhu Yanni. (2018). 'Research on the legal supply mechanism of BRI', *Journal of Northwestern University (Philosophy and Social Science Edition)*, (3), 83–89.

Jin Gang, Shen Kun Rong. (2019). 'The effect of Chinese companies' transportation investment in countries along the Belt and Road: Development effect or debt trap', *China Industrial Economics*, (09), 79–97.[金刚, 沈坤荣. 中国企业对"一带一路"沿线国家的交通投资效应:发展效应还是债务陷阱[J]. 中国工业经济, 2019, (09): 79–97.]

Jin Qiu, (2019). Xi Jinping's keynote speech at the opening ceremony of the second Belt and Road Initiative Forum for International Cooperation, 2019. https://www.ccps.gov.cn/xtt/201904/t20190426_131284.shtml

Kong Qinfeng, Dong Hongwei. (2015). 'Research on trade facilitation level measurement and trade potential of "One Belt One Road" countries', *International Trade Issues*, (12), 158–168. [孔庆峰, 董虹蔚. "一带一路"国家的贸易便利化水平测算与贸易潜力研究[J]. 国际贸易问题, 2015, (12): 158–168.]

Law Business Connection. (2017). 'A first glimpse of the BRI legal risk map,' *The Import and Export Manager*, (5), 66–67.

Li Bin, Yan Xiaochen. (2018). 'New comparative advantages of bilateral trade between China and countries along the "Belt and Road": A public security perspective', *Economic Research*, 53(01), 183–197.[李兵, 颜晓晨. 中国与"一带一路"沿线国家双边贸易的新比较优势——公共安全的视角[J]. 经济研究, 2018, 53(01): 183–197.]

Li Dan. (2019). '"One Belt, One Road": A practical exploration of building a community of human destiny', *Nankai Journal (Philosophy and Social Science Edition)*, (01), 136–145.[李丹. "一带一路":构建人类命运共同体的实践探索[J]. 南开学报(哲学社会科学版), 2019, (01): 136–145.]

Li F. (2017). 'The construction of BRI legal cooperation mechanism and its risk prevention', *Journal of Hebei University of Economics and Business (Comprehensive Edition)*, (1), 20–24.

Li Ming. (2016). 'International Law and "One Belt, One Road" ', *Journal of Law*, 37(01), 11–17. [李鸣. 国际法与"一带一路"研究[J]. 法学杂志, 2016, 37(01): 11–17.]

Li Mo. (2021). White paper on China's International Development Cooperation in the New era (2021, January 10). http://www.scio.gov.cn/zfbps/32832/Document/1696685/1696685.htm.

Li Na. (2018). "Implications of Egyptian arbitral decision recognition and enforcement practice on the rules of legal application of arbitration procedures and BRI Legal Risk Prevention", *Legal Application*, (12), 60–63.

Li Xiangyang. (2017). 'The concept of righteousness and profit in the construction of "One Belt, One Road"', *World Economy and Politics*, (09), 4–14+156.[李向阳. "一带一路"建设中的义利观[J]. 世界经济与政治, 2017, (09): 4–14+156.]

Li Xiangyang. (2018). "The "absence" of regional economic integration in Asia and the development orientation of BRI", *Chinese Social Science*, (8), 33–43.

Li Xiaofan., Jiang Lingduo. (2020). 'The construction of "One Belt and One Road", the opening up of central and western China and regional economic development', *World Economy*, 43(10), 3–27.[李小帆, 蒋灵多. "一带一路"建设、中西部开放与地区经济发展[J]. 世界经济, 2020, 43(10): 3–27.]

Li Yuhong. (2020). 'Exploring international engineering project management in the context of "One Belt, One Road" ', proceedings of the 2020 Wanchi Science Development Forum, Xi'an, Shaanxi, China, F.[李豫红. "一带一路"背景下国际工程项目管理探讨; proceedings of the 2020万知科学发展论坛, 中国陕西西安, F, 2020[C].]

Li Zheng. (2016). 'Risk control chronicle of the Qassim Power Station Project', *International Engineering and Labor*, (07), 84–86.[李铮. 卡西姆电站项目风险控制纪实[J]. 国际工程与劳务, 2016, (07): 84–86.]

Li Y, Yang Y, Li G, Zhang X. (2020). 'Study on Sustainable Development of Microfinance Institutions from the perspective of inclusive finance—Based on MFI data in countries along the Belt and Road', *Emerging Markets Finance and Trade*, 56(13), 3205–3216.

Li Yiran, （2020）. Xinhua Review: China-Africa relations have been in the same boat for 20 years, and Sino-African relations continue to move forward (2020, October 13), https://www.yidaiyilu.gov.cn/xwzx/gnxw/151226.htm.

Li Z, Huang Z, Dong H. (2019). 'The Influential Factors on Outward Foreign Direct Investment: Evidence from the "The Belt and Road"', *Emerging Markets Finance and Trade*, 55(14), 3211–3226.

Lin Yifu. (2017). 'Belt and Road and Free Trade Zone: China's new initiatives and initiatives for opening up to the world', *Journal of Peking University (Philosophy and Social Sciences Edition)*, 54(01), 11–13.[林毅夫. 一带一路与自贸区:中国新的对外开放倡议与举措[J]. 北京大学学报(哲学社会科学版), 2017, 54(01): 11–13.]

Lin, Yifu, Wu and Mao, Xing, Yiqing. (2010). 'The "tidal surge phenomenon" and the formation mechanism of overcapacity', *Economic Studies*, (10).

Liu Chang. (2020). The proposal of the Central Committee of the Communist Party of China on the formulation of the 14th Five-Year Plan for National Economic and Social Development and the Visionary Goals for 2035. (2020, November 3). Xinhua News Agency. http://www.gov.cn/zhengce/2020-11/03/content_5556991.htm.

Liu Chang. (2020). Proposal of the Central Committee of the Communist Party of China on the Formulation of the 14th Five-Year Plan for National Economic and Social Development and the Visionary Goals for 2030 (2020, November 30). http://www.gov.cn/zhengce/2020-11/03/content_5556991.htm.

Liu Chang. (2021). Exclusive interview: Belarus is optimistic about the prospect of cooperation in agricultural products and food export to China— Interview with Belarusian Minister of Agriculture and Food Kropko (2021, January 21)

Liu Guobin. (2016). 'On the financial support role of AIIB in promoting the construction of "One Belt, One Road"', *Northeast Asia Forum*, 25(02), 58–66+128.[刘国斌. 论亚投行在推进"一带一路"建设中的金融支撑作用[J]. 东北亚论坛, 2016, 25(02): 58–66+128.]

Liu H-D, Li W-Y, Chen H. (2016). 'How cultural integration affects bilateral trade between China and countries along the "Belt and Road": An empirical test based on micro trade data from 1995–2013', *International Trade Issues*, (02), 3–13.[刘洪铎, 李文宇, 陈和. 文化交融如何影响中国与"一带一路"沿线国家的双边贸易往来——基于1995–2013年微观贸易数据的实证检验[J]. 国际贸易问题, 2016, (02): 3–13.]

Liu Jingdong. (2017). 'Study on the construction of the rule of law system of BRI', *Political and Legal Forum*, (5), 125–135.

Liu Z-Biao, Wu F-Xiang. (2018). 'Dual embedding of global value chains under the "Belt and Road" initiative', *Chinese Social Sciences*, (08), 17–32.[刘志彪, 吴福象. "一带一路"倡议下全球价值链的双重嵌入[J]. 中国社会科学, 2018, (08): 17–32.]

Lu, Xiaoxiao, Liang, Ying. (2020). 'Infrastructure Construction and Global Value Chain Reconstruction in the Belt and Road', *China Economic Issues*, (01), 11–26.[卢潇潇, 梁颖. "一带一路"基础设施建设与全球价值链重构[J]. 中国经济问题, 2020, (01): 11–26.]

Lv Qiang. (2020). Kenya-Inner Mongolia Railway resumes passenger service (2020, July 19). https://www.yidaiyilu.gov.cn/xwzx/hwxw/138256.htm.

Ma Dan, He Yaxing, Yu Xia. (2021). 'Dual value chains, economic uncertainty and regional trade competitiveness: A perspective of "One Belt, One Road" construction', *China Industrial Economics*, (04), 81–99. [马丹, 何雅兴, 郁霞. 双重价值链、经济不确定性与区域贸易竞争力——"一带一路"建设的视角[J]. 中国工业经济, 2021, (04): 81–99.]

Ma Yun. (2015). 'Risk management in the construction of "One Belt, One Road"', *Review of Political Economy*, 6(04), 189–203.[马昀. "一带一路"建设中的风险管控问题[J]. 政治经济学评论, 2015, 6(04): 189–203.]

Mei Dongzhou, Gong Liutang. (2011). 'Currency mismatch, exchange rate appreciation and economic volatility', *Quantitative Economics and Technical Economics Research*, (6), 2011.

Mei Dongzhou, Gong Liutang. (2011). 'Exchange rate regime choices of emerging market economies', *Economic Research*, (11).

Nash. Jr, J. F. (1950). 'The bargaining problem', *Econometrica*, 18(2), 155–162.

Nie Na. (2016). 'Sources of outbound investment risks and prevention mechanisms for China's participation in the "Belt and Road" project', *Contemporary Economic Management*, 38(09), 84–90.[聂娜. 中国参与共建"一带一路"的对外投资风险来源及防范机制[J]. 当代经济管理, 2016, 38(09): 84–90.]

Ouyang Kang. (2018). '"One Belt, One Road" in the changing landscape of global governance', *Chinese Social Sciences*, (08), 5–16.[欧阳康. 全球治理变局中的"一带一路"[J]. 中国社会科学, 2018, (08): 5–16.]

Pei, Chang-Hong, Yu, Yan. (2015). 'The construction of "One Belt, One Road" and the expansion of China's opening up', *International Business and Economics Exploration*, 31(10): 4–17.[裴长洪, 于燕. "一带一路"建设与我国扩大开放[J]. 国际经贸探索, 2015, 31(10): 4–17.]

Qi Wei. (2016). Xiamen University, Malaysia: The Pearl embedded in "Belt and Road Initiative". (2016, September 27) http://www.scio.gov.cn/ztk/wh/slxy/31209/document/1492497/1492497.htm.

Qian Zhongbin. (2021). *The outline of the 14th Five-Year Plan and Vision 2035 for the National Economic and Social Development of the People's Republic of China* (2021, March 13). http://www.xinhuanet.com/2021-03/13/c_1127205564.htm.

Qiu, Junfeng. (2015). 'International reserve accumulation, real and asset price inflation and Mombasa-Nairobi Standard Gaugeary policy independence', *Economics (Quarterly)*, (2).

Shi Jingxia. (2021). 'The "Belt and Road" initiative and international law: An analysis based on the perspective of International Public Goods Supply', *Chinese Social Sciences*, (01), 156–179+207–208.[石静霞. "一带一路"倡议与国际法——基于国际公共产品供给视角的分析[J]. 中国社会科学, 2021, (01): 156–179+207–208.]

Sun, Hongtao. (2018). "Legal risk management of BRI for Big Petroleum and Petrochemical Engineering Services", *SME Management and Technology* (Zhongguo Xian), (2), 99–100.

Sun Y, Chen L, Sun H, Taghizadeh-Hesary F. (2020). 'Low-carbon financial risk factor correlation in the belt and road PPP project', *Finance Research Letters*, 35, 101–491.

Tao Pingsheng. (2020). 'Compliance, improvement and innovation of international rules for building "One Belt, One Road" in the perspective of global governance', *Management World*, 36(05), 161–171+203+116.[陶平生. 全球治理视角下共建"一带一路"国际规则的遵循、完善和创新[J]. 管理世界, 2020, 36(05): 161–171+203+116.]

Tong, Jia-Dong. (2017). 'Theoretical transcendence of the "Belt and Road" Initiative', *Economic Research*, 52(12): 22–25.[佟家栋. "一带一路"倡议的理论超越[J]. 经济研究, 2017, 52(12): 22–25.]

Voon J P, Xu X. (2020). 'Impact of the Belt and Road Initiative on China's soft power: Preliminary evidence', *Asia-Pacific Journal of Accounting & Economics*, 27(1), 120–131.

Wang Shidan. (2017). "Belt and Road Initiative" financing guiding principles (2017, May 16). http://www.scio.gov.cn/31773/35507/htws35512/Document/1552375/1552375.htm.

Wang Yajun. (2017). 'Theoretical innovation and exemplary value of the "Belt and Road" initiative', *World Economy and Politics*, (03), 4–14+156.[王亚军. "一带一路"倡议的理论创新与典范价值[J]. 世界经济与政治, 2017, (03): 4–14+156.]

Wang Yi. (2019). Speech delivered at the Second International Summit on the BRI, April 26, 2019, https://news.sina.com.cn/c/2019-12-13/doc-iihnzahi7327073.shtml

Wen Hao, Shen Jiben. (2019). 'Policy reflections on the investment and financing model and cooperation mechanism of "One Belt, One Road"', *Macroeconomic Management*, (02), 54–61.[温灏, 沈继奔. "一带一路"投融资模式与合作机制的政策思考[J]. 宏观经济管理, 2019, (02): 54–61.]

Wu Shuyu, Hu Biling. (2019). "China foreign exchange, BRI financial cooperation: Status, assessment and improvement", *China Foreign Exchange*, Issue 10, 2019, <http://www.chinaforex.com.cn/index.php/cms/item-view-id-48478.shtml>

Wu Yan, Qu Song (2017). Chinese investment helps Belarus to develop national industrial culture and integrate into Chinese cultural landscape (2017, August 17). https://www.yidaiyilu.gov.cn/xwzx/hwxw/23046.htm

Xiaobing Y, Xianrui Y, Chenliang L, Zhonghui J. (2020). 'Information diffusion-based risk assessment of natural disasters along the Silk Road Economic Belt in China', *Journal of Cleaner Production*, 244, 118–744.

Yang G, Huang X, Huang J, Chen H. (2020). 'Assessment of the effects of infrastructure investment under the belt and road initiative', *China Economic Review*, 60, 101–418.

Xi Jinping. (2020). Xi Jinping's speech at the general debate of the 75th session of the United Nations General Assembly (2020, September 22), https://www.ccps.gov.cn/xxsxk/zyls/202009/t20200922_143558.shtml.

Xiong Xuanguo (Vice Minister of Justice). (2017). "Promoting the BRI legal services is our common responsibility", *Chinese Lawyer*, (07): 6–7.

Xu S, He XY, Zhong K. (2019). 'The "Belt and Road" initiative and financing constraints of Chinese Enterprises', *China Industrial Economics*, (07), 155–173.[徐思, 何晓怡, 钟凯. "一带一路"倡议与中国企业融资约束[J]. 中国工业经济, 2019, (07): 155–173.]

Yu L, Zhao D, Niu H, Lu F. (2020). 'Does the belt and road initiative expand China's export potential to countries along the belt and road?', *China Economic Review*, 60, 101–419.

Yu S, Qian X, Liu T. (2019). 'Belt and road initiative and Chinese firms' outward foreign direct investment', *Emerging Markets Review*, 41, 100–629.

Zeng Wenge. (2018). 'On the construction and improvement of the land trade facilitation rule system of BRI', *Journal of Law*, (11), 16–22.

Zhai F. (2018). 'China's belt and road initiative: A preliminary quantitative assessment', *Journal of Asian Economics*, 55, 84–92.

Zhang C, Xiao C, Liu H. (2019). 'Spatial Big Data analysis of political risks along the Belt and Road', *Sustainability*, 11(8).

Zhang Jianguang, Zhang Peng. (2017). 'A study on trade efficiency and influencing factors between China and "One Belt One Road" countries', *International Business and Economics Exploration*, 33(08), 4–23.[张剑光, 张鹏. 中国与"一带一路"国家的贸易效率与影响因素研究[J]. 国际经贸探索, 2017, 33(08): 4–23.]

Zhang Wenmu. (2017). '"One Belt, One Road" and the Chinese solution to world governance', *World Economy and Politics*, (08), 4–25+156. [张文木. "一带一路"与世界治理的中国方案[J]. 世界经济与政治, 2017, (08): 4–25+156.]

Zhao Dongqi, Sang Baichuan. (2016). 'International production capacity cooperation under the "Belt and Road" initiative: An empirical analysis based on the international competitiveness of industries', *International Trade Issues*, (10), 3–14. [赵东麒, 桑百川. "一带一路"倡议下的国际产能合作——基于产业国际竞争力的实证分析[J]. 国际贸易问题, 2016, (10): 3–14.]

Zhao, Jingrui, Sun, Hui. (2019). 'A study on the evolution of trade relations between China and countries along the "Belt and Road"', *International Business and Economics Exploration*, 35(11), 36–48. [赵景瑞, 孙慧. 中国与"一带一路"沿线国家贸易关系演进研究[J]. 国际经贸探索, 2019, 35(11): 36–48.]

Zhao, Suisheng, Guo, Dan, Gong, Kasali, Xu, Jia. (2019). 'China's top-level design: Strategic objectives, attractiveness and challenges of the Belt and Road Initiative', *Northeast Asia Forum*, 28(03), 28–39+127. [赵穗生, 郭丹, 宫笠俐, 许佳. 中国的顶层设计:"一带一路"倡议的战略目标、吸引力及挑战[J]. 东北亚论坛, 2019, 28(03): 28–39+127.]

Zhou Guping, Kan Read. (2015). 'Talent support and education path of "One Belt, One Road" strategy', *Educational Research*, 36(10), 4–9+22. [周谷平, 阚阅. "一带一路"战略的人才支撑与教育路径[J]. 教育研究, 2015, 36(10): 4–9+22.]

Zhou Guomei. (2020). 'What efforts has China made to promote the co-construction of green BRI?', *China Environment News*, 12.[周国梅.推动共建绿色"一带一路", 中国做出了哪些努力? [J].中国环境报,2020,12.]

Zhou W, Fang X. (2015). 'The political economy of "One Belt, One Road" strategy', *Marxist Studies*, (10), 62–72. [周文, 方茜. "一带一路"战略的政治经济学思考[J]. 马克思主义研究, 2015, (10): 62–72.]

Zhou, Wuying. (2019). The data of the World Bank's Global Economic Outlook report were transferred from the Economic Information Daily (2019, January 11). http://www.jjckb.cn/2019-01/11/c_137736714.htm?from=singlemessage.

Zhang X, Zhang W, Lee P T-W. (2020). 'Importance rankings of nodes in the China Railway Express network under the Belt and Road Initiative', *Transportation Research Part A: Policy and Practice*, 139, 134–147.

Zhang Yuan. (2018). 'The poverty reduction effects of China's aid and investment in the BRI', *Finance and Trade Economics*, 12, 111–125.

Zong, Hui-Ming, Zheng, Li-Li. (2017). 'Analysis of trade patterns between China and Southeast Asian countries in the context of "One Belt, One Road"', *Economic Geography*, 37(08), 1–9. [宗会明, 郑丽丽. "一带一路"背景下中国与东南亚国家贸易格局分析[J]. 经济地理, 2017, 37(08): 1–9.]

Index

www.ingramcontent.com/pod-product-compliance
Ingram Content Group UK Ltd.
Pitfield, Milton Keynes, MK11 3LW, UK
UKHW021836100926
8556IPUK00004B/368

9 781433 192951